*Student's Solutions Manual for*

# Intermediate Algebra • 5th

**Margaret L. Lial**
*American River College*

**Charles D. Miller**
*American River College*

*Prepared with the assistance of*
**Karen Koopmans**

**Scott, Foresman and Company**
*Glenview, Illinois* *Boston* *London*

Cover: A detail from Spirals II by Pauline Burbidge, Nottingham, England. A quilt, pieced and quilted by machine in cottons, which first appeared in The Art Quilt, published by the Quilt Digest Press, San Francisco.

ISBN: 0-673-18852-3


Printed in the United States of America.

1 2 3 4 5 6 - VIK - 92 91 90 89 88 87

To the student:

This book is to be used with Intermediate Algebra, Fifth Edition, by Margaret Lial and Charles Miller. The solutions to the following are given.

1. Section exercises numbered 1, 5, 9, ...

2. Chapter review exercises numbered 1, 5, 9, ...

3. All chapter test problems

This book should be used as an aid as you work to master your course work. Try to solve the exercises that your instructor assigns before you refer to the solutions in this book. Then, read these solutions to determine where you went wrong.

You may find that some solutions in each set are given in greater detail than others. Thus, if you cannot find an explanation for a difficulty that you met in one exercise, you may find the explanation in a similar solution elsewhere in the set.

When a solution instructs you to "see the answer graph in the textbook," please refer to the graph in the answer section in the back of your textbook.

CONTENTS

## 4 RATIONAL EXPRESSIONS



## 5 RATIONAL EXPONENTS AND RADICALS



## 6 QUADRATIC EQUATIONS AND INEQUALITIES

## 7 THE STRAIGHT LINE



## 8 SYSTEMS OF LINEAR EQUATIONS



## 9 CONIC SECTIONS

## CHAPTER 1 THE REAL NUMBERS

### Section 1.1 (page 7)

1. {2, 3, 4, 5}
See answer graph in the textbook.

5. $\{-\frac{1}{2}, \frac{3}{4}, \frac{5}{3}, \frac{7}{2}\}$
Note that $\frac{5}{3} = 1\frac{2}{3}$, $\frac{7}{2} = 3\frac{1}{2}$. See answer graph in textbook.

9. {z|z is an integer greater than 11} is translated "the set of all numbers z such that z is an integer greater than 11." Thus, we have
{12, 13, 14, 15, ...}
*The three dots, ..., mean to continue on in the same manner*

13. {x|x is an irrational number that is also rational}
An irrational number can never be rational. Thus, there are no elements in this set. Therefore, the answer is the empty set represented by the symbol ∅.

17. {z|z is a whole number multiple of 5}
The whole number multiples of 5 are found by multiplying each whole number by 5. Thus, $5 \cdot 0 = 0$, $5 \cdot 1 = 5$, $5 \cdot 2 = 10$, $5 \cdot 3 = 15$, etc. So we have the set
{0, 5, 10, 15, ...}
for our answer.

21. {January, February, March, April, May, June} can be described as "the set of the first six months of the year."

25. {2, 4, 6, 8}
Since 2, 4, 6, and 8 are natural numbers, whole numbers, and positive integers, then we have the following possibilities for answers: "the set of even natural numbers less than 9" or "the set of even whole numbers between 1 and 9" or "the set of even positive integers less than 9."

29. $-|6| = -(6) = -6$
Notice that the negative sign stays with the answer since it is outside the absolute value sign, so is not affected by it.

33. $-|16| = -(16) = -16$

37. $|-8| - |3| = 8 - 3 = 5$

41. $|-12| + |-3| = 12 + 3 = 15$

45. Since $8 + (-8) = 0$, $-8$ is the additive inverse of 8.

49. Since $0 + 0 = 0$, 0 is its own additive inverse.

53. Since $|10| = 10$ and $10 + (-10) = 0$, $-10$ is the additive inverse of $|10|$.

57. $-|-1| = -(1) = -1$. Now, since $-1 + (1) = 0$, then 1 is the additive inverse of $-1 = -|-1|$.

61. $|-2| + |-3| = 2 + 3 = 5$. Since $5 + (-5) = 0$, then $-5$ is the additive inverse of $5 = |-2| + |-3|$.

65. Integers in the given set are -6, 0, 2, 3, 10/2 (or 5).

69. Every rational number is an integer. False, for example, 1/2 is a rational number that is not an integer.

73. Every whole number is a real number. True. The set of all real numbers is discussed in this section. The set of whole numbers is part of this "larger" set.

77. Some rational numbers are integers. True. For example, $\frac{10}{2}$ is a rational number, which equals 5, an integer.

81. 1, 2, 9, 14 are all natural numbers so $\{1, 2, 9, 14\} \subseteq N$.

85. $\{1, 3, 5, 7, 9, 11, 13 \ldots\} \subseteq W$ because $\{1, 3, 5, 7, 9, 11, 13, \ldots\}$ are the odd whole numbers which are certainly contained in the set of whole numbers $W = \{0, 1, 2, 3, \ldots\}$.

89. $W \not\subseteq N$ because the whole numbers W contain the element 0, and 0 is not contained in the set of natural numbers N.

93. $Q \subseteq R$ Every rational number is a real number so the set of rationals is contained in the set of real numbers.

97. $4 \in \{-2, 0, 2, 4, 6\}$ because 4 is in the set.

101. $\{5\} \not\in \{2, 3, 4, 5, 6\}$ but $5 \in \{2, 3, 4, 5, 6\}$
Notice: none of the elements of $\{2, 3, 4, 5, 6\}$ are sets.

105. $A \neq B$ and $A \subseteq B$ is true if every element of A is an element of B, but there is at least one element of B that is not in A.

109. $|x| = -|x|$ only if $x = 0$. If x is not zero then $|x|$ is positive and $-|x|$ is negative and thus $|x|$ cannot equal $-|x|$.

**Section 1.2 (page 13)**

1. "6 is less than 10" can be written as $6 < 10$.

5. r is not equal to 4.
$r \neq 4$

9. 3 is greater than or equal to 7y.
$3 \geq 7y$ *The symbol* $\geq$ *points to 7y*

13. x is between 5 and 9 means x is larger than 5, or $5 < x$, and x is less than 9, or $x < 9$. So,
$$5 < x < 9.$$

17. a is between 1 and 5, including 1 and excluding 5. "Including 1" requires just the symbol $\leq$
Thus, with a symbol between 1 and 5, we have
$$1 \leq a < 5.$$

21. $4 \leq 4$ is read "4 is less than or equal to 4." Since 4 is equal to 4, then the statement is true.

25. $-4 < 2 < 6$. Since the $<$ symbol always points toward the smaller quantity, the statement is true.

29. $6 > 2$ becomes $2 < 6$
Reverse the numbers, but keep the inequality symbol pointing to 2.

33. $5 < x < 6$ becomes $6 > x > 5$

37. $3(8 + 2) > 6 - 5$
$3(10) > 1$
$30 > 1$ True

41. $4 + |-4| \ngtr |4|$
$4 + 4 \ngtr 4$
$8 \ngtr 4$ False

45. $\{x|x > -1\}$
See answer graph in textbook. Use a parenthesis at -1 since -1 is not an element of the set.

49. $\{x|0 < x < 3\}$
See answer graph in textbook.

53. $\{x|\ -4 < x \leq 2\}$
See answer graph in textbook. Read the inequality as "-4 is less than x and x is less than or equal to 2." Use a parenthesis for "less than" and a bracket for "less than or equal to."

57. If $6 = 5 + 1$, then $5 + 1 = 6$.
Symmetric property

61. If $x + y = 3$ and $5(x + y) = W$, then $5(3) = W$.
Substitution property (Substitute 3 for $(x + y)$.)

65. If $x = 5$ and $5 < 8$, then $x < 8$.
This is not the transitive property because the transitive property of equality uses all equal signs, and the transitive property of inequality uses all inequality signs. We are substituting x for 5 to get the last inequality, $x < 8$. Thus, we are using the substitution property.

69. Under what conditions on x is $\frac{1}{x} < x$ if $x \neq 0$?
When $-1 < x < 0$ and $x > 1$, then $\frac{1}{x} < x$. For example, if $x = \frac{-1}{2}$ then $\frac{1}{\frac{-1}{2}} < \frac{-1}{2}$ is equivalent to $-2 < \frac{-1}{2}$, which is a true statement. So for any numbers between -1 and 0 and for numbers greater than 1, $\frac{1}{x} < x$.

## Section 1.3 (page 18)

1. $4 + 8 = 8 + 4$ *Reverse order of 4 and 8*
Commutative property

5. $(7 + 4) + 9 = (4 + 7) + 9$
*Reverse order of 7 and 4*
Commutative property

9. $y + (-y) = 0$
Inverse property

13. $(m + 6) + 2 = (6 + m) + 2$
*Reverse order of m and 6*
Commutative property

17. $t \cdot 0 = 0$
Multiplication property of zero

21. $-5 + [f + (-f)] = -5 + 0$
Inverse property

25. $-(-v) = v$
Double negative property

29. If $8r + 2 = 18$, then $8r + 2 + (-2) = 18 + (-2)$.
$8r + 2 = 18$
$8r + 2\ (-2) = 18 + (-2)$
*Add (-2) to both sides*
Addition property of equality

33. Switch the numerator and denominator of 5/8 to get its reciprocal 8/5.

37. $.125 = \frac{1}{8}$

The reciprocal is thus 8/1 or 8.

41. $6 + (9 + p) = (6 + 9) + p = 15 + p$

45. $8(2r) = (8 \cdot 2)r = 16r$

49. $5(\frac{1}{5}m) = (5 \cdot \frac{1}{5})m = m$

53. $6(3q + 5) = 6(3q) + 6 \cdot 5$ *Distributive property*

$= (6 \cdot 3)q + 30$ *Associative property*

$= 18q + 30$

57. $(2m + 5p - r)4 = (2m)4 + (5p)4 - r \cdot 4$ *Distributive property*

$= 8m + 20p - 4r$

61. $2x + 3x = (2 + 3)x$

$= 5x$ *Simplify*

65. $-3 + 3 = (-3) + 3$

$= 0$

69. $3a + 5a + 6a = (3 + 5 + 6)a$

$= (14)a$ *Simplify*

$= 14a$

73. $8(2 + 3) = (8)2 + (8)3$

$= 16 + 24$ *Simplify*

$= 40$

77. If $k = 4$,

then $3k = 3 \cdot 4$

$= 12$.

81. Replace x with 5 to show that

$2 + 6x \neq 8x$.

$2 + 6x = 2 + 6 \cdot 5$ *Replace x with 5*

$= 2 + 30$

$= 32$

$8x = 8 \cdot 5$ *Replace x with 5*

$= 40$

Since $32 \neq 40$, $2 + 6x \neq 8x$.

85. Does $a + (b \cdot c) = (a + b)(a + c)$ for all real numbers a, b, and c? No.

There is no distributive property of addition over multiplication.

For example, try

$a = 7$, $b = 5$, and $c = 3$.

$a + (b \cdot c) = 7 + (5 \cdot 3)$ *Substitution*

$= 7 + 15$

$= 22$

$(a + b)(a + c) = (7 + 5)(7 + 3)$ *Substitution*

$= (12)(10)$

$= 120$

$22 \neq 120$

**Section 1.4 (page 28)**

1. $-12 + (-8)$

$= -(12 + 8)$ *Add absolute values*

$= 20$ *Sum is negative since both numbers are negative*

5. $13 - 15 = 13 + (-15)$ *Change signs*

$= -2$ *Add*

9. $-5 + 11 + 3$

$= 6 + 3$ *Add in order from left to right*

$= 9$

13. $-9 - (-11) - (4 - 6)$

$= -9 - (-11) - (-2)$ *Work inside parentheses*

$= -9 + [-(-11)] + [-(-2)]$ *Change signs*

$= -9 + 11 + 2$

$= 4$

17. $-\frac{3}{4} - (\frac{1}{2} - \frac{3}{8})$

$= -\frac{6}{8} - (\frac{4}{8} - \frac{3}{8})$ *Use 8 as a common denominator*

$= -\frac{6}{8} - \frac{1}{8}$ *Simplify inside parentheses*

$= -\frac{7}{8}$

21. $-2 = |-3| + 5 = -2 + 3 + 5$

*Evaluate the absolute value*

$= 6$

25. $8.946 - 3.2574 + (-3.99421)$

$= 8.94600 - 3.25740 - 3.99421$

$= 5.68860 - 3.99421$

$= 1.69439$

29. A = -4 and B = 2.

$-4 - 2 = 6$ *Take the absolute value of the difference*

33. The temperature was -19° and we add 39° to it. Thus, we have -19 + 39 = 20°.

37. $(-12)(-2) = 24$ *The product of two numbers with the same sign is positive*

41. $-\frac{5}{2}(-\frac{12}{25}) = \frac{60}{50}$

$= \frac{6}{5}$ *Write in lowest terms*

45. 9/0 is undefined, since division by 0 is not possible.

49. $\dfrac{\frac{7}{10}}{-\frac{8}{15}} = \frac{7}{10} \cdot (\frac{-15}{8})$ *Multiply the numerator by the reciprocal of the denominator*

$= \frac{-105}{80}$ *Write in lowest terms*

$= \frac{-21}{16}$

53. $(-\frac{6}{7})^3 = (-\frac{6}{7})(-\frac{6}{7})(-\frac{6}{7})$

$= -\frac{216}{343}$

57. $\sqrt{256} = 16$ *Remember that $\sqrt{\ }$ means only the positive square root*

61. $\sqrt[3]{-27} = -3$ *$(-3)(-3)(-3) = -27$ Notice odd roots, such as 3, may be negative Also, odd roots of negative numbers, here -27, exist*

65. $\sqrt{18,499} = 136.011$ *(rounded)*

69. $-6[2 + (-5)] = -6[-3]$ *Simplify inside the brackets*

$= 18$ *Multiply*

73. $-4 - 3(-2) + 5^2$

$= -4 - 3(-2) + 25$ *Simplify the powers*

$= -4 + 6 + 25$ *Do multiplication*

$= 2 + 25$ *Add from left to right*

$= 27$

77. $\dfrac{(-10 + 4)\cdot(-3)}{-7 - 2}$

$= \dfrac{(-6)(-3)}{-9}$ *Work separately above and below the fraction bar*

$= \frac{18}{-9}$ *Simplify the numerator*

$= -2$

81. $2\sqrt{100} - 10 \div 5$

$= 2(10) - 10 \div 5$ *Simplify roots first*

$= 20 - 2$ *Multiply and divide from left to right*

$= 18$

85. $-\frac{4}{5}[6(-4) + (-5)(-5)]$

$= -\frac{4}{5}[-24 + 25]$ *Work inside brackets first*

$= -\frac{4}{5}[1]$ *Simplify inside brackets*

$= -\frac{4}{5}$

89. $\frac{-11.14 - 3.289}{-2.98(-5.234)}$

$= \frac{-14.429}{15.59732}$ *Work separately above and below the fraction bar*

$= -.925$ *(rounded)*

93. $\frac{2c + a}{4b + 6a} = \frac{2(-7) + (-4)}{4(6) + 6(-4)}$

*Substitute -7 for c, -4 for a, and 6 for b*

$= \frac{-14 + (-4)}{24 + (-24)}$ *Multiply in numerator and denominator*

$= \frac{-18}{0}$ *Add*

Since division by 0 is impossible, -18/0 is undefined.

97. $a - b = b - a$

Subtraction is not commutative. For example, if a = 6 and b = 4, we have $6 - 4 = 4 - 6$. Since $2 \neq 2$, the statement is false.

101. $|a - b| = |a| - |b|$ is false. For example, if a = 2 and b = 3 we have

$|2 - 3| = |-1| = 1$

$|2| - |3| = 2 - 3 = -1$

Since $1 \neq -1$, the statement is false.

105. Yes, $\frac{a}{b} = \frac{b}{a}$ for nonzero numbers a and b with equal absolute values. For example, if a = b = 3 then $\frac{a}{b} = \frac{3}{3} = 1$ and $\frac{b}{a} = \frac{3}{3} = 1$. Also if a = 3, and b = -3, then

$\frac{a}{b} = \frac{b}{a} = -1$

**Chapter 1 Review Exercises (page 33)**

1. $\{-3, -5, 1, \frac{1}{4}, \frac{9}{4}\}$

See answer graph in back of textbook.

5. {negative integers greater than or equal to -6}

Given as a list, this set is {-6, -5, -4, -3, -2, -1}.

9. $-|4| = -4$ *Since* $|4| = 4$

13. $-(-9) = 9$ Since $-9 + 9 = 0$, the additive inverse of $-(-9)$ is $-9$.

17. $-9 = \frac{-9}{1}$, $\frac{-4}{3}$, $0 = \frac{0}{1}$, $\frac{5}{3}$, and $\frac{12}{3}$ are rational numbers. That is, they can be expressed as integers divided by integers.

21. "x is between -2 and 1" means that -2 is less than x and x is less than 1.

$-2 < x < 1$

25. $6 < |-2|$ False, since $|-2| = 2$ and 6 is <u>not</u> less than 2.

29. $\{x|x < -4\}$ See answer graph in textbook.

33. The transitive property justifies "if 5 < 9 and 9 < p, then 5 < 9."

37. $12 + (13 + 2) = 12 + (2 + 13)$
We switched 2 and 13 using the commutative property.

41. $\frac{-3}{4}(-\frac{4}{3}) = 1$
Inverse property

45. If $m = 8$, then $2m = 2 \cdot 8$.
Multiplication property of equality

49. $7p + 7r = 7(p + r)$; distributive property

53. $-(-22 = 22$; double negative property

57. The reciprocal of $-\frac{8}{3}$ is $-\frac{3}{8}$ since
since $-\frac{8}{3}(-\frac{3}{8}) = 1$.

61. $-5(11) = -55$ *The product of two numbers with opposite signs is negative*

65. $-\frac{2}{3}[5(-2) + 8 - 16]$
$= -\frac{2}{3}[-10 + 8 - 16]$ *Multiply inside brackets*
$= -\frac{2}{3}[-18]$ *Simplify inside brackets*
$= \frac{36}{3}$ *Multiply*
$= 12$

69. $-\frac{3}{7}\cdot(-\frac{14}{9}) = \frac{42}{63}$
$= \frac{2}{3}$ *Write in lowest terms*

73. $-2[3 + 5(4)] = -2[3 + 20]$ *Multiply inside brackets*
$= -2[23]$ *Add inside brackets*
$= -46$ *Multiply*

77. $\frac{-15}{-3} = 5$

81. $-2 + |-4| - |-3|$
$= -2 + 4 - 3$ *Evaluate absolute value*
$= -1$ *Simplify*

85. $m - z + n$
$= 3 - (-4) + (-5)$ *Substitute*
$= 3 + 4 + (-5)$ *Change signs*
$= 2$ *Add*

89. $(\frac{3}{7})^3 = (\frac{3}{7})(\frac{3}{7})(\frac{3}{7}) = \frac{27}{343}$

93. $\sqrt{196} = 14$ since $\sqrt{\ }$ means the positive square root, and $14 \cdot 14 = 196$.

97. $\sqrt[4]{81} = 3$ since $3 \cdot 3 \cdot 3 \cdot 3 = 81$.

## Chapter 1 Test (page 35)

1. All elements in this set are real numbers.
$\{-\sqrt{7}, -2, -1/2, 0/2, 15/2, 3, 24/2, \sqrt{11}\}$

2. $-\sqrt{7}$, $\sqrt{11}$ are irrational numbers.

3. $-2, -\frac{1}{2}, \frac{0}{2}$ (or 0), $\frac{15}{2}, 3, \frac{24}{2}$ (or 12)
The set of rational numbers includes fractions.

4. $-2, \frac{0}{2}$ (or 0), 3, $\frac{24}{2}$ (or 12)

   Each of these numbers can be written in the form of an integer.

5. $\frac{0}{2}$ (or 0), 3, $\frac{24}{2}$ (or 12)

   The set of whole numbers includes zero.

For Exercises 6-8, see answer graph in textbook.

9. $m + 3 = 3 + m$

   Addition is being reversed.
   Commutative property

10. $(5 + 6) + 1 = 5 + (6 + 1)$

    The numbers remain in the same order on each side but they are grouped differently.
    Associative property of addition

11. $\frac{2}{3} \cdot 1 = \frac{2}{3}$

    Any number times 1 gives that number back again.
    Identity property

12. If $x = 3$ and $y = x + 1$, then $y = 3 + 1$.

    Note that 3 is substituted for x in $y = x + 1$ to give $y = 3 + 1$.
    Substitution property

13. $(a + b) + [-(a + b)] = 0$

    Treat $(a + b)$ as one number, then its additive inverse is $-(a + b)$.
    Inverse property

14. If $x = 4y$, then $x + 6 = 4y + 6$.

    6 is added to both sides.
    Addition property of equality

15. If $x - y = z$, then $3(x - y) = 3z$.

    3 is multiplied times each side.
    Multiplication property of equality

16. $4(k + 3) = 4k + 4 \cdot 3$

    4 is distributed through multiplication over k and 3.
    Distributive property

17. $\frac{3}{4}(\frac{4}{3}m) = (\frac{3}{4} \cdot \frac{4}{3})m$ *Associative property*

    $= 1m$

    $= m$

18. $5(4r + 7s - 2p)$

    $= 5 \cdot 4r + 5 \cdot 7s - 5 \cdot 2p$ *Distributive property*

    $= 20r + 35s - 10p$ *Associative property*

19. $-6 - (-15) + (-9) - 3$

    $= -6 + 15 - 9 - 3$ *Change subtraction to addition*

    $= -3$ *Add*

20. $(-\frac{1}{3})(-\frac{9}{5}) = \frac{9}{15}$

    $= \frac{3}{5}$

21. $-2 + (-3) - |-5| + |-12 + 2|$

    $= -2 + (-3) - 5 + |-10|$ *Evaluate absolute values*

    $= -2 + (-3) - 5 + 10$

    $= 0$ *Add*

22. $6 - 3 \cdot 3 + 5(-4) - \frac{8}{-2}$

    $= 6 - 9 - 20 - (-4)$ *Multiply or divide*

    $= 6 - 9 - 20 + 4$ *Change signs*

    $= -19$ *Add*

23. $6 - 3^3 + 2(5) + (-4)^2$

$= 6 - 27 + 2(5) + 16$ *Evaluate powers*

$= 6 - 27 + 10 + 16$ *Multiply*

$= 5$ *Add*

24. $\frac{10 - 24 + (-6)}{4(-5)} = \frac{-20}{-20}$ *Simplify numerator and denominator*

$= 1$ *Divide*

25. $\frac{-2[3 - (-1 - 2) + 2]}{3(-3) - (-2)}$

$= \frac{-2[3 - (-3) + 2]}{-9 + 2}$ *Work in numerator and denominator separately*

$= \frac{-2[3 + 3 + 2]}{-7}$

$= \frac{-2[8]}{-7}$ *Add inside brackets*

$= \frac{16}{7}$ *Multiply*

26. $\frac{8 \cdot 4 - 3 \cdot 5 - 2(-1) - (-1)}{-3 \cdot 2 + 1}$

$= \frac{32 - 15 + 2 + 1}{-6 + 1}$ *Simplify*

$= \frac{20}{-5}$

$= -4$

27. $8r^2 + m^3 = 8(5)^2 + (-3)^3$ *Substitute 5 for r and -3 for m*

$= 8(25 + (-27)$ *Evaluate powers*

$= 200 - 27$ *Multiply*

$= 173$

28. $-6(2m - 3k) = -6(2(-3) - 3(-4))$

*Substitute*

$= -6(-6 + 12)$ *Simplify*

$= -6(6)$

$= -36$

29. $\frac{8k + 2m^2}{r - 2} = \frac{8(-4) + 2(-3)^2}{5 - 2}$ *Substitute*

$= \frac{-32 + 18}{3}$ *Simplify*

$= -\frac{14}{3}$

30. $\frac{5k + 6r}{k + r - 1} = \frac{5(-4) + 6(5)}{-4 + 5 - 1}$ *Substitute*

$= \frac{-20 + 30}{0}$ *Simplify*

Undefined, since division by 0 is impossible.

## CHAPTER 2 LINEAR EQUATIONS AND INEQUALITIES

### Section 2.1 (page 43)

1. $4x + 3x + 7 + 9$

$= (4 + 3)x + 7 + 9$ *Distributive property*

$= 7x + 16$ *Simplify*

5. $-6p + 11p - 4p + 6 - 5 = p + 1$

*Combine like terms*

9. $-2(m + 1) + 3(m - 4)$

$= -2m - 2 + 3m - 12$ *Distributive property to remove parentheses*

$= -2m + 3m - 2 - 12$ *Combine like terms*

$= m - 14$ *Simplify*

13. $-(2p + 5) + 3(2p + 4) + p$

$= -2p - 5 + 6p + 12 + p$

*Distributive property*

$= -2p + 6p + p - 5 + 12$

*Combine like terms*

$= 5p + 7$ *Simplify*

17. $m + 2(-m + 4) - 3(m + 6)$

$= m - 2m + 8 - 3m - 18$

*Distributive property*

$= m - 2m - 3m + 8 - 18$

*Combine like terms*

$= -4m - 10$ *Simplify*

21. $-(5.923p + 6.476) - (-8.915p + 11.213)$

$= -.5923p - 6.476 + 8.915p - 11.213$

$= -5.923p + 8.915p - 6.476 - 11.213$

$= 2.992p - 17.689$

25. $5 - 8k = -19$

$-8k + 5 - 5 = -19 - 5$ *Subtract 5 from each side*

$-8k = -24$

$\frac{-8k}{-8} = \frac{-24}{-8}$ *Divide each side by -8*

$k = 3$ *Simplify*

Solution set: $\{3\}$

29. $-9y - 4 = 14$

$-9y - 4 + 4 = 14 + 4$ *Add 4 to each side*

$-9y = 18$

$\frac{-9y}{-9} = \frac{18}{-9}$ *Divide each side by -9*

$y = -2$

Solution set: $\{-2\}$

33. $-4 - 3p = 7$

$-4 - 3p + 4 = 7 + 4$ *Add 4*

$-3p = 11$

$\frac{-3p}{-3} = \frac{11}{-3}$ *Divide by -3*

$p = -\frac{11}{3}$ *Simplify*

Solution set: $\{-\frac{11}{3}\}$

37. $7.689z - 4.132 = -34.888$

$7.689z - 4.132 + 4.132 = -34.888 + 4.132$

$7.689z = -30.756$

$\frac{7.689z}{7.689} = \frac{-30.756}{7.689}$

$z = -4$

Solution set: $\{-4\}$

41. $7m - 2m + 4 - 5 = 3m - 5 + 6$

$5m - 1 = 3m + 1$ *Combine like terms*

$5m - 1 - 3m = 3m + 1 - 3m$ *Subtract 3m*

$2m - 1 = 1$ *Combine like terms*

$2m - 1 + 1 = 1 + 1$ *Add 1*

$2m = 2$

$\frac{2m}{2} = \frac{2}{2}$ *Divide by 2*

$m = 1$

Solution set: $\{1\}$

45. $2(k - 4) = 5k + 2$

$2k - 8 = 5k + 2$ *Distributive property*

$2k - 8 - 5k = 5k + 2 - 5k$ *Subtract 5k*

$-3k - 8 = 2$ *Combine like terms*

$-3k - 8 + 8 = 2 + 8$ *Add 8*

$-3k = 10$

$\frac{-3k}{-3} = \frac{10}{-3}$ *Divide by -3*

$k = -\frac{10}{3}$

Solution set: $\{-\frac{10}{3}\}$

49. $2y + 3(y - 4) = 2(y - 3)$

$2y + 3y - 12 = 2y - 6$ *Distributive property*

$5y - 12 = 2y - 6$ *Combine like terms*

$5y - 12 - 2y = 2y - 6 - 2y$ *Subtract 2y*

$3y - 12 = -6$ *Simplify*

$3y - 12 + 12 = -6 + 12$ *Add 12*

$3y = 6$

$\frac{3y}{3} = \frac{6}{3}$ *Divide by 3*

$y = 2$

Solution set: $\{2\}$

53. $5y - (8 - y) = 2[-4 - (3 + 5y) - 13]$

$5y - 8 + y = 2[-20 - 5y]$ *Distributive property*

$6y - 8 = -40 - 10y$ *Distributive property and simplify*

$6y - 8 + 10y = -40 - 10y + 10y$ *Add 10y to both sides*

$16y - 8 = -40$ *Simplify*

$16y - 8 + 8 = -40 + 8$ *Add 8 to each side*

$16y = -32$ *Simplify*

$\frac{16y}{16} = \frac{-32}{16}$ *Divide each side by 16*

$y = -2$ *Simplify*

Solution set: $\{-2\}$

57. $-\frac{5}{9}k = 2$

$9(-\frac{5}{9}k) = 9(2)$ *Multiply by 9 to eliminate fraction*

$-5k = 18$

$\frac{-5k}{-5} = \frac{18}{-5}$ *Divide by -5*

$k = \frac{-18}{5}$

Solution set: $\{\frac{-18}{5}\}$

61. $\frac{m}{2} + \frac{m}{3} = 5$ *Lowest common denominator of 2 and 3 is 6*

$6(\frac{m}{2} + \frac{m}{3}) = 6(5)$ *Multiply by 6 to eliminate fractions*

$6(\frac{m}{2}) + 6(\frac{m}{3}) = 30$ *Distributive property*

$3m + 2m = 30$

$\frac{5m}{5} = \frac{30}{5}$ *Divide by 5*

$m = 6$

Solution set: $\{6\}$

65. $\frac{m-2}{3} + \frac{m}{4} = \frac{1}{2}$

$12\left(\frac{m-2}{3}\right) + 12\left(\frac{m}{4}\right) = 12\left(\frac{1}{2}\right)$ *Multiply by 12*

$4(m - 2) + 3m = 6$

$4m - 8 + 3m = 6$ *Distributive property*

$7m - 8 + 8 = 6 + 8$ *Add 8*

$7m = 14$

$\frac{7m}{7} = \frac{14}{7}$ *Divide by 7*

$m = 2$

Solution set: {2}

69. $9k + 4 - 3k = 2(3k + 4) - 4$

$9k + 4 - 3k = 6k + 8 - 4$

*Distributive property*

$6k + 4 = 6k + 4$ *Simplify*

This equation is an identity because both sides of the equation are the same. Any real number would make the equation true.

Solution set: Set of all real numbers

73. $-11m = 4(m - 3) + 6m = 4m - 12$

$-11m + 4m - 12 + 6m = 4m - 12$

*Distributive property*

$-m - 12 = 4m - 12$

*Simplify*

$-m - 4m - 12 = 4m - 4m - 12$

*Subtract 4m from each side*

$-5m - 12 = -12$ *Simplify*

$-5m - 12 + 12 = -12 + 12$

*Add 12 to each side*

$-5m = 0$

$\frac{-5m}{-5} = \frac{0}{-5}$ *Divide by -5*

$m = 0$

This is a conditional equation since it has one element in the solution set.

Solution set: {0}

77. $-4[6 - (-2 + 3q)] = 3(7 + 4q)$

$-4[6 + 2 - 3q] = 3(7 + 4q)$

*Distributive property*

$-24 - 8 + 12q = 21 + 12q$

*Distributive property*

$-32 + 12q = 21 + 12q$

$-32 + 12q - 12q = 21 + 12q - 12q$

*Subtract 12q*

$-32 = 21$

Since $-32 \neq 21$, the equation is a contradiction.

Solution set: $\emptyset$

81. $3x + 6 = k + 8$

$3(-4) + 6 = k + 8$ *Substitute - 4 for x*

$-12 + 6 = k + 8$

$-6 = k + 8$

$-6 - 8 = k + 8 - 8$ *Subtract 8*

$-14 = k$

85. $y_m = 46.338 - 0.097x$ and

$y_w = 57.484 - 0.196x$

(a) $y_m = 46.338 - 0.097(92)$

*Substitute 92 for x*

$= 46.338 - 8.924$

$= 37.414$

The men's time would be about 37.4 seconds.

$y_w = 57.484 - 0.196(92)$

$= 57.484 - 18.032$

$= 39.452$

The women's time would be about 39.5 seconds.

(b) $y_m = 46.338 - 0.097(96)$

*Substitute 96 for x*

$= 46.338 - 9.312$

$= 37.026$

The men's time would be about 37.0 seconds.

$$y_w = 57.484 - 0.196(96)$$
$$= 57.484 - 18.816$$
$$= 38.668$$

The women's time would be about 38.7 seconds.

89. $y = -3$

$$\frac{y}{y + 3} = \frac{-3}{y + 3}$$

The solution set of the first equation is {-3}. Solve the second equation.

$$\frac{y}{y + 3} = \frac{-3}{y + 3}$$

*The denominators may not be 0, so $y \neq 3$*

$$(y + 3)\frac{y}{y + 3} = (y + 3)\frac{-3}{y + 3}$$

*Multiply by $y + 3$*

$$y = -3$$

But $y \neq -3$, so this equation is a contradiction.

Solution set: $\emptyset$

Since the solution sets are not the same, the equations are not equivalent.

93. $2L - 2W = 2(10) + 2(6)$ *Substitute 10 for L and 6 for W*

$$= 32$$

97. $\frac{5}{9}(F - 32) = \frac{5}{9}(212 - 32)$ *Substitute 212 for F*

$$= \frac{5}{9}(180)$$
$$= 100$$

**Section 2.2 (page 49)**

1. $d = rt$

Solve for r.

$\frac{d}{t} = \frac{rt}{t}$ *Divide each side by t*

$\frac{d}{t} = r$ *Simplify*

5. $p = br$

Solve for b.

$\frac{p}{r} = \frac{br}{r}$ *Divide each side by r*

$\frac{p}{r} = b$ *Simplify*

9. $V = LWH$

Solve for W.

$\frac{V}{LH} = \frac{LWH}{LH}$ *Divide each side by LH*

$\frac{V}{LH} = W$ *Simplify*

13. $C = 2\pi r$

Solve for r.

$\frac{C}{2\pi} = \frac{2\pi r}{2\pi}$ *Divide each side by $2\pi$*

$\frac{C}{2\pi} = r$ *Simplify*

17. $C = \frac{5}{9}(F - 32)$

Solve for F.

$\frac{9}{5}C = (\frac{9}{5})\frac{5}{9}(F - 32)$ *Multiply each side by $\frac{9}{5}$*

$\frac{9}{5}C = F - 32$ *Simplify*

$\frac{9}{5}C + 32 = F - 32 + 32$ *Add 32 to each side*

$\frac{9}{5}C + 32 = F$ *Simplify*

21. Use $I = prt$

Solve for p.

$\frac{I}{rt} = \frac{prt}{rt}$ *Divide by rt*

$\frac{I}{rt} = p$

25. Use $P = a + b + c$.

Solve this formula for c.

$P - a - b = a + b + c - a - b$ *Subtract a and b from each side*

$P - a - b = c$ *Simplify*

$27 - 8 - 12 = c$ *Substitute 27 for P, 8 for a, and 12 for b*

$7 = c$ *Simplify*

The third side is 7 centimeters.

29. Use $A = \frac{1}{2}(B + b)h$.

Solve this formula for b.

$2A = 2 \cdot \frac{1}{2}(B + b)h$ *Multiply by 2*

$2A = (B + b)h$ *Simplify*

$2A = Bh + bh$ *Distributive property*

$2A - Bh + Bh - Bh + bh$ *Subtract Bh*

$2A - Bh = bh$ *Simplify*

$\frac{2A - Bh}{h} = \frac{bh}{h}$ *Divide by h*

$\frac{2A - Bh}{h} = b$ *Simplify*

$\frac{2(18) - (6)(4)}{4} = b$ *Substitute 18 for A, 6 for B, and 4 for h*

$3 = b$ *Simplify*

The other base is 3 meters.

33. Use $d = rt$.

Solve for r.

$\frac{d}{t} = \frac{rt}{t}$ *Divide both sides by t*

$\frac{d}{t} = r$ *Simplify*

$\frac{800}{6} = r$ *Substitute 800 for d and 6 for t*

$133\frac{1}{3} = r$ *Simplify*

The rate is $133\frac{1}{3}$ kilometers per hour.

37. Use $F = \frac{9}{5}C + 32$.

$F = \frac{9}{5}(-40) + 32$ *Substitute -40 for C*

$F = -72 + 32$

$F = -40$

The Fahrenheit temperature is -40°F.

41. Use $A = 2\pi rh + 2\pi r^2$.

Solve for h.

$S - 2\pi r^2 = 2\pi rh$

$\frac{S - 2\pi r^2}{2\pi r} = \frac{2\pi rh}{2\pi r}$

$\frac{S - 2\pi r^2}{2\pi r} = h$

$\frac{32\pi - 2\pi (2)^2}{2\pi (2)} = h$ *Substitute 32π for S and 2 for r*

$\frac{32\pi - 8\pi}{4\pi} = h$

$\frac{24\pi}{4\pi} = h$

$6 = h$

The height is 6 meters.

45. Use $I = prt$.

Solve for $t$.

$$\frac{I}{pr} = t$$

$$\frac{273.13}{(5496.11)(.1148)} = t \quad \textit{Substitute}$$

$$\frac{273.13}{630.95342} =$$

$$t = .43288 \text{ (rounded)}$$

The time is .43288 year. Since 1 year = 365 days, multiply using 365 days.

The time is (365)(.43288) or 158 days.

49.

$$9x = 3y + bx + 2$$

$$9x - bx = 3y + bx - bx + 2$$

*Subtract bx*

$$9x - bx = 3y + 2$$

$$x(9 - b) = 3y + 2 \quad \textit{Distributive property}$$

$$\frac{x(9 - b)}{(9 - b)} = \frac{3y + 2}{9 - b} \quad \textit{Divide by 9 - b}$$

$$x = \frac{3y + 2}{9 - b}$$

53.

$$\frac{6a}{x} - \frac{y}{3} = \frac{2}{3x} \quad \textit{The lowest common denominator is 3x}$$

$$3x(\frac{6a}{x}) - 3x(\frac{y}{3}) = 3x(\frac{2}{3x}) \quad \textit{Multiply by 3x}$$

$$3(6a) - x(y) = 2$$

$$18a - xy = 2$$

$$18a - 18a - xy = 2 - 18a$$

*Subtract 18a*

$$-xy = 2 - 18a$$

$$\frac{-xy}{y} = \frac{2 - 18a}{y}$$

*Divide by y*

$$-x = \frac{2 - 18a}{y}$$

$$x = -(\frac{2 - 18a}{y}) = \frac{18a - 2}{y}$$

57.

$$\frac{1}{p} + \frac{1}{q} = \frac{1}{f}$$

*The lowest common denominator is pqf*

$$pqf(\frac{1}{p}) + pqf(\frac{1}{q}) = pqf(\frac{1}{f})$$

*Multiply by pqf*

$$qf + pf = pq$$

$$f(q + p) = pq$$

*Distributive property*

$$\frac{f(q + p)}{q + p} = \frac{pq}{q + p}$$

*Divide by q + p*

$$f = \frac{pq}{q + p}$$

61. Use the unearned interest formula.

$$u = f \cdot \frac{k(k + 1)}{n(n + 1)}$$

$$= 450 \cdot \frac{9(9 + 1)}{24(24 + 1)}$$

*Substitute 450 for f, q for k, and 24 for m*

$$= 450 \cdot \frac{90}{600}$$

$$= 67.5$$

The amount of unearned interest is \$67.50.

65. The product of 8 and -2 is (8)(-2) = -16. 9 is added to -16. The result is -16 + 9 = -7.

69. The product of -2 and 4 is (-2)(4) = -8. The product of -9 and -3 is (-9)(-3) = 27. The sum of the two products is -8 + 27 = 19.

**Section 2.3 (page 58)**

1. A number increased by 8

$$\begin{array}{ccc} \downarrow & \downarrow & \downarrow \\ x & + & 8 \end{array}$$

Answer: $x + 8$

5. Product means multiply, so "the product of a number and 5" is 5x.

9. 12 decreased by twice a number

$$\begin{array}{ccc} \downarrow & \downarrow & \downarrow \\ 12 & - & 2x \end{array}$$

Answer: $12 - 2x$

13. The ratio of a and b is $\frac{a}{b}$.

So the ratio of a number and 6 is $\frac{x}{6}$.

17. Let x be the number.

The sum of a number and 11 is 18

$$\begin{array}{ccccc} \downarrow & \downarrow & \downarrow & \downarrow & \downarrow \\ x & + & 11 & = & 18 \end{array}$$

Solve the equation.

$$x + 11 = 18$$

$$x + 11 - 11 = 18 - 11 \quad \textit{Subtract 11}$$

$$x = 7$$

The number is 7.

21. Let x be the number.

A number multiplied by 6 decreased by 4 is 28.

$$\begin{array}{ccccc} \downarrow & \downarrow & \downarrow & \downarrow & \downarrow \\ 6x & - & 4 & = & 28 \end{array}$$

Solve this equation.

$$6x - 4 = 28$$

$$6x - 4 + 4 = 28 + 4 \quad \textit{Add 4}$$

$$\frac{6x}{6} = \frac{32}{6} \quad \textit{Divide by 6}$$

$$x = \frac{32}{6} = \frac{16}{3}$$

The number is 16/3.

25. Let x be the first even integer. Then, (x + 2) and (x + 4) represent the next two consecutive even integers. The sum of 3 consecutive even integers is 276.

$$\begin{array}{ccc} \downarrow & \downarrow & \downarrow \\ x + (x + 2) + (x + 4) & = & 276 \end{array}$$

Solve this equation.

$$x + (x + 2) + (x + 4) = 276$$

$$3x + 6 = 276 \quad \textit{Simplify the left side}$$

$$3x + 6 - 6 = 276 - 6 \quad \textit{Subtract 6}$$

$$3x = 270 \quad \textit{Simplify}$$

$$\frac{3x}{3} = \frac{270}{3} \quad \textit{Divide by 3}$$

$$x = 90$$

$$x + 2 = 90 + 2 = 92$$

$$x + 4 = 90 + 4 = 94$$

The first even integer is x = 90, the second is 92, the third is 94.

29. Let x be the first integer. Then, the first 3 consecutive integers are x, x + 1, x + 2.

12 added to the second integer is $1\frac{1}{2}$ times the third integer.

$$(x + 1) + 12 = 1\frac{1}{2} \cdot (x + 2)$$

Solve this equation.

$$x + 1 + 12 = 1\frac{1}{2}(x + 2)$$

$$x + 13 = \frac{3}{2}(x) + \frac{3}{2}(2)$$ *Distributive property*

$$x + 13 = \frac{3}{2}x + 3$$

$$2(x) + 2(13) = 2(\frac{3}{2}x) + 2(3)$$ *Multiply by 2*

$$2x + 26 = 3x + 6$$

$$2x - 3x + 26 = 3x - 3x + 6$$ *Subtract 3x*

$$-x + 26 = 6$$

$$-x + 26 - 26 = 6 - 26$$ *Subtract 26*

$$-x = -20$$

$$x = 20$$

$$x + 1 = 20 + 1 = 21$$

$$x + 2 = 20 + 2 = 22$$

The first integer is 20,
the second is 21,
the third is 22.

33. Let x be the number of votes received by the second candidate. Then, the first has

twice the votes of the second less 24.

$$2 \cdot x - 24$$

The votes of the first and the votes of the second were 960.

$$2x - 24 + x = 960$$

Solve this equation.

$$2x - 24 + x = 960$$

$$3x - 24 = 960$$ *Combine*

$$3x - 24 + 24 = 960 + 24$$ *Add 24*

$$\frac{3x}{3} = \frac{984}{3}$$ *Divide by 3*

$$x = 328$$

$$2x - 24 = 2(328) - 24$$

$$= 632$$

The first candidate has 632 votes and the second candidate has 328 votes.

37. Let x represent the amount invested at 10%. The balance, 60,000 - x, is invested at 8%. The interest earned annually at 10% is found using the formula.

$$I = p\ r\ t$$

$$= (x)(10\%)(1) = .10x$$

For the balance at 8%:

$$I = p\ r\ t$$

$$= (60{,}000 - x)(8\%)(1)$$

$$= .08(60{,}000 - x)$$

Interest at 10% + interest at 8% = total interest

$$.10x + .08(60{,}000 - x) = 5600$$

Solve the equation.

$$.10x + .08(60{,}000 - x) = 5600$$

$$.10x + 4800 - .08x = 5600$$ *Distributive property*

$$.02x = 5600 - 4800$$

$$.02x = 800$$

$$x = 40{,}000$$

$$60{,}000 - x = 20{,}000$$

The amount invested at 10% is $40,000; the amount invested at 8% is $20,000.

41. Let $x$ represent the number of liters in the 10% solution. The following table represents the problem.

| Strength | Liters of Solution | Liters of Pure Alcohol |
|---|---|---|
| 10% | x | .10x |
| 50% | 30 | .50(30) = 15 |
| 20% | x + 30 | .20(x + 30) |

Use the last column of the table for the equation.

$$.10x + 15 = .20(x + 30)$$
$$.10x + 15 = .20x + 6$$
$$-.10x = -9 \quad \textit{Subtract 15 and .20x}$$
$$x = 90$$

90 liters of the 10% solution must be used.

45. Let $x$ be the number of hours the cars traveled. The distance the first car traveled ($d = rt$) is $50x$; the distance of the second is $55x$. Since they were traveling in opposite directions, the sum of their distances equals 315 miles. Write the equation.

$$50x + 55x = 315$$
$$105x = 315$$
$$x = 3$$

They will be 315 miles apart in 3 hours.

49. Let $x$ represent the sale of goods.

| The sale of goods | plus | 6% sales tax | totals | \$1590. |
|---|---|---|---|---|
| ↓ | ↓ | ↓ | ↓ | ↓ |
| x | + | .06x | = | 1590 |

$$1.06x = 1590$$
$$x = 1500$$

The total amount of cash less the sale of goods equals the amount of tax.

$$1590 - 1500 = 90$$

The amount of tax is \$90.

53. Let $x$ represent the width and $2x$ is the length (twice the width). Use $P = 2W + 2L$ and add one more width to cut the area into two parts. Thus we have the equation:

$$2W + 2L + W = 105$$
$$2x + 2(2x) + x = 105$$
$$7x = 105$$
$$x = 15$$
$$2x = 30$$

The width is 15 meters and the length is 30 meters.

57. Let $x$ be the number of liters drained from the radiator. Make a table.

| Strength | Amount of solution | Amount of pure |
|---|---|---|
| 20% | 20 - x | .20(20 - x) |
| 100% | x | 1x - x |
| 40% | 20 | .40(20) |

From the right hand column,

$$.20(20 - x) + x = .40(20)$$
$$4 - .20x + x = 8$$
$$4 + .80x = 8$$
$$.80x = 4$$
$$x = 5$$

5 liters of pure antifreeze are needed.

61. Let $x$ represent the amount of time Steve travels. David travels 1/2 hour less or $(x - .5)$.

Use $d = rt$: Steve's distance is $50x$.
David's distance is $60(x - .5)$.

The sum of the distances is 80 miles.

$$50x + 60(x - .5) = 80$$
$$50x + 60x - 30 = 80$$
$$110x = 110$$
$$x = 1$$

Steve meets David one hour after Steve leaves.

65. Let x represent the number of heads of lettuce in a crate. The number of heads sold are x - .10x or .90x.

Income - Cost = Profit

↓ ↓ ↓

$.40(.90x) - 10.40 = .20x$

$.36x - .20x = 10.40$

$.16x = 10.40$

$x = 65$

There are 65 heads of lettuce in a crate.

69. $\{x \mid x > 3\}$

See answer graph in textbook.

73. $\{x \mid 3 \le x < 9\}$

See answer graph in textbook.

**Section 2.4 (page 68)**

The answer graphs for Exercises 1-41 may be found in the textbook.

1. $4x > 8$

$\frac{4x}{4} > \frac{8}{4}$ *Divide both sides by 4*

$x > 2$ *Simplify*

Solution set: $(2, +\infty)$

5. $3r + 1 \ge 16$

$3r + 1 - 1 \ge 16 - 1$ *Subtract 1 from each side*

$3r \ge 15$ *Simplify*

$\frac{3r}{3} \ge \frac{15}{3}$ *Divide both sides by 3*

$r \ge 5$ *Simplify*

Solution set: $[5, +\infty)$

9. $-\frac{3}{4}r \ge 21$

$4(-\frac{3}{4}r) \ge 4(21)$ *Multiply both sides by 4*

$-3r \ge 84$

$\frac{-3r}{-3} \le \frac{84}{-3}$ *Divide both sides by -3 and reverse the inequality symbol*

Solution set: $(-\infty, 28]$

13. $-1.3m > 3.9$

$\frac{-1.3m}{-1.3} < \frac{3.9}{-1.3}$ *Divide both sides by -1.3 and reverse inequality symbol*

$m < -3$

Solution set: $(-\infty, -3)$

17. $-r \le -7$

$\frac{-r}{-1} \ge \frac{-7}{-1}$ *Divide both sides by -1 and reverse inequality symbol*

$r \ge 7$

Solution set: $[7, +\infty)$

21. $-3 < x - 5 < 6$

$-3 + 5 < x - 5 + 5 < 6 + 5$

*Add 5 to each of the three parts*

$2 < x < 11$

Solution set: $(2, 11)$

25. $-19 \le 3x - 5 \le -9$

$-14 \le 3x \le -4$ *Add 5 to each part*

$\frac{-14}{3} \le x \le \frac{-4}{3}$ *Divide each part by 3*

Solution set: $[\frac{-14}{3}, \frac{-4}{3}]$

29. $4.2817z \geq -13.27327$

$\frac{4.2817z}{4.2817} \geq \frac{-13.27327}{4.2817}$ *Divide both sides by 4.2817*

$z \geq -3.1$ *Simplify*

Solution set: $[-3.1, +\infty)$

33. $y + 4(2y - 1) \geq 5y$

$y + 8y - 4 \geq 5y$ *Distributive property*

$9y - 5y \geq 4$

$4y \geq 4$

$\frac{4y}{4} \geq \frac{4}{4}$ *Divide by 4*

$y \geq 1$

Solution set: $[1, +\infty)$

37. $-3(z - 6) > 2z - 5$

$-3z + 18 > 2z - 5$ *Distributive property*

$-3z - 2z > -5 - 18$

$-5z > -23$

$\frac{-5z}{-5} < \frac{-23}{-5}$ *Divide by -5 and reverse inequality symbol*

$z < \frac{23}{5}$

Solution set: $(-\infty, \frac{23}{5})$

41. $-\frac{1}{4}(p + 6) + \frac{3}{2}(2p - 5) < 0$

$(4) - \frac{1}{4}(p + 6) + (4)\frac{3}{2}(2p - 5) < (4)0$

*Multiply each term by 4*

$-(p + 6) + 6(2p - 5) < 0$

$-p - 6 + 12p - 30 < 0$

*Distributive property*

$11p < 36$

$\frac{11p}{11} < \frac{36}{11}$

*Divide by 11*

$p < \frac{36}{11}$

Solution set: $(-\infty, \frac{36}{11})$

45. Let x represent her score on the final. The total points needed for a B is 82% of 1100 = (.82)(1100) or 902 points. Given that she has earned 815 points so far, we have the following inequality.

| total points so far | plus | score on final | should be at least | the lowest number of points for a B |
|---|---|---|---|---|
| ↓ | ↓ | ↓ | ↓ | ↓ |
| 815 | + | x | $\geq$ | 902 |

Solve this inequality.

$815 + x \geq 902$

$x \geq 87$

She should earn 87 points or higher on the final for a B.

49. $C = 20x + 100$; $R = 24x$

To make a profit we need $R > C$.

$24x > 20x + 100$ *Substitute for R and C*

$4x > 100$

$x > 25$

26 units is the smallest number that must be sold since more than 25 units must be sold to produce a profit.

53. $5 > 2y + 4 > 12$

Since 5 is not greater than 12, this inequality does not make sense.

57. Six times a number is between -12 and 12. Let x represent the number. -12 is less than 12 and 6x is between these numbers.

$-12 < 6x < 12$

$\frac{-12}{6} < \frac{6x}{6} < \frac{12}{6}$ *Divide by 6*

$-2 < x < 2$

The unknown numbers are all the numbers between -2 and 2, that is, $(-2, 2)$.

61.

| One third of a number | is added to 6 | giving a result of at least 3 | |
|---|---|---|---|
| ↓ | ↑ | ↓ | ↓ |
| $\frac{1}{3}x$ | $+6$ | $\geq$ | 3 |

$$\frac{1}{3}x + 6 \geq 3$$

$$3(\frac{1}{3}x) + 3(6) \geq 3(3)$$ *Multiply by 3*

$$x + 18 \geq 9$$

$$x \geq -9$$

The numbers are all the numbers greater than or equal to -9, that is, $[-9, +\infty)$

65. $|-3| + |-2| = 3 + 2 = 5$

69. $(-5) - (-9) + |-7| = -5 + 9 + 7$
$= 11$

**Section 2.5 (page 73)**

1. $|x| = 8$ gives the two equations $x = 8$ and $x = -8$. The solution set is $\{8, -8\}$.

5. $|q| = -5$
Absolute value can never be negative so there are no solutions.
Solution set: $\emptyset$

9. $|t| - 5 = 8$
$|t| - 5 + 5 = 8 + 5$ *Add 5 on each side*
$|t| = 13$
$t = 13$ or $t = -13$
Solution set: $\{-13, 13\}$

13. $|x + 2| = 6$
We have two equations:
$x + 2 = 6$ or $x + 2 = -6$
$x = 4$ or $x = -8$
Solution set: $\{-8, 4\}$

17. $|2x + 1| = 9$
$2x + 1 = 9$ or $2x + 1 = -9$
$2x = 8$ $\quad 2x = -10$
$x = 4$ or $x = -5$
*Solve each equation*
Solution set: $\{-5, 4\}$

21. $|2y + 5| = 12$
$2y + 5 = 12$ or $2y + 5 = -12$
$2y = 7$ $\quad 2y = -17$
$y = \frac{7}{2}$ or $y = -\frac{17}{2}$
Solution set: $\{\frac{7}{2}, -\frac{17}{2}\}$

25. $\left|\frac{1}{2}x + 3\right| = 4$
$\frac{1}{2}x + 3 = 4$ or $\frac{1}{2}x + 3 = -4$
$\frac{1}{2}x = 1$ $\quad \frac{1}{2}x = -7$
$x = 2$ or $x = -14$
*Solve each equation*
Solution set: $\{-14, 2\}$.

29. $\left|\frac{1}{8}p + 6\right| = -2$
An absolute value is never negative. So there is no solution for p. The solution set is $\emptyset$.

33. $|x + 2| = |2x + 2|$
$x + 2 = 2x + 2$ or $x + 2 = -(2x + 2)$
$-x = 0$ $\quad x + 2 = -2x - 2$
$x = 0$ or $3x = 4$
$x = -\frac{4}{3}$
Solution set: $\{-\frac{4}{3}, 0\}$

37. $\left|m - \frac{1}{2}\right| = \left|\frac{1}{2}m - 2\right|$
$m - \frac{1}{2} = \frac{1}{2}m - 2$
$\frac{1}{2}m = -\frac{3}{2}$
$m = -3$
or

$$m - \frac{1}{2} = -(\frac{1}{2}m - 2)$$
$$m - \frac{1}{2} = \frac{1}{2}m + 2$$
$$\frac{3}{2} = \frac{5}{2}$$
$$m = \frac{5}{3}$$

Solution set: $\{-3, \frac{5}{3}\}$

41. $|2p - 6| = |2p + 5|$

$2p - 6 = 2p + 5$ or $-(2p - 6) = 2p + 5$

$-6 = 5$     $-4p = -1$

*False*     $p = \frac{1}{4}$

The first equation has no solution since $-6 = 5$ is a false statement.

Therefore, the solution set is $\{\frac{1}{4}\}$.

45. Let x be the number. The problem translates to the following equation:

The absolute value of 4 added to a number is 8.

$|4 + x| = 8$

Solve the equation.

$|4 + 8| = 8$

$4 + x = 8$ or $4 + x = -8$

$x = 4$ or $x = -12$

The number is 4 or -12.

49. $|x - 3| = |3 - x|$

Since $|3 - x| = |(-1)(x - 3)| = |x - 3|$ is true for all real numbers x, the solution set is all the real numbers.

53. $|2x + 1| = 3x - 1$

$2x + 1 = 3x - 1$ or $2x + 1 = -(3x - 1)$

$-x = -2$     $2x + 1 = -3x + 1$

$x = 2$ or $5x = 0$

$x = 0$

By substituting $x = 0$ into the equation, we have $|2\cdot 0 + 1| = 3\cdot 0 - 1$ or $|1| = -1$. Zero cannot be a solution because absolute value is never negative.

Solution set: $\{2\}$

57. $x + |x| = 2$

If $x > 0$, then $|x| = x$.

$x + |x| = 2$

$x + x = 2$    *Replace $|x|$ with $x$*

$2x = 2$

$x = 1$

If $x < 0$, then $|x| = -x$.

$x + |x| = 2$    *Replace $|x|$ with $-x$*

$x - x = 2$

$0 = 2$    *False*

There are no $x < 0$ which satisfy $x + |x| = 12$.

Solution set: $\{1\}$

61. $\{x|x \geq -2\}$

See the answer graph in your textbook.

65. $x \leq 2$, $x \geq -1$ is satisfied by all values of x that are both greater than or equal to -1 and less than or equal to 2.

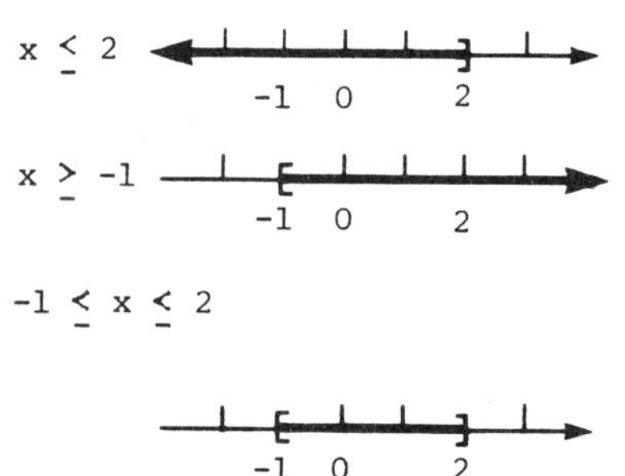


Solution set: $[-1, 2]$

**Section 2.6 (page 78)**

1. The intersection of A and B is the set of elements that are common to A and B.
$A \cap B = \{a, b, c, d, e, f\} \cap \{a, c, e\}$
$= \{a, c, e\}$

5. The elements common to both sets B and C is a.
$B \cap C = \{a\}$

9. The union of B and C is the set of all elements in either B or C or both.
$B \cup C = \{a, c, e\} \cup \{a, f\}$
$= \{a, c, e, f\}$

13. $B \cup \emptyset = B$ since $\emptyset$ contains no elements. So $B \cup \emptyset = \{a, c, e\}$.

17. $B \cup (C \cup D) = \{a, c, e\} \cup \{a, d, f\}$
$= \{a, c, d, e, f\}$

21. $x > 3$ and $x < 7$ are all real numbers greater than 3 and less than 7. So the solution set is the intersection of $(3, +\infty)$ and $(-\infty, 7)$.

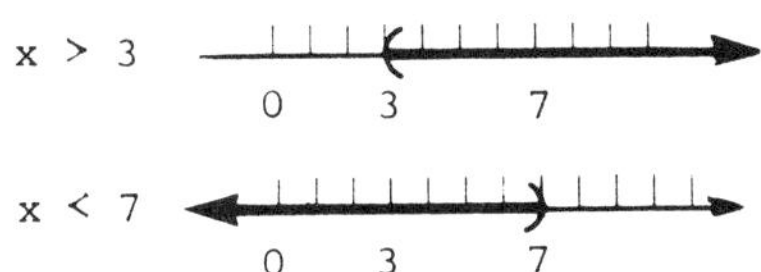


x > 3 and x < 7

Solution set: $(3, 7)$

25. $x \leq -8$ or $x \leq -12$
$x \leq -8$ includes $x \leq -12$.

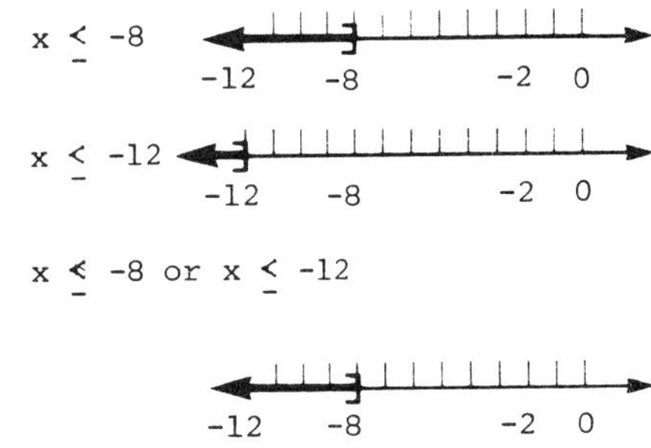


Solution set: $(-\infty, -8]$

29. $x \geq 3$ or $x \leq 5$
The set of points where x is greater than or equal to 3 or less than or equal to 5 comprises the entire number line. Thus, x is any real number.

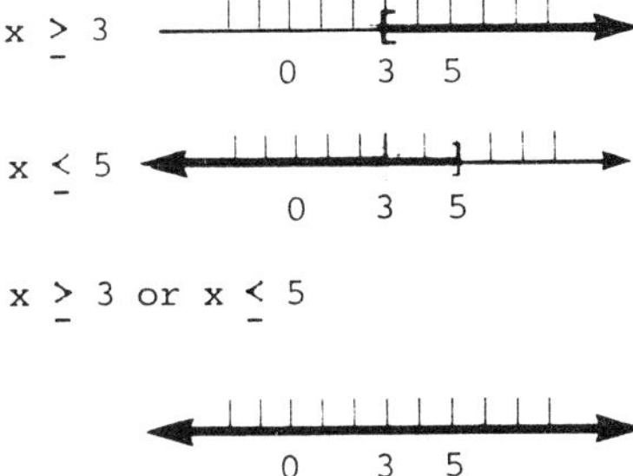


Solution set: $(-\infty, +\infty)$

33. $3x < -3$ and $x + 2 > 0$
First, solve each inequality.
$3x < -3$ $\quad$ $x + 2 > 0$
$x < -1$ $\quad$ $x > -2$
So, we wish to find the set where
$x < -1$ and $x > -2$.
These sets intersect or overlap where x is between -2 and -1.

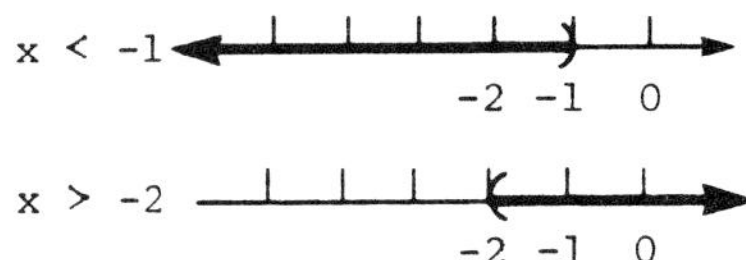


x < -1 and x > -2

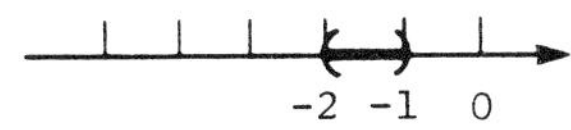


Solution set: $(-2, -1)$

37. $6x - 8 \le 16$ and $4x - 1 \le 15$

$6x \le 24$ and $4x \le 16$

$x \le 4$ and $x \le 4$

These inequalities have the same solution set and the same graph.

Solution set: $(-\infty, 4]$

41. Solve each inequality.

$2x + 3 > 7$ or $4x - 1 < 3$

$2x > 4$ $4x < 4$

$x > 2$ or $x < 1$

We wish to find the set where x is greater than 2 or x is less than 1.

$x > 2$

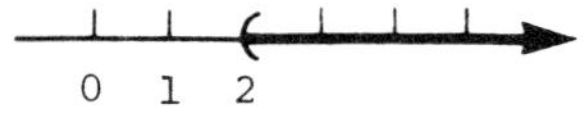


$x < 1$

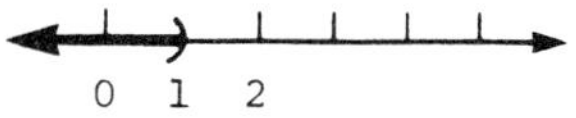


$x > 2$ or $x < 1$

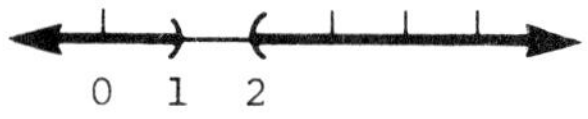


Solution set: $(-\infty, 1) \cup (2, +\infty)$

45. Solve each inequality.

$-5x + 2 \le 17$ or $2x + 1 \le 9$

$-5x \le 15$ $2x \le 8$

$x \ge -3$ or $x \le 4$

We wish to find the set where x is greater than or equal to -3 or x is less than or equal to 4. This is the entire number line.

$x \ge -3$

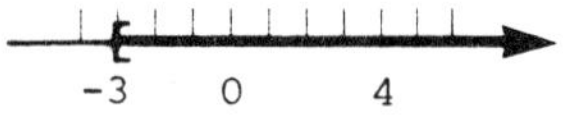


$x \le 4$

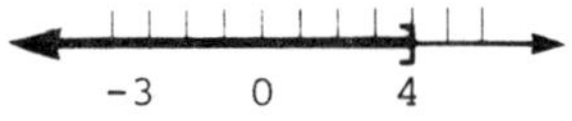


$x \ge -3$ or $x \le 4$

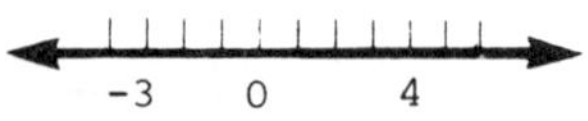


Solution set: $(-\infty, +\infty)$

49. $[4, +\infty) \cap (-\infty, 12]$

x is greater than or equal to 4 <u>and</u> x is less than or equal to 12. The set of real numbers between (and including) 4 and 12 make up the solution.

Solution set: $[4, 12]$

53. $(-\infty, 1] \cup [4, +\infty)$

This set includes the real numbers less than or equal to 1 <u>or</u> the real numbers greater than or equal to 4. There are no real numbers common to both original sets.

Solution set: $(-\infty, 1] \cup [4, +\infty)$

57. $(-1, 4) \cap (2, 7)$

This set includes the real numbers between -1 and 4 and between 2 and 7. The real numbers that are common to both original sets (their intersection) are between 2 and 4.

Solution set: $(2, 4)$

61. $X \cap Y = Y \cap X$

Always true because the elements that are common to both sets are the same. Intersection is commutative.

65. $(X \cap Y) \cup Z = X \cap (Y \cup Z)$

Not always true

Let $x = \{1\}$, $Y = \{0\}$, and $Z = \{2\}$. Then $(X \cap Y) \cup Z = \{2\}$ and $X \cap (Y \cup Z) = \emptyset$.

69. $B \cup B = B$ is always true.

73. $|x| = 5$

$x = 5$ or $x = -5$

Solution set: $\{5, -5\}$

77. $|2r - 3| = 11$

$2r - 3 = 11$ or $-(2r - 3) = 11$

$2r = 14$ $\quad -2r + 3 = 11$

$r = 7$ $\quad -2r = 8$

$r = -4$

Solution set: $\{-4, 7\}$

**Section 2.7 (page 84)**

1. $|k| < 7$

$-7 < k < 7$

Solution set: $(-7, 7)$

5. $|z + 1| \leq 4$

$-4 \leq z + 1 \leq 4$ *Change the absolute value to an equivalent inequality*

$-5 \leq z \leq 3$ *Subtract 1 from both parts of the inequality*

Solution set: $[-5, 3]$

9. $|2z - 5| < 10$

$-10 < 2z - 5 < 10$

$-5 < 2z < 15$ *Add 5*

$-\frac{5}{2} < \frac{2z}{2} < \frac{15}{2}$ *Divide by 2*

$-\frac{5}{2} < z < \frac{15}{2}$

Solution set: $(-\frac{5}{2}, \frac{15}{2})$

13. $|2k - 1| \leq -6$

There are no numbers whose absolute value is less than -6.

Solution set: $\emptyset$

17. $|.0385z - .1745| \leq .172$

$-.172 \leq .0385z - .1745 \leq .172$

$.0025 \leq .0385z \leq .3465$

*Add .1745 to each part*

$.064935 \leq z \leq 9$ *Divide by .0385*

Solution set: $[.06, 9]$

*(.06 is rounded.)*

21. $|p| \geq 12$

$p < -12$ or $p > 12$

Solution set: $(-\infty, -12) \cup (12, +\infty)$

25. $|3a - 2| > 8$

$3a - 2 < -8$ or $3a - 2 > 8$

$3a < -6$ $\quad 3a > 10$

$a < -2$ $\quad a > \frac{10}{3}$

Solution set: $(-\infty, -2) \cup (\frac{10}{3}, +\infty)$

29. $|2.75y + 1.62| \geq 12.62$

$2.75y + 1.62 \leq -12.62$

$2.75y \leq -14.24$

$y \leq -5.2$ *(rounded)*

or

$2.75y + 1.62 \geq 12.62$

$2.75y \geq 11$

$y \geq 4$

Solution set: $(-\infty, -5.2] \cup [4, +\infty)$

*(5.2 is rounded.)*

33. $|s - 12| \geq 6$ is equivalent to $s - 12 \leq -6$ or $s - 12 \geq 6$.

$s - 12 \leq -6$ or $s - 12 \geq 6$

$s \leq 6$ $\quad s \geq 18$

Solution set: $(-\infty, 6] \cup [18, +\infty)$

37. $|5t + 2| > 8$

$5t + 2 < 8$ or $5t + 2 > 8$

$5t < -10$ $\quad 5t > 6$

$t < -2$ $\quad t > \frac{6}{5}$

Solution set: $(-\infty, -2) \cup (\frac{6}{5}, +\infty)$

41. $|4n - 3| \geq -5$ is true for all numbers since absolute value is never negative.

Solution set: $(-\infty, +\infty)$

45. Let x be the number.

$|x| < 10$

$-10 < x < 10.$

Solution set: $(-10, 10)$

49. Let x be the number.

$|x + 4| - 6 \geq -3$

$|x + 4| \geq 3$

$x + 4 \leq -3$ or $x + 4 \geq 3$

$x \leq -7$ $\quad$ $x \geq -1$

Solution set: $(-\infty, -7] \cup [-1, +\infty)$

53. k is at least 4 units from 3, or k's distance from 3 is at least 4.

$|k - 3| \geq 4$

57. $|x - 4| < 2x$ where x must not be negative since absolute value is never negative.

$-2x < x - 4 < 2x$

Rewrite as two inequalities:

$-2x < x - 4$ <u>and</u> $x - 4 < 2x.$

Solve each inequality separately.

$-2x - x < x - 4 - x$

and $x - 4 - x < 2x - x$

$-3x < -4$ and $-4 < x$

$\frac{-3x}{-3} > \frac{-4}{-3}$

$x > \frac{4}{3}$ and $x > -4$

For x to be greater than both 4/3 and -4, x must be greater than 4/3.

Solution set: $(4/3, +\infty)$

61. $2^3 = (2)(2)(2) = 8$

65. $(-6)^3 = (-6)(-6)(-6) = -216$

69. $(\frac{2}{3})^3 = \frac{2^3}{3^3} = \frac{8}{27}$

**Supplementary Exercises on Linear Equations and Inequalities (page 85)**

1.
$$4z + 1 = 53$$
$$4z + 1 - 1 = 53 - 1$$
$$4z = 52$$
$$z = 13$$
Solution set: $\{13\}$

5.
$$3p + 7 = 9 - 8p$$
$$3p + 7 + 8p = 9 - 8p + 8p$$
$$11p + 7 = 9$$
$$11p + 7 - 7 = 9 - 7$$
$$11p = 2$$
$$p = \frac{2}{11}$$
Solution set: $\{\frac{2}{11}\}$

9.
$$4(a - 11) + 3a = 2a + 11$$
$$4a - 44 + 3a = 2a + 11$$
$$7a - 44 = 2a + 11$$
$$7a - 44 - 2a = 2a + 11 - 2a$$
$$5a - 44 = 11$$
$$5a - 44 + 44 = 11 + 44$$
$$5a = 55$$
$$a = 11$$
Solution set: $\{11\}$

13. $|x| < 3$

$-3 < x < 3$

Solution set: $(-3, 3)$

See answer graph in textbook.

17. $|r - 2| = |4r + 1|$

$r - 2 = 4r + 1$ or $r - 2 = -(4r + 1)$

$-3 = 3r$ $\quad$ $r - 2 = -4r - 1$

$-1 = r$ $\quad$ $5r = 1$

$r = \frac{1}{5}$

Solution set: $\{-1, \frac{1}{5}\}$

In Exercises 21-49, see answer graph in textbook.

21. $|q| > 1$ is equivalent to
$q < -1$ or $q > 1$
Solution set: $(-\infty, -1) \cup (1, +\infty)$

25. $\frac{1}{4}p < -3$
$4(\frac{1}{4}p) < 4(-3)$
$p < -12$
Solution set: $(-\infty, -12)$

29. $|r - 1| < 5$ is equivalent to
$-5 < r - 1 < 5$ *Add 1 to each part*
$-4 < r < 6$
Solution set: $(-4, 6)$

33. $|2p - 3| > 1$ is equivalent to
$2p - 3 < -1$ or $2p - 3 > 1$
$2p < 2$ $\quad$ $2p > 4$
$p < 1$ or $p > 2$

Solution set: $(-\infty, 1) \cup (2, +\infty)$

37. $5z - (3 + z) \geq 6z + 2$
$5z - 3 - z \geq 6z + 2$
$4z - 3 \geq 6z + 2$
$4z - 3 - 4z \geq 6z + 2 - 4z$
$-3 \geq 2z + 2$
$-3 - 2 \geq 2z + 2 - 2$
$-5 \geq 2z$
$\frac{1}{2}(-5) \geq \frac{1}{2}(2z)$
$-\frac{5}{2} \geq z$ (or $z \leq \frac{-5}{2}$)

Solution set: $(-\infty, -\frac{5}{2}]$

41. $-1 \leq 4a + 3 \leq 5$
$-4 \leq 4a \leq 2$ *Divide by 4*
$-1 \leq a \leq \frac{1}{2}$
Solution set: $[-1, \frac{1}{2}]$

45. $|3r - 1| \geq 4$
$3r - 1 \leq -4$ or $3r - 1 \geq 4$
$3r \leq -3$ $\quad$ $3r \geq 5$
$r \leq -1$ or $r \geq \frac{5}{3}$
Solution set: $(-\infty, -1] \cup [\frac{5}{3}, +\infty)$

49. $-6 \leq \frac{3}{2} - x \leq 2$
$-6 - \frac{3}{2} \leq -x \leq 2 - \frac{3}{2}$
$-\frac{15}{2} \leq -x \leq \frac{1}{2}$ *Multiply by -1; reverse signs*
$\frac{15}{2} \geq x \geq -\frac{1}{2}$
Solution set: $[-\frac{1}{2}, \frac{15}{2}]$

53. $8q - (2 - q) = 3(1 + 3q) - 5$
$8q - 2 + q = 3 + 9q - 5$
$9q - 2 = 9q - 2$
This is true for all q.
Solution set: $(-\infty, +\infty)$

57. $|2 + 5q| < 3$
$-3 < 2 + 5q < 3$
$-5 < 5q < 1$ *Subtract 2*
$-1 < q < \frac{1}{5}$ *Divide by 5*
Solution set: $(-1, \frac{1}{5})$
See the graph in your textbook.

**Chapter 2 Review Exercises (page 88)**

1. $12p - 8p + 3p - 9 = 7p - 9$

5. $-(y + 5) - 7y - (3y - 9) - 3(4 - y) - y$
$= -y - 5 - 7y - 3y + 9 - 12 + 3y - y$
$= -y - 7y - 3y + 3y - y - 5 + 9 - 12$
$= -9y - 8$

9. $-r + 9 = 4r - 7$
$-r + 9 + r = 4r - 7 + r$
$9 = 5r - 7$
$9 + 7 = 5r - 7 + 7$
$16 = 5r$
$\frac{16}{5} = r$
Solution set: $\{\frac{16}{5}\}$

13. $5(k - 2) + 8k = 16$
$5k - 10 + 8k = 16$
$13k - 10 = 16$
$13k - 10 + 10 = 16 + 10$
$13k = 26$
$k = 2$
Solution set: $\{2\}$

17. $-3.94k - 2.814 = 1.62k - 7.262$
$-3.94k - 2.814 + 3.94k = 1.62k - 7.262 + 3.94k$
$-2.814 = 5.56k - 7.262$
$-2.814 + 7.262 = 5.56k - 7.262 + 7.262$
$4.448 = 5.56k$
$.8 = k$
Solution set: $\{.8\}$

21. $\frac{m - 2}{4} + \frac{m + 1}{2} = 1$
$m - 2 + 2(m + 1) = 4$ *Multiply by 4*
$m - 2 + 2m + 2 = 4$
$3m = 4$
$m = \frac{4}{3}$
Solution set: $\{\frac{4}{3}\}$

25. $-2r + 6(r - 1) + 3r - (4 - r) = -(r + 5) - 5$
$-2r + 6r - 6 + 3r - 4 + r = -r - 5 - 5$
*Distributive property*
$9r = 0$
$r = 0$
There is only one element in the solution set so it is a conditional equation.
Solution set: $\{0\}$

29. $M = \frac{1}{4}(x + 2y)$
$4M = x + 2y$ *Multiply by 4*
$4M - 2y = x$

33. Let H be the height of the box. Now $V = LWH$ where $L = 6$, $W = 5$, and $V = 180$, so $H = \frac{V}{LW}$
$= \frac{180}{(6)(5)}$
$= 6.$
The box is 6 feet in height.

37. The formula is $C = \frac{5}{9}(F - 32)$.
$F = 68 \quad C = \frac{5}{9}(68 - 32)$
$= \frac{5}{9}(36)$
$= 20$
The temperature is 20°C.

41. "Half a number, added to 5" is written $\frac{1}{2}x + 5$.

45. Let x be the number.
$2x - 3 = 11$
$2x = 14$
$x = 7$
The number is 7.

49. Let x be the number. Then
$x - .35x = 260$
$1x - .35x = 260$
$.65x = 260$
$x = 400$
The number is 400.

53. Let x represent the money invested at 7%. (42,000 - x) is the money invested at 10%. Use I = prt where t = 1.
$.07x + .10(42{,}000 - x) = 3690$
$.07x + 4200 - .10x = 3690$
$-.03x = -510$
$x = 17{,}000$
She invested $17,000 at 7% and $25,000 at 10%.

57. $3m < 24$
$\frac{1}{3}(3m) < \frac{1}{3}(24)$
$m < 8$
Solution set: $(-\infty, 8)$

61. $-5x - 4 \geq 10$
$-5x - 4 + 4 \geq 10 + 4$
$-5x \geq 14$
$\frac{-5x}{-5} \leq \frac{14}{-5}$ *Reverse direction of inequality*
$x \leq -\frac{14}{5}$
Solution set: $(-\infty, -\frac{14}{5}]$

65. $-(4 + 2m) - 3 \leq 5m + 2$
$-4 - 2m - 3 \leq 5m + 2$
$-7 - 2m \leq 5m + 2$
$-7 - 2m + 2m \leq 5m + 2 + 2m$
$-7 \leq 7m + 2$
$-7 - 2 \leq 7m + 2 - 2$
$-9 \leq 7m$
$-\frac{9}{7} \leq m$ or $m \geq -\frac{9}{7}$
Solution set: $[-\frac{9}{7}, +\infty)$

69. $\frac{9y + 5}{-3} > 2$
$-3(\frac{9y + 5}{-3}) < (-3)(2)$ *Reverse inequality symbol*
$9y + 5 < -6$
$9y + 5 - 5 < -6 - 5$
$9y < -11$
$y < \frac{-11}{9}$
Solution set: $(-\infty, -\frac{11}{9})$

73. $\frac{5}{3}(m - 2) + \frac{2}{5}(m + 1) > 1$
$25(m - 2) + 6(m + 1) > 15$ *Multiply by 15*
$25m - 50 + 6m + 6 > 15$
$31m > 59$
$m > \frac{59}{31}$
Solution set: $(\frac{59}{31}, +\infty)$

77. Let x be the number. One third of a number is between 1 and 4.
$1 < \frac{1}{3}x < 4$
$3 < x < 12$ *Multiply by 3*
The number is between 3 and 12, that is in the interval (3, 12).

81. $|-5 + 2x| = 1$

$-5 + 2x = 1$ or $-5 + 2x = -1$

$2x = 6$ $\qquad$ $2x = 4$

$x = 3$ or $x = 2$

Solution set: $\{2, 3\}$

85. $|4a + 2| + 7 = 3$

$4a + 2 = -4$

Absolute value is never negative.

Solution set: $\emptyset$

89. The absolute value of the sum of a number and 6 is 11.

Let x be the number. Then

$|x + 6| = 11$

$x + 6 = 11$ or $x + 6 = -11$

$x = 5$ or $x = -17$

The number is -17 or 5.

93. $x + 4 > 12$ and $x - 2 < 1$

$x > 8$ and $x < 3$

No solution.

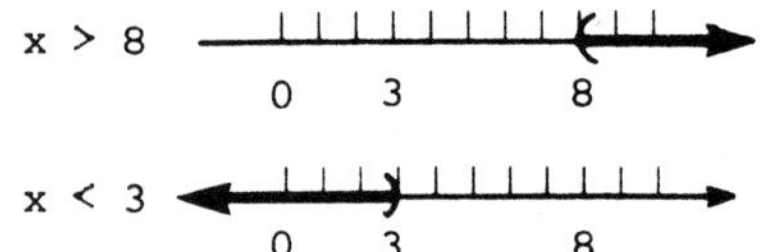


$x > 8$ and $x < 3$

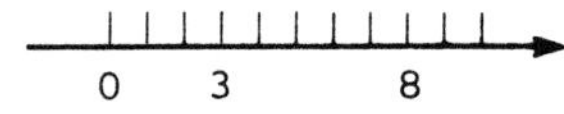


Solution set: $\emptyset$

97. $A \cap B$

$= \{a, b, c, d\} \cap \{a, c, e, f\}$

$= \{a, c\}$

101. $(-3, +\infty) \cap (-\infty, 4)$

$(-3, +\infty)$ includes all real numbers larger than -3.

$(-\infty, 4)$ includes all real numbers less than 4.

The numbers common to both sets are larger than -3 and less than 4:

$-3 < x < 4$

Solution set: $(-3, 4)$

105. $|p| < 4$

$-4 < p < 4$

Solution set: $(-4, 4)$

See answer graph in textbook.

109. $|x + 1| \geq 9$

$x + 1 \leq -9$ or $x + 1 \geq 9$

$x \leq -10$ $\qquad$ $x \geq 8$

Solution set: $(-\infty, 10] \cup [8, +\infty)$

See answer graph in the textbook.

113. Let x be the number. "The absolute value of a number is at least 5" is written

$|x| \geq 5$

$x \leq -5$ or $x \geq 5$.

The number is less than or equal to -5 or greater than or equal to 5, that is, $(-\infty, -5] \cup [5, +\infty)$.

117. $x < 3$ and $x \geq -2$

We need the real numbers that are common to both sets: the numbers greater or equal to -2 and less than 3.

$-2 \leq x < 3$

Solution set: $[-2, 3)$

121. $|5r - 1| > 14$

$5r - 1 > 14$ or $5r - 1 < -14$

$5r - 1 > 14$ $\quad$ $5r - 1 < -14$

$5r > 15$ $\quad$ $5r < -13$

$r > 3$ or $r < -\frac{13}{5}$

Solution set: $(-\infty, -\frac{13}{5}) \cup (3, +\infty)$

125. $\frac{3y}{5} - \frac{y}{2} = 3.$ *The lowest common denominator is 10*

$10(\frac{3y}{5}) - 10(\frac{y}{2}) = 10(3)$

$6y - 5y = 30$

$y = 30$

Solution set: {30}

**Chapter 2 Test (page 92)**

1. $5r - 3 + 2r = 3(r - 2) + 11$

$7r - 3 = 3r - 6 + 11$

$7r - 3 = 3r + 5$

$7r - 3 - 3r = 3r + 5 - 3r$

$4r - 3 = 5$

$4r - 3 + 3 = 5 + 3$

$4r = 8$

$r = 2$

Solution set: {2}

2. $9 - (3 + 4z) - 5z = -4(z - 5) - 9$

$9 - 3 - 4z - 5z = -4z + 20 - 9$

$6 - 9z = -4z + 11$

$6 - 9z + 9z = -4z + 11 + 9z$

$6 = 5z + 11$

$6 - 11 = 5z + 11 - 11$

$-5 = 5z$

$-1 = z$

Solution set: {-1}

3. $\frac{2p - 1}{3} + \frac{p + 1}{4} = \frac{7}{4}$

$12(\frac{2p - 1}{3}) + 12(\frac{p + 1}{4}) = 12(\frac{7}{4})$

$4(2p - 1) + 3(p + 1) = 3(7)$

$8p - 4 + 3p + 3 = 21$

$11p - 1 = 21$

$11p = 22$

$p = 2$

Solution set: {2}

4. $3p - (2 - p) + 4p = 7p - 2(-p)$

$3p - 2 + p + 4p = 7p - 2 + p$

$8p - 2 = 8p - 2$

This equation is an identity; the solution set is the set of all real numbers.

Solution set: $(-\infty, +\infty)$

5. $S = vt - 16t^2$

$S + 16t^2 = vt - 16^2 + 16t^2$

$S + 16t^2 = vt$

$\frac{S + 16t^2}{t} = v$

6. $I = 12$, $p = 300$, $r = 8\%$, $t = ?$

Solve $I = prt$ for t: $\frac{I}{pr} = t.$

$\frac{12}{300(.08)} = t$

$= \frac{12}{24} = \frac{1}{2}$ *The time is $\frac{1}{2}$ of a year or 6 months*

7. Let x be the width of the rectangle. So the length of the rectangle is 2x + 1. Use 2W + 2L = P.

$2(x) + 2(2x + 1) = 56$

$2x + 4x + 2 = 56$

$6x + 2 = 56$

$6x = 54$

$x = 9$

Answer: The width is 9 meters.

8. Let x be the smaller of the two integers. This means the larger integer is x + 72. The sum is 112.

$$x + (x + 72) = 112$$
$$2x + 72 = 112$$
$$2x = 40$$
$$x = 20$$

The integers are 20 and 92.

9. Let x represent the money invested at 8% and 2x the money invested at 10%.

$$.08x + .10(2x) = 1400$$
$$.28x = 1400$$
$$x = 5000$$

\$5000 was invested at 8%.
\$10,000 was invested at 10%.

10. $-\frac{3}{4}r > -6$

$$-\frac{4}{3}(-\frac{3}{4})r < -\frac{4}{3}(-6)$$
$$r < 8$$

Solution set: $(-\infty, 8)$

See the graph in your textbook.

11. $2 - 3(p - 1) < 5p$

$$2 - 3p + 3 < 5p$$
$$5 - 3p < 5p$$
$$5 < 8p$$
$$\frac{5}{8} < p$$

Solution set: $(\frac{5}{8}, +\infty)$

See the graph in your textbook.

12. $6z - (4 + 5z) \leq 8 - 3z$

$$6z - 4 - 5z \leq 8 - 3z$$
$$z - 4 \leq 8 - 3z$$
$$4z \leq 12$$
$$z \leq 3$$

Solution set: $(-\infty, 3]$

See the graph in your textbook.

13. $-1 < \frac{2}{3}a - 2 < 2$

$$1 < \frac{2}{3}a < 4$$
$$\frac{3}{2} < a < 6$$

Solution set: $(\frac{3}{2}, 6)$

See the graph in your textbook.

14. Let x be John's weight. Then $x + 40 \leq 182$ and so $x \leq 142$. John doesn't weigh more than 142 pounds.

15. Let x be the student's grade on the fourth test as a percent. The average of the four exams equals

$$\frac{83 + 76 + 79 + x}{4} = \frac{238 + x}{4}$$

To get a B the student must have

$$\frac{238 + x}{4} \geq 80.$$
$$238 + x \geq 320$$
$$x \geq 82$$

The minimum grade is 82%.

16. $|x - 3| = 3$

$$x - 3 = 3 \quad \text{or} \quad x - 3 = -3$$
$$x = 6 \quad \text{or} \quad x = 0$$

Solution set: $\{0, 6\}$

17. $|2k - 3| = 7$

$$2k - 3 = 7 \quad \text{or} \quad 2k - 3 = -7$$
$$k = 5 \quad \text{or} \quad 2k = -4$$
$$k = -2$$

Solution set: $\{-2, 5\}$

18. $|m - 1| = 2m + 5$

$$m - 1 = 2m + 5 \quad \text{or} \quad m - 1 = -(2m + 5)$$
$$-6 = m \quad \text{or} \quad m - 1 = -2m - 5$$
$$3m = -4$$
$$m = -\frac{4}{3}$$

Solution set: $\{-6, -\frac{4}{3}\}$

For Exercises 19-25, see answer graphs in textbook.

19. $3x - 2 < 10$ and $-2x < 10$

$3x < 12$ $\qquad -\frac{1}{2}(-2x) > -\frac{1}{2}(10)$

$x < 4$ $\qquad x > -5$

Solution set: $(-5, 4)$

20. $-4x \leq -20$ or $4x - 2 < 10$

$-\frac{1}{4}(-4x) \geq -\frac{1}{4}(-20)$ $\qquad 4x < 12$

$x \geq 5$ or $x < 3$

Solution set: $(-\infty, 3) \cup [5, +\infty)$

21. $[-2, +\infty) \cap (-\infty, 3$

$x \geq -2$ and $x < 3$

$-2 \leq x < 3$

Solution set: $[-2, 3)$

22. (a) $A \cap B$

$= \{1, 2, 5, 7\} \cap \{1, 5, 9, 12\}$

$= \{1, 5\}$

(b) $A \cup B$

$= \{1, 2, 5, 7\} \cup \{1, 5, 9, 12\}$

$= \{1, 2, 5, 7, 9, 12\}$

23. $|2k + 3| \leq 11$

$-11 \leq 2k + 3 \leq 11$

$-14 \leq 2k \leq 8$

$-7 \leq k \leq 4$

Solution set: $[-7, 4]$

24. $|y + 5| > 2$ is equivalent to

$y + 5 < -2$ or $y + 5 > 2.$

$y + 5 < -2$ $\qquad y + 5 > -3$

$y < -7$ or $y > -3$

Solution set: $(-\infty, -7) \cup (-3, +\infty)$

25. $3r + 2 \geq 6$ is equivalent to

$3r + 2 \leq -6$ or $3r + 2 \geq 6$

$3r + 2 \leq -6$ or $3r + 2 \geq 6$

$3r \leq -8$ $\qquad 3r \geq 4$

$r \leq -\frac{8}{3}$ or $\qquad r \geq \frac{4}{3}$

Solution set: $(-\infty, -\frac{8}{3}] \cup [\frac{4}{3}, +\infty)$

## CHAPTER 3 EXPONENTS AND POLYNOMIALS

### Section 3.1 (page 101)

1. In $5^7$, the exponent is 7 and the base is 5.

5. In $(-9)^4$, the exponent is 4 and, since -9 is in parentheses, (-9) is the base.

9. In $p^{-7}$, the exponent is -7 and the base is p.

13. $-3q^{-4}$ has exponent -4 and since -3q is not in parentheses, the exponent affects only q. Thus, the base is q.

17. $5^4 = 5\cdot 5\cdot 5\cdot 5 = 625$

21. $(-2)^5 = (-2)(-2)(-2)(-2)(-2) = -32$

25. $-(-3)^4 = -[(-3)(-3)(-3)(-3)] = -[81] = -81$

29. $\frac{1}{5^{-2}} = \frac{1}{\frac{1}{5^2}} = \frac{1}{\frac{1}{25}} = 25$

33. $\frac{2^{-3}}{3^{-2}} = \frac{3^2}{2^3} = \frac{3\cdot 3}{2\cdot 2\cdot 2} = \frac{9}{8}$

37. $(\frac{2}{3})^{-2} = \frac{1}{(\frac{2}{3})^2} = \frac{1}{\frac{2^2}{3^2}} = \frac{1}{\frac{4}{9}} = \frac{9}{4}$

41. $(4.6175)^{-2} = \frac{1}{(4.6175)^2} = \frac{1}{(4.6175)(4.6175)} = \frac{1}{21.3211306}$
$= .046901$ *(rounded)*

45. $2^6\cdot 2^{10} = 2^{6+10}$ *Product rule*
$= 2^{16}$ *Simplify*

49. $\frac{3^5}{3^2} = 3^{5-2}$ *Quotient rule*
$= 3^3 = 27$

53. $\frac{3^{-5}}{3^{-2}} = 3^{-5-(-2)}$ *Quotient rule*
$= 3^{-5+2} = 3^{-3} = \frac{1}{3^3} = \frac{1}{27}$

57. $t^5t^{-12} = t^{5+(-12)}$ *Product rule*
$= t^{-7} = \frac{1}{t^7}$

61. $a^{-3}a^2a^{-4} = a^{-3+2+(-4)}$ *Product rule*
$= a^{-5}$
$= \frac{1}{a^5}$

65. $\frac{r^3r^{-4}}{r^{-2}r^{-5}} = \frac{r^{3+(-4)}}{r^{-2+(-5)}}$ *Product rule*
$= \frac{r^{-1}}{r^{-7}}$
$= r^{-1-(-7)}$
$= r^{-1+7} = r^6$ *Quotient rule*

69. $-4(2x^3)(3xy^2)$
$= (-4\cdot 2\cdot 3)(x^3\cdot x^1)(y^2)$
$= -24x^4y^2$

73. $0.02 = 0.02_\wedge$ *Caret to right of 2, count from right to left 2 places to decimal*
$0.02 = 2 \times 10^{-2}$ *Right to left, exponent negative*

77. $93{,}000{,}000 = 9_{\wedge}3{,}000{,}000$ *Caret to 7 places right of the 9. Count from caret left to right to decimal point*

$= 9.3 \times 10^7$

81. $1.52 \times 10^{-2}$

$= .0152$ *Move decimal point 2 places to the left for a negative exponent*

85. $4.69 \times 10^{-2}$

$= .0469$ *Move the decimal point 2 places to the left*

89. $a^r a^{r+1} a^{r+2}$

$= a^{r+r+1+r+2}$

$= a^{3r+3}$

93. $(x + y)^{-1} = x^{-1} + y^{-1}$

$(2 + 3)^{-1} = 2^{-1} + 3^{-1}$ *Substitute 2 for x, 3 for y*

$5^{-1} = \frac{1}{2} + \frac{1}{3}$

$\frac{1}{5} \neq \frac{5}{6}$

The statement is false.

97. First find how many seconds in a year: $365 \times 24 \times 60 \times 60$

$= 31{,}536{,}000$ seconds in a year.

If light travels $1.86 \times 10^5$ miles per second, then there are

$(1.86 \times 10^5)(3.1536 \times 10^7)$

$= 5.8657 \times 10^{12}$

miles in a light year (rounded).

101. $\frac{9^{-4}}{9^{-4}} = 9^{-4 - (-4)} = 9^{-4 + 4}$

$= 9^0 = 1$

**Section 3.2 (page 106)**

1. $(4^2)^5 = 4^{2 \cdot 5}$ *Power rule*

$= 4^{10}$ *Simplify*

5. $(6^{-3})^{-2} = 6^{(-3)(-2)}$ *Power rule*

$= 6^6$ *Simplify*

9. $(-23)^0 = 1$ *Zero exponent*

13. $3^0 - 4^0 = 1 - 1 = 0$

17. $(\frac{3}{4})^{-2} = (\frac{4}{3})^2$ *4/3 is reciprocal of 3/4*

$= \frac{4^2}{3^2}$

21. $(3^{-2} \cdot 4^{-1})^3 = 3^{(-2)(3)} \; 4^{(-1)(3)}$ *Power rule*

$= 3^{-6} \cdot 4^{-3}$

$= \frac{1}{3^6} \cdot \frac{1}{4^3}$ *Negative exponents*

$= \frac{1}{3^6 \cdot 4^3}$

25. $(z^3)^{-2} z^2 = z^{(3)(-2)} z^2$ *Power rule*

$= z^{-6} z^2$

$= z^{-6 + 2}$ *Product rule*

$= z^{-4}$

$= \frac{1}{z^4}$ *Negative exponent*

29. $(3a^{-2})^3 (a^3)^{-4} = 27a^{-6} a^{-12}$ *Power rule*

$= 27a^{-18} = \frac{27}{a^{18}}$ *Negative exponet*

33. $(2^1p^2q^{-3})^2(4^1p^{-3}q^1)^2$

$= 2^{(1)(2)}p^{(2)(2)}q^{(-3)(2)} \cdot 4^{(1)(2)}p^{(-3)(2)}q^{(1)(2)}$ *Power rule*

$= 2^2p^4q^{-6} \cdot 4^2p^{-6}q^2$

$= (2^2 \cdot 4^2)(p^4 \cdot p^{-6})(q^{-6}q^2)$

$= (4 \cdot 16)(p^{4+(-6)})(q^{-6+2})$ *Product rule*

$= 64p^{-2}q^{-4}$

$= 64 \cdot \frac{1}{p^2} \cdot \frac{1}{q^4}$ *Negative exponents*

$= \frac{64}{p^2q^4}$

37. $\frac{4a^5(a^{-1})^3}{(a^{-2})^{-2}} = \frac{4a^5 \cdot a^{(-1)(3)}}{a^{(-2)(-2)}}$ *Power rule*

$= \frac{4a^5 \cdot a^{-3}}{a^4}$

$= \frac{4a^{5+(-3)}}{a^4}$ *Product rule*

$= \frac{4a^2}{a^4}$

$= 4a^{2-4}$ *Quotient rule*

$= 4a^{-2}$

$= \frac{4}{a^2}$

41. $\frac{(2k)^2m^{-5}}{(km)^{-3}} = \frac{2^2k^2m^{-5}}{k^{-3}m^{-3}}$ *Power rule*

$= 4k^{2-(-3)}m^{-5-(-3)}$ *Quotient rule*

$= 4k^5m^{-2}$

$= \frac{4k^5}{m^2}$

45. $\frac{(2k)^2k^3}{k^{-1}k^{-5}}(5k^{-2})^{-3} = \frac{2^2k^2k^3}{k^{-1}k^{-5}} \cdot \frac{5^{-3}k^6}{1}$ *Power rule*

$= \frac{2^2k^{2+3+6}}{5^3k^{(-1)+(-5)}}$ *Product rule*

$= \frac{4k^{11}}{125k^{-6}}$

$= \frac{4k^{11-(-6)}}{125}$ *Quotient rule*

$= \frac{4k^{17}}{125}$

49. $(\frac{2p}{q^2})^3(\frac{3p^4}{q^{-4}})^{-1} = \frac{2^3p^3}{q^6} \cdot \frac{3^{-1}p^{-4}}{q^4}$ *Power rule*

$= \frac{2^3p^{-1}}{3^1q^{10}}$ *Product rule*

$= \frac{8}{3pq^{10}}$

53. $\frac{9 \times 10^2}{3 \times 10^6} = \frac{9}{3} \times 10^{2-6} = 3 \times 10^{-4} = .0003$

57. $\frac{.002 \times 3900}{.000013} = \frac{(2 \times 10^{-3}) \times (3.9 \times 10^3)}{1.3 \times 10^{-5}}$

$= \frac{7.8 \times 10^0}{1.3 \times 10^{-5}} = 6 \times 10^5$

$= 600{,}000$

61. $\frac{840{,}000 \times .03}{0.00021 \times 600} = \frac{(8.4 \times 10^5) \times (3 \times 10^{-2})}{(2.1 \times 10^{-4}) \times (6 \times 10^2)}$

$= \frac{(8.4)(3)}{(2.1)(6)} \times \frac{10^{5+(-2)}}{10^{-4+2}}$

$= 2 \times \frac{10^3}{10^{-2}}$

$= 2 \times 10^5$

$= 200{,}000$

65. $\frac{6863 \times 142{,}000}{.00965 \times .0843} =$

$$\frac{6.863 \times 10^3 \times 1.42 \times 10^5}{9.65 \times 10^{-3} \times 8.43 \times 10^{-2}}$$

$$= \left(\frac{6.863 \times 1.42}{9.65 \times 8.43}\right) \times \left(\frac{10^3 \times 10^5}{10^{-3} \times 10^{-2}}\right)$$

$$= .1198 \times \frac{10^8}{10^{-5}} \text{ (rounded)}$$

$$= .1198 \times 10^{13}$$

$$= 1.198 \times 10^{12}$$

$$= 1{,}198{,}000{,}000{,}000$$

69. To find the number of bacteria that would fit into the earth, divide the volume of earth by the volume of a bacterium.

$$\frac{5 \times 10^{14}}{2.5 \times 10^{-16}} = 2 \times 10^{30}$$

The earth could contain $2 \times 10^{30}$ bacteria.

73. $\frac{(-3y^3x^3)(-4y^4x^2)(x^2)^{-4}}{18x^3y^2(y^3)^3(x^3)^{-2}}$

$$= \frac{(-3y^3x^3)(-4y^4x^2)(x^{-8})}{18x^3y^2y^9x^{-6}} \quad \textit{Power rule}$$

$$= \frac{12x^{3+2-8}y^{3+4}}{18x^{3-6}y^{2+9}} \quad \textit{Product rule}$$

$$= \frac{2x^{-3}y^7}{3x^{-3}y^{11}}$$

$$= \frac{2x^{-3-(-3)}}{3y^{11-7}}$$

$$= \frac{2}{3y^4}$$

77. $x^{-m} \cdot x^{3+4m}(x^{-2})^m$

$$= x^{-m} \cdot x^{3+4m} \cdot x^{-2m} \quad \textit{Power rule}$$

$$= x^{-m+3+4m-2m} \quad \textit{Product rule}$$

$$= x^{3+m}$$

81. $\frac{(a^{4+k}a^{3k})^{-1}}{(a^2)^{-k}}$

$$= \frac{(a^{4+4k})^{-1}}{a^{-2k}} \quad \textit{Product rule inside parentheses}$$

$$= \frac{a^{-1(4+4k)}}{a^{-2k}} \quad \textit{Power rule}$$

$$= \frac{a^{-4-4k}}{a^{-2k}}$$

$$= a^{-4-4k+2k} \quad \textit{Quotient rule}$$

$$= a^{-4-2k}$$

85. $9x + 5x - x = 13x$

89. $7x - (5 + 5x)$

$$= 7x - 5 - 5x \quad \textit{Distributive property}$$

$$= 2x - 5$$

**Section 3.3 (page 113)**

1. The numerical factor in $9k$ is 9, so the coefficient is 9. The exponent of $k$ or $k^1$ is 1, so the degree is 1.

5. The numerical factor in $\pi k^3$ is $\pi$, so the coefficient is $\pi$. The exponent of the variable in $\pi k^3$ is 3, so the degree is 3.

9. Since $-y^8 = (-1)y^8$, the coefficient is -1 and the degree is 8.

13. $8x + 9x^2 + 3x^3 = 3x^3 + 9x^2 + 8x$

*The largest exponent is 3 and the smallest is 1 since $8x = 8x^1$*

17. $13y^3 - 9y^4 + 8y^2 - 7y + 2 =$

$-9y^4 + 13y^3 + 8y^2 - 7y + 2$

*The largest exponent is 4 and the smallest is 0 since $2 = 2x^0$*

21. $-8a^2 - 9a + 2$ has three terms, so it is a trinomial.

*The largest exponent is 2 so it has the degree 2*

25. -2 has one term of degree 0 since $2 = 2x^0$. Thus, it is a monomial of degree 0.

29. The largest exponent in $12k^5 - 9k^2$ is 5. Thus, the degree of the polynomial is 5.

33. 8 is the coefficient of some variable to the power zero. For example, $8 = 8x^0$. Thus, the degree of 8 is zero.

37.
$$\begin{array}{r} 21p - 8 \\ \underline{-9p + 4} \\ 12p - 4 \end{array}$$
*Add vertically*

41.
$$\begin{array}{r} -6y^3 + 8y + 5 \\ \underline{9y^3 + 4y - 6} \\ 3y^3 + 12y - 1 \end{array}$$
*Add vertically*

45.
$$\begin{array}{rrrr} 6y^3 & - 9y^2 & & + 8 \\ 4y^3 & + 2y^2 & + 5y & \\ \hline 10y^3 & - 7y^2 & + 5y & + 8 \end{array}$$

*Add vertically Make sure you have terms of the same degree in each vertical column*

49. $6p^2 - 11p + 8p^2 - 21p$

$= 6p^2 + 8p^2 - 11p - 21p$

*Group terms of same degree*

$= (6 + 8)p^2 - (11 + 21)p$

*Distributive property*

$= 14p^2 - 32p$

53. $6z^2 - 12z + 8z^3 - 5z^2 - 9$

$= 6z^2 - 5z^2 - 12z + 8z^3 - 9$

$= (6 - 5)z^2 - 12z + 8z^3 - 9$

*Distributive property*

$= z^2 - 12z + 8z^3 - 9$

$= 8z^3 + z^2 - 12z - 9$

*Write in descending powers*

57. $[(4x - 8) - (-1 + x)] - (11x + 5)$

$= 4x - 8 + 1 - x - 11x - 5$

*Work inside innermost grouping symbols outward*

$= (4 - 1 - 11)x + (-8 + 1 - 5)$

*Distributive property*

$= -8x - 12$

61. $(y^2 + 3y) + [2y - (5y^2 + 3y + 4)]$

$= y^2 + 3y + 2y - 5y^2 - 3y - 4$

*Work inside grouping symbols*

$= (1 - 5)y^2 = (3 + 2 - 3)y - 4$

$= -4y^2 + 2y - 4$

65. $(8a + 3a^2) - (2a - a^2) - (4a + a^2)$

$= 8a + 3a^2 - 2a + a^2 - 4a - a^2$

$= (8 - 2 - 4)a + (3 + 1 - 1)a^2$

$= 2a + 3a^2$

$= 3a^2 + 2a$ *In descending powers*

69. $P(x) = x - 10$

(a) $P(-1 = -1 - 10$ *Substitute -1 for x*

(b) $P(2) = 2 - 10$ *Substitute 2 for x*

$= -8$

73. $P(x) = 2x^4 + 3x^2 - 5x$

(a) $P(-1) = 2(-1)^4 + 3(-1)^2 - 5(-1)$

$= 2 \cdot 1 + 3 \cdot 1 - (-5)$

$= 2 + 3 + 5 = 10$

(b) $P(2) = 2(2)^4 + 3(2)^2 - 5(2)$

$= 2 \cdot 16 + 3 \cdot 4 - 10$

$= 32 + 12 - 10$

$= 34$

77. $-5y + 3a$ *Place like terms in columns*

$-3y + 2a$

$\underline{4y + 6a}$

$-4y + 11a$ *Add*

81. First,

$(3m^2 - 5n^2 + 2n) + [-(3m^2 - 4n^2)]$

$= 3m^2 - 5n^2 + 2n - 3m^2 + 4n^2$

$= -n^2 + 2n.$

So we have the following subtraction.

$(-4m^2 + 3n^2 - 5n) - (-n^2 + 2n)$

$= -4m^2 + 3n^2 + n^2 - 5n - 2n$

*Change signs, add, combine like terms*

$= -4m^2 + 4n^2 - 7n$

85. $[(6xy - 10z^2) + (3z^2 - 2xy)]$

$= 6xy - 10z^2 + 3z^2 - 2xy$

*Simplify inside the brackets*

$= -7z^2 + 4xy$

$[(5xy + 3z^2) - (-2xy - z^2)]$

$= 5xy + 3z^2 + 2xy + z^2$

*Simplify inside the brackets*

$= 4z^2 + 7xy$

$(-7z^2 + 4xy) - (4z^2 + 7xy)$

$= -7z^2 + 4xy - 4z^2 - 7xy$

*Subtract the results*

$= -11z^2 - 3xy$

89. $(3a^{2x} + a^x - 4) + (5a^{3x} - a^{2x} + 2a^x + 6)$

$= 3a^{2x} + a^x - 4 + 5a^{3x} - a^{2x} + 2a^x + 6$

*Add like terms Do not use the product rule--this is addition!*

$= 5a^{3x} + 2a^{2x} + 3a^x + 2$

93. $(5p^{2k} + 3p^k - 4) + [(6p^k - 3) - (2p^{2k} + p^k)]$

$= 5p^{2k} + 3p^k - 4 + [6p^k - 3 - 2p^{2k} - p^k]$

$= 5p^{2k} + 3p^k - 4 + 5p^k - 2p^{2k} - 3$

$= 3p^{2k} + 8p^k - 7$

97. $10z(5z^4) = 10z^1 5z^4 = 10 \cdot 5 \cdot z^{1+4}$

$= 50z^5$

101. $-2a^4(3a^5b^2) = -2 \cdot 3a^4a^5b^2$

$= -6a^4 + 5b^2$

$= -6a^9b^2$

**Section 3.4 (page 120)**

1. $3p(7p - 4k = 3p(7p) + 3p(-4k)$ *Distributive property*

$= 21p^2 - 12kp$

5. $-6a^5(-2a^2 + 4a - 5)$

$= (-6a^5)(-2a^2) + (-6a^5)(4a) + (-6a^5)(-5)$

$= 12a^7 - 24a^6 + 30a^5$

9. $10rt(-3r^2t^2 + 2rt^2 - 5r^2t)$

$= 10rt(-3r^2t^2) + (10rt)(2rt^2) + (10rt)(-5r^2t)$

$= -30r^3t^3 + 20r^2t^3 - 50r^3t^2$

13. $(5m + 1)(9m - 4)$

$= (5m)(9m) + (5m)(-4) +$

F O

$((1)(9m) + (1)(-4)$ *Use FOIL method*

I L

$= 45m^2 - 20m + 9m - 4$

$= 45m^2 - 11m - 4$

17. $(-y + 1)(3y + 1)$

$= (-y)(3y) + (-y)(1) +$

$(1)(3y) + (1)(1)$ *Use FOIL method*

$= -3y^2 - y + 3y + 1$

$= -3y^2 + 2y + 1$

21. $(-8m + 5p)(3m - 5p)$

$= (-8m)(3m) + (-8m)(-5p) +$

$(5p)(3m) + (5p)(-5p)$ *Use FOIL method*

$= -24m^2 + 40mp + 15mp - 25p^2$

$= -24^2 + 55mp - 25p^2$

25. $(3w + \frac{1}{4}z)(w - 2z)$

$= (3w)(w) + (3w)(-2z) +$

$(\frac{1}{4}z)(w) + (\frac{1}{4}z)(-2z)$ *Use FOIL method*

$= 3w^2 - 6wz + \frac{1}{4}wz - \frac{1}{2}z^2$

$= 3w^2 - \frac{23}{4}wz - \frac{1}{2}z^2$

29.
$$\begin{array}{rl} 3k - 2 & \\ \underline{5k + 1} & \\ 3k - 2 & \leftarrow 1(3k - 2) \\ \underline{15k^2 - 10k} & \underline{\leftarrow 5k(3k - 2)} \\ 15k^2 - 7k - 2 & \end{array}$$

33.
$$\begin{array}{rl} 2z^3 - 5z^2 + 8z - 1 & \\ \underline{4z - 3} & \\ -6z^3 + 15z^2 - 24z + 3 & \textit{Multiply by -3} \\ \underline{8z^4 - 20z^3 + 32z^2 - 4z} & \textit{Multiply by 4z} \\ 8z^4 - 26z^3 + 47z^2 - 28z + 3 & \end{array}$$

37. $(4z - t)(4z + t)$

$= (4z)^2 - t^2$ *Use the product of the sum and difference of two terms*

$= 16z^2 - t^2$

41. $(5r - 3w)(5r + 3w)$

$= (5r)^2 - (3w)^2$ *Use the product of the sum and difference of two terms*

$= 25r^2 - 9w^2$

45. $(\frac{1}{2}y - \frac{2}{3}z)(\frac{1}{2}y + \frac{2}{3}z)$

$= (\frac{1}{2}y)^2 - (\frac{2}{3}z)^2$ *Product of the sum and difference of two terms*

$= \frac{1}{4}y^2 - \frac{4}{9}z^2$

49. $6t(t - 2s)(t + 2s)$

$= 6t[(t)^2 - (2s)^2]$ *Product of the sum and difference of two terms*

$= 6t]t^2 - 4s^2]$

$= (6t)(t^2) - (6t)(4s^2)$

$= 6t^3 - 24s^2t$

53. $(3k - 2p)^2$

$= (3k)^2 - 2(3k)(2p) + (2p)^2$ *Square of a binomial*

$= 9k^2 - 12kp + 4p^2$

57. $(3ax - 5by)^2$

$= (3ax)^2 - 2(3ax)(5by) + (5by)^2$ *Square of a binomial*

$= 9a^2x^2 - 30abxy + 25b^2y^2$

61. $[(5x + 1) + 6y]^2$

$= (5x + 1)^2 + 2(5x + 1)(6y) + (6y)^2$ *Square of a binomial*

$= [(5x)^2 + 2(5x)(1) + (1)^2] + 12y(5x + 1) + 36y^2$ *Square of a binomial*

$= 25x^2 + 10x + 1 + 60xy + 12y + 36y^2$

65. $[(2a + b) = 3][(2a + b) + 3]$

$= (2a + b)^2 - (3)^2$ *Product of the sum and difference of two terms*

$= [(2a)^2 + 2(2a)(b) + (b)^2] - 9$ *Square of a binomial*

$= 4a^2 + 4ab + b^2 - 9$

69. $[5t - (3 + 2s)][5t + (3 + 2s)]$

$= (5t)^2 - (3 + 2s)^2$ *Product of the sum and differences of two terms*

$= 25t^2 - [(3)^2 + 2(3)(2s) + (2s)^2]$ *Square of a binomial*

$= 25t^2 - [9 + 12s + 4s^2]$

$= 25t^2 - 9 - 12s - 4s^2$

73. $(4z - x)(z^3 - 4z^2x + 2zx^2 - x^3)$

$= (4z)(z^3 - 4z^2x + 2zx^2 - x^3) + (-x)(z^3 - 4z^2x + 2zx^2 - x^3)$

$= 4z^4 - 16z^3x + 8z^2x^2 - 4zx^3 - z^3x + 4z^2x^2 - 2zx^3 + x^4$

$= 4z^4 - 17z^3x + 12z^2x^2 - 6zx^3 + x^4$

77. $(a + b)(a + 2b)(a - 3b)$

$= [(a + b)(a + 2b)](a - 3b)$

$= [a^2 + 3ab + 2b^2](a - 3b)$

$$\begin{array}{rrrr} & a^2 & + 3ab & + 2b^2 \\ & & a & - 3b \\ \hline & -3a^2b & - 9ab^2 & - 6b^3 \\ a^3 & + 3a^2b & + 2ab^2 & \\ \hline a^3 & & - 7ab^2 & - 6b^3 \end{array}$$ *Multiply vertically*

81. $(q - 2)^4$

$= [(q - 2)(q - 2)][(q - 2)(q - 2)]$

$= [q^2 - 4q + 4][q^2 - 4q + 4]$

$= (q^2)[q^2 - 4q + 4] + (-4q)[q^2 - 4q + 4] + 4[q^2 - 4q + 4]$ *Multiply each term from the first bracket times the second bracket*

$= q^4 - 4q^3 + 4q^2 - 4q^3 + 16q^2 - 16q + 4q^2 - 16q + 16$

$= q^4 - 8q^3 + 24q^2 - 32q + 16$

85. $(x + y)^4 = x^4 + y^4$

$(2 + 3)^4 = 2^4 + 3^4$

$(5)^4 = 2^4 + 3^4$

$625 = 16 + 81$

$625 \neq 97$

$(x + y)^4 = [(x + y)^2]^2$

$= (x^2 + 2xy + y^2)^2$

$$
\begin{array}{r}
x^2 + 2xy + y^2 \\
x^2 + 2xy + y^2 \\
\hline
x^2y^2 + 2xy^3 + y^4 \\
2x^3y + 4x^2y^2 + 2xy^3 \quad \\
x^4 + 2x^3y + x^2y^2 \quad\quad\quad\quad \\
\hline
x^4 + 4x^2y + 6x^2y^2 + 4xy^3 + y^4
\end{array}
$$

89. $z^{m-4}(3z^{m+2} - 4z^{-m})$

$= 3z^{m-4+m+2} - 4z^{m-4+(-m)}$ *Product rule*

$= 3z^{2m-2} - 4z^{-4}$

93. $(8k^m - 3y^{3m})(2k^{2m} + 7y^m)$

$= (8k^m)(2k^{2m}) + (8k^m)(7y^m) + (-3y^{3m})(2k^{2m}) + (-3y^{3m})(7y^m)$

$= 16k^{3m} + 56k^my^m - 6y^{3m}k^{2m} - 21y^{4m}$

97. $(4p^n - q^n)(4p^n + q^n)$

$= (4p^n)^2 - (q^n)^2$ *Product of the sum and difference of two terms*

$= 16p^{2n} - q^{2n}$

101. $9 \cdot 6 + 9 \cdot r = 9(6 + r)$ *The common factor is 9*

105. $x(x + 1) + 4(x + 1)$

$= (x + 4)(x + 1)$ *The common factor is $(x + 1)$*

109. $2p^4q^{-2}(5pq)^{-1}$

$= \dfrac{2p^4q^{-2}}{5pq}$

$= \dfrac{2p^3}{5p^3}$

**Section 3.5 (page 126)**

1. $15k + 30 = 15 \cdot k + 15 \cdot 2$

*15 is the greatest common factor*

$= 15(k + 2)$

5. $8m^2p - 4mp^2$

$= 4mp \cdot 2m + 4m(-p)$ *4mp is the greatest common factor*

$= 4mp(2m - p)$

9. $8m^4 + 6m^3 - 4m^2$

$= 2m^2 \cdot 4m^2 + 2m^2 \cdot 3m - 2m^2 \cdot 2$

*$2m^2$ is the greatest common factor*

$= 2m^2(4m^2 + 3m - 2)$

13. $6r^2s - 3rs^2 - 9r^3s$

$= 3rs \cdot 2r - 3rs \cdot s - 3rs \cdot 3r^2$

*3rs is the greatest common factor*

$= 3rs(2r - s - 3r^2)$

17. $12km^3 - 24k^3m^2 + 36k^2m^4$

$= 12km^2 \cdot m + 12km^2(-2k^2) + 12km^2(3km^2)$

*$12km^2$ is the greatest common factor*

$= 12km^2(m - 2k^2 + 3km^2)$

21. $16z^2n^6 + 64zn^7 - 32z^3n^3$

$= 16zn^3 \cdot zn^3 + 16zn^3 \cdot 4n^4 - 16zn^3 \cdot 2z^2$

$= 16zn^3(zn^3 + 4n^4 - 2z^2)$

25. $9x^2y^2z^4 - 18x^3y^2z$

$= 9x^2y^2z \cdot z^3 - 9x^2y^2z \cdot 2x$

$= 9x^2y^2z(z^3 - 2x)$

29. $14a^3b^2 + 7z^2b - 21a^5b^3 + 42ab^4$

$= 7ab \cdot 2a^2b + 7ab \cdot a + 7ab(-3a^4b^2) + 7ab \cdot 6b^3$

*7ab is the greatest common factor*

$= 7ab(2a^2b + a - 3a^4b^2 + 6b^3)$

33. $(m - 9)(m + 1) + (m - 9)(m + 2)$

*The greatest common factor is (m - 9)*

$= (m - 9)[(m + 1) + (m + 2)]$

$= (m - 9)(2m + 3)$

*Simplify inside the brackets*

37. $(r - 6)(2r + 1) - (r - 6)(r + 3)$

*The greatest common factor is (r - 6)*

$= (r - 6)[(2r + 1) - (r + 3)]$

$= (r - 6)[2r + 1 - r - 3]$

*Work inside brackets*

$= (r - 6)(r - 2)$ *Simplify inside brackets*

41. $4(3 - x)^2 - (3 - x)^3 + 3(3 - x)$

*(3 - x) is the greatest common factor*

$= (3 - x)[4(3 - x) - (3 - x)^2 + 3]$

*Simplify in brackets*

$= (3 - x)[12 - 4x - 9 + 6x - x^2 + 3]$

*Change signs*

$= (3 - x)[6 + 2x - x^2]$

45. $5(m + p)^3 - 10(m + p)^2 - 15(m + p)^4$

*The common factor is $5(m + p)^2$*

$= 5(m + p)^2[(m + p) - 2 - 3(m + p)^2]$

*Simplify inside the brackets*

$= 5(m + p)^2[m + p - 2 - 3(m^2 + 2mp + p^2)]$

$= 5(m + p)^2[m + p - 2 - 3m^2 - 6mp - 3p^2]$

49. $-2x^5 + 6x^3 + 4x^2$

$= 2x^2(-x^3 + 3x + 2)$ *Factor out $2x^2$*

$= -2x^2(x^3 - 3x - 2)$ *Factor out $-2x^2$*

53. $pq + 3rq + pm + 3rm$

$= (pq + 3rq) + (pm + 3rm)$

$= q(p + 3r) + m(p + 3r)$

$= (p + 3r)(q + m)$ *Factor out (p + 3r)*

57. $p^2 + pq - 3py - 3yq$

$= p(p + q) - 3y(p + q)$

*Factor out p from the first two terms and -3y from the second two terms*

$= (p + q)(p - 3y)$ *Factor out (p + q)*

61. $x^2 - 3x + 2x - 6$

$= x(x - 3) + 2(x - 3)$

$= (x - 3)(x + 2)$ *Factor out (x - 3)*

65. $21y^2 + 14y - 15y - 10$

$= 7y(3y + 2) + 5(-3y - 2)$

$= 7y(3y + 2) - 5(3y + 2)$

$= (3y + 2)(7y - 5)$

69. $14m^2 + 21mq - 2mq - 3q^2$

$= 7m(2m + 3q) + q(-2m - 3q)$

*Factor out 7m and q*

$= 7m(2m + 3q) - q(2m + 3q)$

*Factor out -1 from the second two terms*

$= (2m + 3q)(7m - q)$

73. $1 - a + ab - b$

$= (1 - a) + b(a - 1)$

$= 1(1 - a) - b(-a + 1)$

*-(a + 1) = a - 1*

$= (1 - a)(1 - b)$ *Factor out (1 - a)*

77. $p^{6m} - 2p^{4m}$

$= p^{4m}(p^{2m} - 2)$

81. $y^{r+5} + y^{r+4} + y^{r+2}$

$= y^{r+2}(y^3 + y^2 + 1)$

85. $r^p m^p + q^p m^p - r^p z^p - q^p z^p$

$= m^p(r^p + q^p) + z^p(-r^p - q^p)$

$= m^p(r^p + q^p) - z^p(r^p + q^p)$

$= (r^p + q^p)(m^p - z^p)$

89. $3p^{-3} + 2p^{-2} - 4p^{-1}$

$= p^{-3}\cdot 3 + p^{-3}(2p) + p^{-3}(-4)$

$= p^{-3}(3 + 2p - 4)$

93. $(8r + 5s)(2r - 3s)$

$= (8r)(2r) + (8r)(-3s) +$
F O

$(5s)(2r) + (5s)(-3s)$
I L

$= 16r^2 - 24rs + 10rs - 15s^2$

$= 16r^2 - 14rs - 15s^2$

97. $(2r - 5s)(7r + 3s)$

$= (2r)(7r) + (2r)(3s) +$

$(-5s)(7r) + (-5s)(3s)$

$= 14r^2 + 6rs - 35rs - 15s^2$

$= 14r^2 - 29rs - 15s^2$

**Section 3.6 (page 133)**

1. $c^2 + 4c - 5$

We need two numbers with a product of -5 and a sum of 4. Use 5 and -1 since (5)(-1) = -5 and 5 - 1 = 4.

$c^2 + 4c - 5 = (c + 5)(c - 1)$

5. $a^2 - a - 12$

We need two integers whose sum is -1 (coefficient of the middle term) and whose product is -12. Use -4 and 3, since -4 + 3 = -1, and $-4\cdot 3 = -12$.

$a^2 - a - 12 = (a - 4)(a + 3)$

9. $x^2 - 3x - 40$

Two numbers with a product of -40 and a sum of -3 are -8 and 5.

$x^2 - 3x - 40 = (x - 8)(x + 5)$

13. $y^2 - 3yx - 10x^2$

We need to find two integers whose sum is -3 and whose product is -10. -5 and +2 are the integers, since $-5 + 2 = -3$ and $5 \cdot -2 = 10$.

$y^2 - 3yx - 10x^2 = (y - 5x)(y + 2x)$

17. $5y^2 + y - 6$

Look for two integers with a product of $5(-6) = -30$ and a sum of 1. Use 6 and -5.

$= 5y^2 + 6y - 5y - 6$

$= y(5y + 6) - 1(5y + 6)$ *Factor by grouping*

$= (5y + 6)(y - 1)$ *Factor out $(5y + 6)$*

21. $8y^2 + 13y - 6$ *Look for a product of $8(-6) = -48$ and sum of 13*

$= 8y^2 + 16y - 3y - 6$ *Use 16 and -13*

$= 8y(y + 2) - 3(y + 2)$ *Factor by grouping*

$= (y + 2)(8y - 3)$ *Factor out $(y + 2)$*

25. $35p^2 - 4p - 15$ *Look for a product of $35(-15) = -525$ and sum of -4*

$= 35p^2 - 25p + 21p - 15$ *Use -25 and 21*

$= 5p(7p - 5) + 3(7p - 5)$ *Factor by grouping*

$= (7p - 5)(5p + 3)$

29. $4k^2 - 12ak + 9a^2$ *Look for a product of $4 \cdot 9 = 36$ and sum of -12*

$= 4k^2 - 6ak - 6ak + 9a^2$ *Use -6 and -6*

$= 2k(2k - 3a) - 3a(2k - 3a)$

*Factor by grouping*

$= (2k - 3a)(2k - 3a)$

$= (2k - 3a)^2$

33. $8m^2 - 14mp - 39p^2$ *Look for a product of $8(-39) = -312$ and sum of -14*

$= 8m^2 - 26mp + 12mp - 39p^2$

*Use -26 and 12*

$= 2m(4m - 13p) + 3p(4m - 13p)$

$= (4m - 13p)(2m + 3p)$

37. $12m^2 + 14m - 40$

$= 2(6m^2 + 7m - 20)$ *Factor out 2*

To factor $(6m^2 + 7m - 20)$, look for two integers with a product of $6(-20) = -120$ and a sum of 7. Use 15 and -8.

$= 2(6m^2 + 15m - 8m - 40)$

$= 2[3m(2m + 5) - 4(2m + 5)]$

*Factor by grouping*

$= 2(2m + 5)(3m - 4)$

41. $6a^3 + 12a^2 - 90a$

$= 6a(a^2 + 2a - 15)$ *Factor out 6a Then, factor $a^2 + 2a - 15$*

$= 6a(a - 3)(a + 5)$ *Since $-3 \cdot 5 = -15$ and $-3 + 5 = 2$*

45. $2x^3y^3 - 48x^2y^4 + 288xy^5$

$= 2xy^3(x^2 - 24xy + 144y^2)$

*Factor out $2xy^3$*

$= 2xy^3(x - 12y)^2$ *Since $(-12)^2 = 144$ and $2(-12) = -24$*

49. $z^4 - 7z^2 - 30$

$= x^2 - 7x - 30$ *Let $x = z^2$*

$= (x - 10)(x + 3)$ *Since $-10.3 = -30$ and $-10 + 3 = -7$*

$= (z^2 - 10)(z^2 + 3)$ *Substitute $z^2$ for $x$*

53. $12p^4 + 28p^2r - 5r^2$

$= 12y^2 + 28yr - 5r^2$ *Let $y = p^2$*

Then, look for two integers with a product of $12(-5) = -60$ and a sum of 28.

$= 12y^2 + 30\ yr - 2yr - 5r^2$ *Use 30 and -2*

$= 6y(2y + 5r) - r(2y + 5r)$

$= (2y + 5r)(6y - r)$

$= (2p^2 + 5r)(6p^2 - r)$ *Substitute $p^2$ for $y$*

57. $(3q + 7h)(q - 2h)$

$= 3q^2 - 3qh + 7qh - 14h^2$ *Use the FOIL method*

$= 3q^2 + 4qh - 14h^2$

61. $6(p + 3)^2 + 13(p + 3) + 5$

$= 6x^2 + 13x + 5$ *Let $x = p + 3$*

$= 6x^2 + 3x + 10x + 5$

$= 3x(2x + 1) + 5(2x + 1)$

$= (2x + 1)(3x + 5)$

$= [2(p + 3) + 1][3(p + 3) + 5]$ *Substitute $(p + 3)$ for $x$*

$= [2p + 6 + 1][3p + 9 + 5]$ *Simplify*

$= (2p + 7)(3p + 14)$

65. $a^2(a + b)^2 - ab(a + b)^2 - 6b^2(a + b)^2$

$= (a + b)^2(a^2 - ab - 6b^2)$ *Factor out $(a + b)^2$*

$= (a + b)^2(a - 3b)(a + 2b)$

69. $z^2(z - x) - 2z(x - 2) - 2x^2(z - x)$

$= z^2(z - x) - zx\ [-(z - x) - 2x^2(z - x)]$ *$x - z = -(z - x)$*

$= z^2(z - x) + zx(z - x) - 2x^2(z - x)$ *Simplify*

$= (z - x)(z^2 + zx - 2x^2)$ *Factor out $(z - x)$*

$= (z - x)(z + 2x)(z - x)$

$= (z - x)^2(z + 2x)$

73. $6z^{4r} - 5z^{2r} - 4$

$= 6x^2 - 5x - 4$ *Let $x = z^{2r}$*

$= 6x^2 - 8x + 3x - 4$

$= 2x(3x - 4) + (3x - 4)$

$= (3x - 4)(2x + 1)$

$= (3z^{2r} - 4)(2z^{2r} + 1)$ *Substituting $z^{2r}$ for $x$*

77. $-5° = -(5°) = -1$

81. $(2m - 5)(2m + 5) = (2m)^2 - 5^2$ *Product of the sum and difference of two terms*

$= 4m^2 - 25$

85. $(y - 2)(y^2 + 2y + 4) = y^3 - 8$ *By vertical multiplication shown below*

$$\begin{array}{r} y^2 + 2y + 4 \\ y - 2 \\ \hline -2y^2 - 4y - 8 \\ y^3 + 2y^2 + 4y \qquad \\ \hline y^3 \qquad\qquad\quad - 8 \end{array}$$

**Section 3.7 (page 137)**

1. $x^2 - 25 = x^2 - 5^2$ *Differences of two squares*

$= (x + 5)(x - 5)$

5. $16y^2 - 81q^2 = (4y)^2 - (9q)^2$

$= (4y + 9q(4y - 9q)$

9. $a^4 - 4b^4 = (a^2)^2 - (2b^2)^2$

$= (a^2 + 2b^2)(a^2 - 2b^2)$

13. In $a^2 - 10a + 25$, we see that $a^2 = (a)^2$, $25 = (5)^2$, and $-10a = 2(a)(-5)$. Thus, we have the square of the binomial $(a - 5)$, or a perfect square trinomial.

$a^2 - 10a + 25 = (a - 5)^2$

17. $25x^2y^2 - 20xy + 4$

$= (5xy)^2 - (2)(5xy)(2) + (2)^2$ *Perfect square trinomial*

$= [(5xy) - 2]^2$

$= (5xy - 2)^2$

21. $8a^3 + 1$

$= (2a)^3 + (1)^3$ *Sum of two cubes*

$= (2a + 1)[(2a)^2 - 2a + 1^2]$

$= (2a + 1)(4a^2 - 2a + 1)$

25. $64x^3 + 125y^3$

$= (4x)^3 + (5y)^3$ *Sum of two cubes*

$= (4x + 5y)(16x^2 - 20xy + 25y^2)$

29. $1000 + 27r^3s^3$

$= 10^3 + (3rs)^3$ *Sum of two cubes*

$= (10 + 3rs)[10^2 - 10(3rs + (3rs^2]$

$= (10 + 3rs)(100 - 30rs + 9r^2s^2)$

33. $(x + y)^2 - 16$

$= (x + y)^2 - 4^2$ *Difference of two squares*

$= [(x + y) - 4][(x + y) + 4]$

$= (x + y - 4)(x + y + 4)$

37. $m^2 - (3p - 5)^2$

$= [m - (3p - 5)][m + (3p - 5)]$ *Difference ot two squares*

$= (m - 3p + 5)(m + 3p - 5)$

41. $(a + b)^2 + 2(a + b) + 1$

$= x^2 + 2x + 1$ *Let $a + b = x$*

$= (x + 1)^2$

$= (a + b + 1)^2$ *Replace $x$ with $a + b$*

45. $p^2 - 6p + 9 - r^2$

$= (p - 3)^2 - r^2$ *Difference of squares*

$= [(p - 3) + r][(p - 3) - r]$

$= (p - 3 + r)(p - 3 - r)$

49. $r^2 - 16s^2 + 24s - 9$

$= r^2 - (16a^2 - 24s + 9)$

$= r^2 - (4s - 3)^2$ *Difference of squares*

$= [r + (4s - 3)][r - (4s - 3)]$

$= (r + 4s - 3)(r - 4s + 3)$

53. $(p - 5)^3 + 125$

$= (p - 5)^3 + 5^3$ *Sum of two cubes*

$= [(p - 5) + 5][(p - 5)^2 - 5(p - 5) + (5)^2]$

$= p(p^2 - 10p + 25 - 5p + 25 + 25)$

$= p(p^2 - 15p + 75)$

57. The middle term of $p^2 + 6p + c$ is $6p$. Since the coefficient of $p^2$ is 1, we have $2 \cdot p \cdot$(second factor) $= 6p$ or the second factor $= 3$.

In factored form: $(p + 3)^2$ so $c = 9$.

61. $16q^2 + bq + 25 = (4q)^2 + 2(4q)(5) + (5)^2$

Thus, $bq = 2 \cdot 4q \cdot 5 = 40q$

$b = 40$.

Check: $16q^2 + 40q + 25 = (4q + 5)^2$

Note: $-40$ is also a correct answer, giving $(4q - 5)^2$.

65. $64r^{8z} - 1$

$= [(8r^{4z})^2 - (1)^2]$

$= (8r^{4z} - 1)(8r^{4z} + 1)$ *Difference of two squares*

69. $9a^{4z} - 30a^{2z} + 25$

$= 9x^2 - 30x + 25$ *Let $x = a^{2z}$*

$= (3x)^2 - 2(3x)(5) + 5^2$ *Perfect square trinomial*

$= (3x - 5)^2$

$= (3a^{2z} - 5)^2$ *Replace $x$ with $a^{2z}$*

73. $27z^{12y} + 125x^{6y}$

$= (3z^{4y})^3 + (5x^{2y})^3$ *Factor as the sum of two cubes*

$= (3z^{4y} + 5x^{2y})(9z^{8y} - 15z^{4y}x^{2y} + 25x^{4y})$

77. $xy + 2y + 4x + 8$

$= y(x + 2) + 4(x + 2)$ *Factor by grouping*

$= (y + 4)(x + 2)$

81. $6r^2 + 19rz - 7z^2$ *Find two integers with product of $6(-7) = -42$ and sum of 19*

$= 6r^2 + 21rz - 2rz - 7z^2$

*Use 21 and -2*

$= 3r(2r + 7z) - z(2r + 7z)$

*Factor by grouping*

$= (3r - z)(2r + 7z)$

**Section 3.8 (page 141)**

1. $100a^2 - 9b^2$

$= (10a)^2 - (3b)^2$ *Difference of squares*

$= (10a + 3b)(10a - 3b)$

5. $3a^2pq + 3abpq - 90b^2pq$

$= 3pq(a^2 + ab - 30b^2)$ *Factor out the greatest common factor*

$= 3pq(a + 6b)(a - 5b)$

9. $6b^2 - 17b - 3$

$= 6b^2 + b - 18b - 3$

$= b(6b + 1) - 3(6b + 1)$

$= (6b + 1)(b - 3)$

13. $2p^2 + 11pq + 15q^2$

$= 2p^2 + 6pq + 5pq + 15q^2$

$= 2p(p + 3q) + 5q(p + 3q)$

$= (p + 3q)(2p + 5q)$

17. In $54m^3 - 2000$

$= 2(27m^3 - 1000)$ *Factor out 2*

$= 2[(3m^3 - (10)^3]$ *Difference of two cubes*

$= 2[3m - 10][(3m)^2 + (3m)(10) + (10)^2]$

$= 2(3m - 10)(9m^2 + 30m + 100)$

21. $kq - 9q + kr - 9r$

$= q(k - 9) + r(k - 9)$ *Factor by grouping*

$= (q + r)(k - 9)$

25. $8p^3 - 125 = (2p)^3 - (5)^3$

*Factor as the difference of 2 cubes*

$= (2p - 5)[(2p)^2 + (2p)(5) + (5)^2]$

$= (2p - 5)(4p^2 + 10p + 25)$

29. $p^3 + 64 = p^3 + 4^3$ *Sum of two cubes*

$= (p + 4)(p^2 - 4p + 4^2)$

$= (p + 4)(p^2 - 4p + 16)$

33. $12z^3 - 6z^2 + 18z$

$= 6z(2z^2 - z + 3)$ *Factor out 6z*

$(2z^2 - z + 3)$ *Cannot be factored further*

37. $1000z^3 + 512$

$= 8(125z^3 + 64)$

$= (5z)^3 + (4)^3$ *Sum of two cubes*

$= 8[5z + 4][(5z)^2 - (5z)(4) + (4)^2]$

$= 8(5z + 4)(25z^2 - 20z + 16)$

41. $24p^3q + 52p^2q^2 + 20pq^3$

$= 4pq(6p^2 + 13pq + 5q^2)$

$= 4pq(6p^2 + 10pq + 3pq + 5q^2)$

$= 4pq[2p(3p + 5q) + q(3p + 5q)]$

$= 4pq(3p + 5q)(2p + q)$

45. $50p^2 - 162 = 2(25p^2 - 81)$ *Factor out 2*

$= 2(5p + 9)(5p - 9)$

*Difference of 2 squares*

49. $21a^2 - 5ab - 4b^2$

$= 21a^2 + 7ab - 12ab - 4b^2$

$= 7a(3a + b) - 4b(3a + b)$

$= (3a + b)(7a - 4b)$

53. $(p + 8q)^2 - 10(p + 8q) + 25$

*Let $(p + 8q) = x$*

$= x^2 - 10x + 25$

$= (x - 5)^2$

$= [(p + 8q) - 5]^2$ *Replace $x$ with $p + 8q$*

$= (p + 8q - 5)^2$

57. $z^{4x} - w^{2x}$

$= (z^{2x})^2 - (w^x)^2$ *Difference of two squares*

$= (z^{2x} - w^x)(z^{2x} + w^x)$

61. $3m^{2k} - 7m^k - 6$

$= 3m^{2k} + 2m^k - 9m^k - 6$

$= m^k(3m^k + 2) - 3(3m^k + 2)$

$= (3m^k + 2)(m^k - 3)$

65. $15c^{4y} + 3c^{2y} - 6c^y$

$= 3c^y(5c^{3y} + c^y - 2)$ *Factor out $3c^y$*

69. $\frac{3^6 \cdot 3^{-1}}{3^3} = 3^{6-1-3} = 3^2 = 9$

73. $5r - 2 = 8$

$5r = 10$ *Add 2*

$r = 2$ *Divide by 5*

Solution set: $\{2\}$

77. $-9r = 0$

$r = 0$ *Divide by -9*

Solution set: $\{0\}$

**Section 3.9 (page 145)**

1. $(x - 2)(x - 3) = 0$ implies

$x - 2 = 0$ or $x - 3 = 0$

$x = 2$ or $x = 3$.

The solution set is $\{2, 3\}$.

5. $r(r - 4)(2r + 5) = 0$ implies

$r = 0$ or $r - 4 = 0$ or $2r + 5 = 0$

$r = 0$ or $r = 4$ or $2r = -5$

$r = -\frac{5}{2}$.

The solution set is $\{0, 4, -\frac{5}{2}\}$.

9. $y^2 + 9y + 14 = 0$

$(y + 7)(y + 2) = 0$ *Factor*

$y + 7 = 0$ or $y + 2 = 0$

$y = -7$ or $y = -2$

The solution set is $\{-7, -2\}$.

13. $3x^2 - 11x - 20 = 0$

$(3x + 4)(x - 5) = 0$ *Factor*

$3x + 4 = 0$ or $x - 5 = 0$

$3x = -4$ or $x = 5$

$x = \frac{-4}{3}$

The solution set is $\{\frac{-4}{3}, 5\}$.

17. $2a^2 - 8a = 0$

$2a(a - 4) = 0$ *Factor*

$2a = 0$ or $a - 4 = 0$

$a = 0$ or $a = 4$

The solution set is $\{0, 4\}$.

21. $-2a^2 + 8 = 0$

$-2(a^2 - 4) = 0$

$-2(a + 2)(a - 2) = 0$

$a + 2 = 0$ or $a - 2 = 0$

$a = -2$ or $a = 2$

Solution set: $\{-2, 2\}$.

25. $(x + 4)(x - 6) = -16$

$x^2 - 2x - 24 = -16$ *Multiply out the left side*

$x^2 - 2x - 8 = 0$ *Add 16*

$(x - 4)(x + 2) = 0$

$x - 4 = 0$ or $x + 2 = 0$

$x = 4$ or $x = -2$

The solution set is $\{-2, 4\}$.

29. $2x(x + 2) = x(x - 3) - 12$

$2x^2 + 4x = x^2 - 3x - 12$ *Multiply*

$x^2 + 7x + 12 = 0$ *Bring to left side*

$(x + 3)(x + 4) = 0$ *Factor*

$x + 3 = 0$ or $x + 4 = 0$

$x = -3$ or $x = -4$

The solution set is $\{-3, -4\}$.

33. Let x be one of the numbers. The other, then, is 12 - x.

$$x^2 + (12 - x)^2 = 90$$
$$x^2 + 144 - 24x + x^2 = 90$$
$$2x^2 - 24x + 54 = 0$$
$$2(x^2 - 12x + 27) = 0$$
$$2(x - 9)(x - 3) = 0$$

Thus, $x - 9 = 0$ or $x - 3 = 0$

$x = 9$ or $x = 3$.

The two numbers are 9 and 3.

37. Let x be one odd integer, then x + 2 is the next consecutive odd integer.

$$x(x + 2) = 99$$
$$x^2 + 2x = 99$$
$$x^2 + 2x - 99 = 0$$
$$(x - 9)(x + 11) = 0$$

$x - 9 = 0$ or $x + 11 = 0$

$x = 9$ or $x = -11$

If $x = 9$, $x + 2 = 11$.

If $x = -11$, $x + 2 = -9$.

The numbers are 9 and 11 or -9 and -11.

41. Let x represent the width and x + 4 the length. The area of a rectangle is A = LW.

$$A = LW.$$
$$140 = (x + 4)(x)$$
$$140 = x^2 + 4x$$
$$0 = x^2 + 4x - 140$$
$$0 = (x + 14)(x - 10)$$

$x + 14 = 0$ or $x - 10 = 0$

$x = -14$ or $x = 10$

A rectangle cannot have negative width so $x \neq -14$. The width is 10 meters and the length is 14 meters.

45.

$$5(3a - 1)^2 + 3 = -16(3a - 1)$$
$$(45a^2 - 30a + 5) + 3 = -48a + 16$$ *Multiply on each side*
$$45a^2 + 18a - 8 = 0$$
$$(15a) - 4)(3a + 2)$$ *Factor*

$15a - 4 = 0$ or $3a + 2 = 0$

$15a = 4$ $\quad 3a = -2$

$a = \frac{4}{15}$ $\quad a = -\frac{2}{3}$

Solution set: $\{\frac{4}{15}, -\frac{2}{3}\}$

49.

$$(2k - 3)^2 = 16k^2$$
$$4k^2 - 12k + 9 = 16k^2$$ *Multiply out*
$$-12k^2 - 12k + 9 = 0$$
$$-3(4k^2 + 4k - 3) = 0$$ *Factor out -3*
$$-3(2k - 1)(2k + 3) = 0$$ *Factor*

$2k - 1 = 0$ or $2k + 3 = 0$

$2k = 1$ $\quad 2k = -3$

$k = \frac{1}{2}$ $\quad k = \frac{-3}{2}$

Solution set: $\{\frac{1}{2}, -\frac{3}{2}\}$

53.

$$\frac{-100a^5b^2}{75a^6b^5} = \frac{-100a^{5-6}b^{2-5}}{75}$$
$$= \frac{-4}{3}a^{-1}b^{-3}$$
$$= \frac{-4}{3ab^3}$$

57.

$$\frac{11}{15} = \frac{11}{15}\cdot\frac{5}{5} = \frac{55}{75}$$

**Chapter 3 Review Exercises (page 147)**

1. The exponent in $9^3$ is 3 and the base is 9.

5. $3^5 = 3 \cdot 3 \cdot 3 \cdot 3 \cdot 3 = 243$

9. $(\frac{3}{4})^{-3} = (\frac{4}{3})(\frac{4}{3})(\frac{4}{3}) = \frac{64}{27}$

13. $\frac{m^{-2}m^{-5}}{m^{-8}m^{-4}} = \frac{m^{-7}}{m^{-12}}$ *Simplify*

$= m^{-7-(-12)}$ *Quotient rule*

$= m^5$

17. $3450 = 34.5 \times 10^3$

21. $3.8 \times 10^{-3}$

$= .0038$ *Move decimal 3 places to the left*

25. $-9^{\circ} = -(9^{\circ}) = -1$

29. $(\frac{5z^{-3}}{z^{-1}})(\frac{5}{z^2}) = \frac{25z^{-3}}{z^{-1}z^2} = \frac{25z^{-3}}{z^{-1+2}} =$

$\frac{25}{z^3 z^1} = \frac{25}{z^4}$

33. $k^r \cdot k^{2r-3} = k^{r+(2r-3)}$ *Product rule*

$= k^{3r-3}$

37. $\frac{6 \times 10^4}{3 \times 10^8} = \frac{6}{3} \times 10^{4-8}$

$= 2 \times 10^{-4}$

41. In $14p^5$, 14 is the coefficient.

45. $9k + 11k^3 - 3k^2$

(a) $11k^3 - 3k^2 + 9k$ Descending powers.

(b) Trinomial: 3 terms.

(c) Degree 3: the highest degree of the exponents is 3 in $11k^3$.

49. $P(x) = -5x + 11$

(a) $P(-4) = -5(-4) + 11$

$= 31$

(b) $P(2) = -5(2) + 11$

$= 1$

53. $8p + 11p - 6p = 13p$

57.
$$\begin{array}{r} -5x^2 + 7x - 3 \\ \underline{6x^2 + 4x + 9} \\ x^2 + 11x + 6 \end{array}$$

61. $(12m^5 + 7m^3 + 8m^2) + (4m^3 - 9m)$

$= 12m^5 + 7m^3 + 8m^2 + 4m^3 - 9m$

$= 12m^5 + 11m^3 + 8m^2 - 9m$

*Combine like items*

65. $(6y)(2y^2) = 12y^3$

69.
$$\begin{array}{r} 2p^2 + 5p - 3 \\ \underline{p + 4} \\ 8p^2 + 20p - 12 \\ \underline{2p^3 + 5p^2 - 3p \phantom{- 12}} \\ 2p^3 + 13p^2 + 17p - 12 \end{array}$$

73. $(m - 7)(m + 3)$

$= (m)(m) + (m)(3) + (-7)(m) + (-7)(3)$

*Use FOIL method*

$= m^2 - 4m - 21$

77. $(3r^2 + 5)(3r^2 - 5)$

$= (3r^2)^2 - (5)^2$ *Product of sum and difference of two terms*

$= 9r^4 - 25$

81. $9y^q(2y^{q-1} - 5y^{-q})$

$= 18y^{q+(q-1)} - 45y^{q+(-q)}$

$= 18y^{2q-1} - 45y^0$

$= 18y^{2q-1} - 45$

85. $30p + 25$

$= 5(6p + 5)$

89. $12a^2b + 18ab^3 - 24a^3b^2$

$= 6ab(2a + 3b^2 - 4a^2b)$

93. $2m + 6 - am - 3a$

$= 2(m + 3) + (-a)(m + 3)$

$= (m + 3)(2 - a)$

97. $r^2 + 15r + 54$

$= (r + 9)(r + 6)$, since $9 \cdot 6 = 54$ and $9 + 6 = 15$.

101. In $12p^2 + p - 20$, after trying repeated combinations, we see that $12p^2 = 3p \cdot 4p$ and $-20 = 4 \cdot -5$. Putting these into binomial factors, we have

$$(3p + 4)(4p - 5).$$

105. $r^4 + r^2 - 6 = (r^2 + 3)(r^2 - 2)$, since $3(-2) = -6$ and $3 - 2 = 1$.

109. $9p^2 - 49$

$= (3p)^2 - (7)^2$ *Difference of two squares*

$= (3p - 7)(3p + 7)$

113. $9k^2 - 12k + 4$

$= (3k - 2)^2$ *Perfect square trinomial*

117. $121a^{10m} - 144b^{6m}$

$= (11a^{5m})^2 - (12b^{3m})^2$ *Factor the difference of 2 squares*

$= (11a^{5m} + 12b^{3m})(11a^{5m} - 12b^{3m})$

121. $(5x - 2)(x + 1) = 0$

$5x - 2 = 0$ or $x + 1 = 0$

$5x = 2$ or $x = -1$

$x = \frac{2}{5}$

The solution set is $\{\frac{2}{5}, -1\}$.

125. $6r^2 + 7r = 3$

$6r^2 + 7r - 3 = 0$

$(2r + 3)(3r - 1) = 0$ *Factor*

$2r + 3 = 0$ or $3r - 1 = 0$

$2r = -3$ or $3r = 1$

$r = -\frac{3}{2}$ $r = \frac{1}{3}$

The solution set is $\{-\frac{3}{2}, \frac{1}{3}\}$.

129. $(2x + 1)(x - 2) = -3$

$2x^2 - 3x - 2 = -3$

$2x^2 - 3x + 1 = 0$

$(2x - 1)(x - 1) = 0$

$2x - 1 = 0$ or $x - 1 = 0$

$2x = 1$ or $x = 1$

$x = \frac{1}{2}$

The solution set is $\{1, \frac{1}{2}\}$.

133. Let x be the number.

$$3x^2 + 5x = 100$$
$$3x^2 + 5x - 100 = 0$$
$$(3x + 20)(x - 5) = 0 \quad \textit{Factor}$$
$$3x + 20 = 0 \quad \text{or} \quad x - 5 = 0$$
$$3x = -20 \qquad x = 5$$
$$x = \frac{-20}{3}$$

The number may be 5 or $\frac{-20}{3}$.

137. $(4x + 1)(2x - 3)$
$$= (4x)(2x) + (4x)(-3) + (1)(2x) + (1)(-3)$$
$$= 8x^2 - 12x + 2x - 3$$
$$= 8x^2 - 10x - 3$$

141. $(-5) + 11w) + (6 + 5w) + (-15 - 8w^2)$
$$= -5 + 6 - 15 + 11w + 5w - 8w^2$$
$$= -14 + 16w - 8w^2$$
$$= -8w^2 + 16w - 14$$

145. $\dfrac{(5z^2x^3)^2(2zx^2)^{-1}}{(-10zx^{-3})^{-2}(3z^{-1}x^{-4})^2}$
$$= \frac{5^2z^4x^6 2^{-1}z^{-1}x^{-2}}{(-10)^{-2}z^{-2}x^6 3^2 z^{-2}x^{-8}}$$
$$= \frac{(25)(100)z^3x^4}{2\cdot 9z^{-4}x^{-2}}$$
$$= \frac{(50)(25)}{9}z^7x^6$$
$$= \frac{1250}{9}z^7x^6$$

149. $8 - a^3 = (2)^3 - (a)^3$ *Factor the difference of cubes*
$$= (2 - a)[(2)^2 + 2a + (a)^2]$$
$$= (2 - a)(4 + 2a + a^2)$$

**Chapter 3 Test (page 150)**

1. $(\frac{3}{2})^3 = \frac{3^3}{2^3} = \frac{27}{8}$

2. $-(-2)^{-6} = \frac{-1}{(-2)^6} = \frac{-1}{64}$

3. $-4^2 + (-4)^2 = -16 + 16 = 0$

4. $8^{-2}\cdot 8^5\cdot 8^{-6} = 8^{-2+5+(-6)}$
$$= 8^{-3}$$
$$= \frac{1}{8^3} \text{ or } \frac{1}{512}$$

5. $\dfrac{y^{3r}\cdot y^{2r}}{y^{2-r}} = \dfrac{y^{3r+2r}}{y^{2-r}}$
$$= \frac{y^{5r}}{y^{2-r}}$$
$$= y^{5r-2+r}$$
$$= y^{6r-2}$$

6. $(-8)^0 = 1$

7. $\dfrac{(2y)^3y^5}{y^{-2}y^{-6}} = \dfrac{2^3y^3y^5}{y^{-2}y^{-6}}$
$$= \frac{8y^{3+5}}{y^{-2+(-6)}}$$
$$= \frac{8y^8}{y^{-8}}$$
$$= 8y^{8-(-8)}$$
$$= 8y^{16}$$

8. $(3a^{-2}b^{-2})^{-3} \cdot (2^{-2}ab^{-3})^2$

$= 3^{-3}a^6b^6 \cdot 2^{-4}a^2b^{-6}$

$= \frac{1}{3^3} \cdot \frac{1}{2^4} \cdot a^{6+2}b^{6-6}$

$= \frac{1}{27 \cdot 16} a^8b^0$

$= \frac{1}{432} a^8$

9. $\frac{3^{-2}r^{-3}(r^{-2})^{-1}}{5(r^2)^{-3}} = \frac{3^{-2}r^{-3} \cdot r^2}{5r^{-6}}$

$= \frac{3^{-2}r^{-1}}{5r^{-6}}$

$= \frac{3^{-2}r^{-1-(-6)}}{5}$

$= \frac{3^{-2}r^5}{5}$

$= \frac{1}{3^2} \cdot \frac{r^5}{5}$

$= \frac{r^5}{9 \cdot 5} = \frac{r^5}{45}$

10. $.00974 = 9.74 \times 10^{-3}$ *Count 3 places right to left to decimal, negative exponent*

11. $3.41 \times 10^{-3} = .00341$ *Move decimal 3 places to the left for a negative exponent*

12.
$$\begin{array}{r} 5m^2 - 9m + 6 \\ \underline{-8m^2 + 4m - 9} \\ -3m^2 - 5m - 3 \end{array}$$

13. $(3k^3 - 5k^2 + 8k - 2) - (4k^3 + 11k + 7)$

$= 3k^3 - 5k^2 + 8k - 2 - 4k^3 - 11k - 7$

$= -k^3 - 5k^2 - 3k - 9$

14. $-2r - [(r - 4) - (5 - 3r)] + (9 - 4r)$

$= -2r - [r - 4 - 5 + 3r] + 9 - 4r$

$= -2r - [4r - 9] + 9 - 4r$

$= -2r - 4r + 9 + 9 - 4r$

$= -10r + 18$

15.
$$\begin{array}{r} 3k^2 + 8k - 9 \\ \underline{k + 4} \\ 12k^2 + 32k - 36 \\ \underline{3k^3 + 8k^2 - 9k \quad\quad} \\ 3k^3 + 20k^2 + 23k - 36 \end{array}$$

16. $[2y + (3z - x)][2y - (3z - x)]$

$= (2y)^2 - (3z - x)^2$

$= 4y^2 - (9z^2 - 6zx + x^2)$

*Product of sum and difference of two terms*

$= 4y^2 - 9z^2 + 6zx - x^2$

17. $P(x) = -x^2 - 12x + 6$

$P(-4) = -(-4)^2 - 12(-4) + 6$

*Substitute -4 for x*

$= -16 + 48 + 6$

$= 38$

18. $ax + ay + bx + by$

$= a(x + y) + b(x + y)$

$= (x + y)(a + b)$

19. $3x^2 + 6xy^2 + 24xy$

$= 3x(x + 2y^2 + 8y)$

20. $k^2 + 10k + 24 = (k + 6)(k + 4)$

21. $2p^2 - 5pq + 3q^2$

$= 2p^2 - 3pq - 2pq + 3q^2$

$= (p(2p - 3q) - q(2p - 3q)$

$= (2p - 3q)(p - q)$

22. $x^2 + 100 = x^2 + 10^2$

The sum of two squares cannot be factored.

23. $x^4 - 9x^2 + 20$ *Let $y - x^2$*

$= y^2 - 9y + 20$

$= (y - 5)(y - 4)$

$= (x^2 - 5)(x^2 - 4)$ *Replace y with $x^2$*

$= (x + 2)(x - 2)(x^2 - 5)$

24. $4x^{2p} + 36x^py^q + 81y^{2q}$

Write $4x^{2p} = 2x^p \cdot 2x^p$

and $81y^{2q} = 9y^q \cdot 9y^q$

so the factorization is

$(2x^p + 9y^q)^2$.

25. $100d^2 - 9b^6 = (10d)^2 - (3b^3)^2$

*Difference of the two squares*

$= (10d + 3b^3)(10d - 3b^3)$

26. $27y^3 - 125z^6 = (3y)^3 - (5z^2)^3$

*Factor as the difference of two cubes*

$= [3y - 5z^2][(3y)^2 + (3y)(5z^2) + (5z^2)^2]$

$= (3y - 5z^2)(9y^2 + 15yz^2 + 25z^4)$

27. $3a^2 + 13a = 30$

$3a^2 + 13a - 30 = 0$ *Subtract 30 from both sides*

$(3a - 5)(a + 6) = 0$ *Factor*

$3a - 5 = 0$ or $a + 6 = 0$

$3a = 5$ or $a = -6$

$a = \frac{5}{3}$

Thus, the solutions are $a = \frac{5}{3}$ or $a = -6$.

The solution set is $\{\frac{5}{3}, -6\}$.

28. $8r^2 = 14r - 3$

$8r^2 - 14r + 3 = 0$

$(4r - 1)(2r - 3) = 0$ *Factor*

$4r - 1 = 0$ or $2r - 3 = 0$

$4r = 1$ or $2r = 3$

$r = \frac{1}{4}$ or $r = \frac{3}{2}$

The solution set is $\{\frac{1}{4}, \frac{3}{2}\}$.

29. $16x^2 - 32x = 0$ *Factor out 16x as a common factor*

$16x(x - 2) = 0$

$16x = 0$ or $x - 2 = 0$

$x = 0$ or $x = 2$

The solution set is $\{0, 2\}$.

30. Let W = width, then length = W + 2 since the length is 2 more meters than the width, and the area is 63 meters. Use A = LW.

$63 = (W + 2)W$ *Substitute 63 for A and W + 2 for L*

$63 = W^2 + 2W$

$0 = W^2 + 2W - 63$

$= (W + 9)(W - 7)$, so,

$W + 9 = 0$ or $W - 7 = 0$

$W = -9$ or $W = 7$

Reject -9, since width is positive. Thus, the width is 7 meters and the length is W + 2 = 7 + 2 = 9 meters.

**CHAPTER 4 RATIONAL EXPRESSIONS**

**Section 4.1 (page 157)**

1. $\frac{m}{m-4}$ is not defined when the denominator $m - 4 = 0$. When $m = 4$, this rational expression is undefined.

5. $\frac{5p + 9}{4}$

Since the variable does not appear in the denominator, no number can replace the variable and make the expression undefined.

9. $\frac{z^2 - 2}{z^2 - 16} = \frac{z^2 - z}{(z + 4)(z - 4)}$ *Factor the denominator*

$z + 4 = 0$ or $z - 4 = 0$ *Set each factor equal to 0*

$z = -4$ $z = 4$

Both -4 and 4 make this rational expression undefined.

13. $\frac{8z - 2}{z^2 + 16}$

$z^2 + 16 = 0$ *Set denominator equal to 0*

$z^2 = -16$

The square root of -16 is not a real number, so there are no real numbers which make this expression undefined.

In Exercises 17-49 to write each expression in its lowest terms, divide the numerator and denominator by the greatest common factor.

17. $\frac{3p^5}{9p^2} = \frac{3p^2 \cdot p^3}{3p^2 \cdot 3} = \frac{p^3}{3}$ *Divide numerator and denominator by the greatest common factor: $3p^2$*

21. $\frac{2x - 2}{3x - 3} = \frac{2(x - 1)}{3(x - 1)}$ *Factor*

$= \frac{2}{3}$ *Divide numerator and denominator by $x - 1$*

25. $\frac{2y - y^2}{4y - y^2} = \frac{y(2 - y)}{y(4 - y)}$ *Factor*

$= \frac{2 - y}{4 - y}$

29. $\frac{r^2 - r - 20}{r^2 + r - 30} = \frac{(r - 5)(r + 4)}{(r - 5)(r + 6)}$ *Factor*

$= \frac{r + 4}{r + 6}$

33. $\frac{a^3 - b^3}{a - b} = \frac{(a - b)(a^2 + ab + b^2)}{(a - b)}$ *Factor*

$= a^2 + ab + b^2$

37. $\frac{xy - yw + xz - zw}{xy + yw + xz + zw}$

$= \frac{y(x - w) + z(x - w)}{y(x + w) + z(x + w)}$

$= \frac{(y + z)(x - w)}{(y + z)(x + w)}$ *Factor*

$= \frac{x - w}{x + w}$

41. $\frac{6 - m}{m - 6} = \frac{-1(m - 6)}{(m - 6)}$ *Factor out -1 in numerator*

$= -1$

45. $\frac{(3 - m)(4 - n)}{(m - 3)(4 + n)} = \frac{-(m - 3)(4 - n)}{(m - 3)(4 + n)}$ *Factor*

$= \frac{-(4 - n)}{4 + n}$ or $\frac{n - 4}{4 + n}$

49. $\frac{x^2 - x}{1 - x} = \frac{x(x - 1)}{1 - x}$ *Factor the numerator*

$= \frac{x(x - 1)}{-1(x - 1)}$ *Factor -1 out of the denominator*

$= \frac{x}{-1}$ or $-x$

53. $\frac{6}{z - 2}$; $5z - 10$

$\frac{6}{z - 2} = \frac{6}{z - 2} \cdot \frac{5}{5} = \frac{30}{5z - 10}$

*Multiply numerator and denominator by 5*

57. $\frac{2x}{x - 4}$; $4 - x$

$\frac{2x}{x - 4} = \frac{2x}{x - 4} \cdot \frac{-1}{-1} = \frac{-2x}{4 - x}$ *Multiply by -1*

61. $\frac{2z}{z - 6}$; $2z^2 - 10z - 12$

$\frac{2z^2 - 10z - 12}{z - 6}$ *Divide denominator by original denominator*

$= \frac{2(z^2 - 5z - 6)}{z - 6}$

$= \frac{2(z - 6)(z + 1)}{z - 6}$ *Factor*

$= 2(z + 1)$

Multiply numerator and denominator by $2(z + 1)$.

$\frac{2z}{z - 6} \cdot \frac{2(z + 1)}{2(z + 1)} = \frac{4z(z + 1)}{2(z - 6)(z + 1)}$

$= \frac{4z(z + 1)}{2z^2 - 10z - 12}$

or $\frac{4z^2 + 4z}{2z^2 - 10z - 12}$

65. $\frac{4}{m + 3p}$; $m^3 + 27p^3$

$\frac{m^3 + 27p^3}{m + 3p}$ *Denominator divided by original denominator*

$= \frac{(m + 3p)(m^2 - 3mp + 9p^2)}{m + 3p}$

$= m^2 - 3mp + 9p^2$

Multiply numerator and denominator by $m^2 - 3mp + 9p^2$.

$\frac{4}{m + 3p} \cdot \frac{m^2 - 3mp + 9p^2}{m^2 - 3mp + 9p^2}$

$= \frac{4(m^2 - 3mp + 9p^2)}{m^3 + 27p^3}$ or $\frac{4m^2 - 12mp + 36p^2}{m^3 + 27p^3}$

69. $\frac{x + 5}{5} = x$

If $x = 2$, then $\frac{x + 5}{5} = \frac{2 + 5}{5} = \frac{7}{5}$.

$\frac{7}{5}$ does not equal 2.

73. $\frac{z^{3d} + 5z^d}{2z^{4d} + z^{3d}}$

$= \frac{z^d(z^{2d} + 5)}{z^d(2z^{3d} + z^{2d})}$ *Factor out $z^d$, the greatest common factor*

$= \frac{z^{2d} + 5}{2z^{3d} + z^{2d}}$

77. $\frac{a^{2p} - 1}{1 - a^{2p}} = \frac{(-1)(a^{2p} - 1)}{(-1)(1 - a^{2p})}$

$= \frac{(-1)(a^{2p} - 1)}{a^{2p} - 1}$ *Simplify the denominator*

$= -1$

81. $\dfrac{6x + 15 - 9x^{-1}}{x(2x - 1)}$

$= \dfrac{(6x + 15 - \frac{9}{x})\cdot x}{x(2x - 1)\cdot x}$

$= \dfrac{6x^2 + 15x - 9}{x^2(2x - 1)}$

$= \dfrac{3(2x^2 + 5x - 3)}{x^2(2x - 1)} = \dfrac{3(2x - 1)(x + 3)}{x^2(2x - 1)}$

$= \dfrac{3(x + 3)}{x^2}$

$= \dfrac{3x + 9}{x^2}$

89. $-3 \div (-\frac{6}{5}) = -3 \cdot -\frac{5}{6}$

$= \frac{15}{6}$

$= \frac{5}{2}$

**Section 4.2 (page 162)**

1. $\dfrac{m^3}{2} \cdot \dfrac{4m}{m^4} = \dfrac{4m^4}{2m^4}$ *Multiply*

$= 2$ *Divide top and bottom by $2m^4$*

5. $\dfrac{6y^5x^6}{y^3x^4} \cdot \dfrac{y^4x^2}{3y^5x^7} = \dfrac{6y^9x^8}{3y^8x^{11}}$ *Multiply*

$= \dfrac{2y}{x^3}$ *Write in lowest terms*

9. $\dfrac{(-3mn)^2 \cdot (4m^2n)^3}{16m^2n^4(mn^2)^3} \div \dfrac{24(m^2n^2)^4}{(3m^2n^3)^2}$

$= \dfrac{9m^2n^2 \cdot 64m^6n^3}{16m^2n^4m^3n^6} \div \dfrac{24m^8n^8}{9m^4n^6}$ *Power rule*

$= \dfrac{576m^8n^5}{16m^5n^{10}} \cdot \dfrac{9m^4n^6}{24m^8n^8}$ *Multiply by reciprocal of divisor*

$= \dfrac{5184m^{12}n^{11}}{384m^{13}n^{18}}$

$= \dfrac{27}{2mn^7}$

13. $\dfrac{2r}{8r + 4} \cdot \dfrac{14r + 7}{3}$

$= \dfrac{28r^2 + 14r}{24r + 12}$

$= \dfrac{7r(4r + 2)}{6(4r + 2)}$ *Factor*

$= \dfrac{7r}{6}$

17. $z^2 - 1 \cdot \dfrac{1}{1 - z}$

$= (z - 1)(z + 1)\cdot\dfrac{1}{1 - z}$ *Factor*

$= (z - 1)(z + 1)\cdot\dfrac{-1(1)}{-1(1 - 2)}$

$= (z - 1)(z + 1)\cdot\dfrac{-1}{(z - 1)}$

$= -(z + 1)$ *Write in lowest terms*

21. $\dfrac{6r - 5s}{3r + 2s} \cdot \dfrac{6r + 4s}{5s - 6r}$

$= \dfrac{6r - 5s}{3r + 2s} \cdot \dfrac{2(3r + 2s)}{(-1)(6r - 5s)}$ *Factor*

$= \dfrac{2(6r - 5s)(3r + 2s)}{-(1)(3r + 2s)(6r - 5s)}$ *Multiply*

$= -2$ *Write in lowest terms*

25. $\frac{2r^2 + 5r - 3}{r^2 - 1} \cdot \frac{r^2 - 2r + 1}{2r^2 - 7r + 3}$

$= \frac{(2r - 1)(r + 3)}{(r + 1)(r - 1)} \cdot \frac{(r - 1)(r - 1)}{(2r - 1)(r - 3)}$ *Factor*

$= \frac{(2r - 1)(r + 3)(r - 1)(r - 1)}{(2r - 1)(r - 3)(r + 1)(r - 1)}$ *Multiply*

$= \frac{(r + 3)(r - 1)}{(r - 3)(r + 1)}$ *Write in lowest terms*

29. $\frac{6a^2 + a - 1}{6a^2 + 5a + 1} \div \frac{3a^2 + 2a - 1}{3a^2 + 4a + 1}$

$= \frac{6a^2 + a - 1}{6a^2 + 5a + 1} \cdot \frac{3a^2 + 4a + 1}{3a^2 + 2a - 1}$ *Invert and change to multiplication*

$= \frac{(3a - 1)(2a + 1)}{(3a + 1)(2a + 1)} \cdot \frac{(3a + 1)(a + 1)}{(3a - 1)(a + 1)}$ *Factor*

$= \frac{(3a - 1)(2a + 1)(3a + 1)(a + 1)}{(3a - 1)(2a + 1)(3a + 1)(a + 1)}$ *Multiply*

$= 1$

The numerator and denominator have the same factors, so dividing each by these factors gives 1 (not 0).

33. $\frac{6k^2 + kr - 2r^2}{6k^2 - 5kr + r^2} \div \frac{3k^2 + 17kr + 10r^2}{6k^2 + 13kr - 5r^2}$

$= \frac{(2k - r)(3k + 2r)}{(2k - r)(3k - r)} \div \frac{(3k + 2r)(k + 5r)}{(3k - r)(2k + 5r)}$ *Factor*

$= \frac{(2k - r)(3k + 2r)}{(2k - r)(3k - r)} \cdot \frac{(3k - r)(2k + 5r)}{(3k + 2r)(k + 5r)}$ *Multiply by reciprocal*

$= \frac{2k + 5r}{k + 5r}$ *Write in lowest terms*

37. $\frac{am - an + bm - bn}{am + an - bm - bn} \div \frac{am - an - 3bm + 3bn}{am + an - 3bm - 3bn}$

$= \frac{am - an + bm - bn}{am + an - bm - bn} \cdot \frac{am + an - 3bm - 3bn}{am - an - 3bm + 3bn}$ *Change to multiplication*

$= \frac{a(m - n) + b(m - n)}{a(m + n) - b(m + n)} \cdot \frac{a(m + n) - 3b(m + n)}{a(m - n) - 3b(m - n)}$

$= \frac{(a + b)(m - n)}{(a - b)(m + n)} \cdot \frac{(a - 3b)(m + n)}{(a - 3b)(m - n)}$ *Factor*

$= \frac{a + b}{a - b}$ *Write in lowest terms*

41. $\frac{x^2 + 2x + 1}{x - 2} \cdot \frac{x^2 - 5x + 6}{x + 1} \cdot \frac{5x - 20}{x - 3}$

$= \frac{(x + 1)(x + 1)}{x - 2} \cdot \frac{(x - 2)(x - 3)}{x + 1} \cdot \frac{5(x - 4)}{x - 3}$ *Factor*

$= \frac{5(x + 1)(x - 4)(x + 1)(x - 2)(x - 3)}{(x + 1)(x - 2)(x - 3)}$ *Multiply*

$= 5(x + 1)(x - 4)$

45. $\frac{6k^2 - 13k - 5}{k^2 + 7k} \cdot \frac{k^3 + 6k^2 - 7k}{2k - 5} \cdot \frac{k^2 - 5k + 6}{3k^2 - 8k - 3}$

$= \frac{(3k + 1)(2k - 5)}{k(k + 7)} \cdot \frac{k(k + 7)(k - 1)}{(2k - 5)} \cdot \frac{(k - 3)(k - 2)}{(3k + 1)(k - 3)}$ *Factor*

$= (k - 1)(k - 2)$

49. $\frac{x^2 - 4x + 4 - y^2}{x^3y - 2x^2y - x^2y^2} \cdot \frac{x^4y}{x - 2 - y}$

$= \frac{(x - 2)^2 - y^2}{x^2y(x - 2 - y)} \cdot \frac{x^4y}{(x - 2 - y)}$

*Numerator of first fraction is difference of squares*

$= \frac{(x - 2 + y)(x - 2 - y)}{x^2y(x - 2 - y)} \cdot \frac{x^4y}{(x - 2 - y)}$ *Factor*

$= \frac{x^2(x - 2 + y)}{x - 2 - y}$ *Write in lowest terms*

53. $\frac{3}{8} + \frac{1}{8} = \frac{4}{8} = \frac{1}{2}$

57. $\frac{5}{12} - \frac{3}{16} = \frac{20}{48} - \frac{9}{48} = \frac{11}{48}$

**Section 4.3 (page 168)**

1. $k, 5k^2$

Least common denominator:

$5k^2$.

5. $4z, z - 1$

Least common denominator:

$4z(z - 1)$.

9. $a^2 - 16, (a - 4)^2$

$a^2 - 16 = (a - 4)(a + 4)$

Least common denominator:

$(a - 4)^2(a + 4)$.

13. $x^2 + 7x + 10, x^2 - 25$

$x^2 + 7x + 10 = (x + 2) \cdot (x + 5)$

$x^2 - 25 = (x + 5)(x - 5)$

Least common denominator:

$(x + 2)(x + 5)(x - 5)$.

17. $p^2 + 6p, p^2 + 7p + 6, 9p^2$

$p^2 + 6p = p(p + 6)$

$p^2 + 7p + 6 = (p + 6)(p + 1)$

$9p^2 = 9p \cdot p$

Least common denominator:

$9p^2(p + 6)(p + 1)$.

21. $\frac{1}{m + 1} + \frac{m}{m + 1} = \frac{1 + m}{m + 1}$ *Add the numerators*

$= \frac{m + 1}{m + 1}$

$= 1$

25. $\frac{a^2}{a + b} - \frac{b^2}{a + b} = \frac{a^2 - b^2}{a + b}$ *Subtract*

$= \frac{(a - b)(a + b)}{(a + b)}$ *Factor the numerator*

$= a - b$

29. $\frac{9}{r} + \frac{5}{2r} = \frac{2}{2} \cdot \frac{9}{r} + \frac{5}{2r}$ *Write each fraction with the least common denominator*

$= \frac{18}{2r} + \frac{5}{2r}$ *Simplify*

$= \frac{18 + 5}{2r}$ *Add numerators*

$= \frac{23}{2r}$ *Simplify*

33. $\frac{3}{4x} + \frac{5}{3x} = \frac{3}{4x} \cdot \frac{3}{3} + \frac{5}{3x} \cdot \frac{4}{4}$

$= \frac{9}{12x} + \frac{20}{12x}$

$= \frac{9 + 20}{12x}$

$= \frac{29}{12x}$

37. $\frac{1}{m - 1} + \frac{1}{m} = \frac{1}{m - 1} \cdot \frac{1}{m} + \frac{1}{m} \cdot \frac{(m - 1)}{(m - 1)}$

*Write each fraction with the least common denominator of m(m - 1)*

$= \frac{m}{m(m - 1)} + \frac{m - 1}{m(m - 1)}$ *Simplify*

$= \frac{m + m - 1}{m(m - 1)}$ *Add*

$= \frac{2m - 1}{m(m - 1)}$ *Simplify the numerator*

41. $\frac{5}{m-4} + \frac{2}{4-m} = \frac{5}{m-4} - \frac{2}{m-4}$

$= \frac{5-2}{m-4}$ *Subtract numerators*

$= \frac{3}{m-4}$ *Simplify*

45. $\frac{7}{3a+9} - \frac{5}{4a+12}$

$= \frac{7}{3(a+3)} - \frac{5}{4(a+3)}$ *Factor each denominator*

$= \frac{4}{4}\cdot\frac{7}{3(a+3)} - \frac{3}{3}\cdot\frac{5}{4(a+3)}$

*Write with common denominator*

$= \frac{28}{12(a+3)} - \frac{15}{12(a+3)}$ *Simplify*

$= \frac{13}{12(a+3)}$ *Subtract*

49. $\frac{-4}{a^2 - ab - 6b^2} - \frac{1}{a^2 + ab - 2b^2}$

$= \frac{-4}{(a-3b)(a+2b)} - \frac{1}{(a-b)(a+2b)}$

*Factor each denominator*

$= \frac{-4}{(a-3b)(a+2b)}\cdot\frac{a-b}{a-b} - \frac{1}{(a-b)(a+2b)}\cdot\frac{a-3b}{a-3b}$

*Write with common denominator*

$= \frac{-4(a-b)}{(a-b)(a-3b)(a+2b)} - \frac{a-3b}{(a-b)(a-3b)(a+2b)}$

*Simplify*

$= \frac{-4(a-b) - (a-3b)}{(a-b)(a-3b)(a+2b)}$ *Subtract*

$= \frac{-4a + 4b - a + 3b}{(a-b)(a-3b)(a+2b)}$

$= \frac{7b - 5a}{(a-b)(a-3b)(a+2b)}$

*Simplify the numerator*

53. $\frac{r-3s}{r^2 - 4s^2} + \frac{r-3}{r^2 - 4rs + 4s^2}$

$= \frac{r-3s}{(r-2s)(r+2s)} + \frac{r-s}{(r-2s)(r-2s)}$

*Factor*

$= \frac{r-2s}{r-2s}\cdot\frac{r-3s}{(r-2s)(r+2s)} + \frac{r+2s}{r+2s}\cdot\frac{r-s}{(r-2s)(r-2s)}$

*Write with common denominator*

$= \frac{(r-2s)(r-3s)}{(r-2s)^2(r+2s)} + \frac{(r+2s)(r-s)}{(r-2s)^2(r+2s)}$

*Simplify*

$= \frac{(r-2s)(r-3s) + (r+2s)(r-s)}{(r-2s)^2(r+2s)}$

*Add numerators*

$= \frac{r^2 - 5rs + 6s^2 + r^2 + rs - 2s^2}{(r-2s)^2(r+2s)}$

$= \frac{2r^2 - 4rs + 4s^2}{(r-2s)^2(r+2s)}$ *Simplify*

57. $\frac{3y}{y^2 + yz - 2z^2} + \frac{4y-1}{y^2 - z^2}$

$= \frac{3y}{(y+2z)(y-z)} + \frac{4y-1}{(y+z)(y-z)}$

*Factor each denominator*

$= \frac{3y}{(y+2z)(y-z)}\cdot\frac{y+z}{y+z} + \frac{4y-1}{(y+z)(y-z)}\cdot\frac{y+2z}{y+2z}$

*Write with common denominator*

$= \frac{3y(y+z) + (4y-1)(y+2z)}{(y+2z)(y-z)(y+z)}$

*Add*

$= \frac{3y^2 + 3yz + 4y^2 + 8yz - y - 2z}{(y+2z)(y-z)(y+z)}$

$= \frac{7y^2 + 11yz - y - 2z}{(y+2z)(y-z)(y+z)}$

61. $\frac{5}{2r^2 - 3rs - 2s^2} - \frac{3}{3r^2 - 7rs + 2s^2} + \frac{4}{6r^2 + rs - s^2}$

$= \frac{5}{(2r + s)(r - 2s)} - \frac{3}{(3r - s)(r - 2s)} + \frac{4}{(2r + s)(3r - s)}$

$= \frac{5}{(2r + s)(r - 2s)} \cdot \frac{3r - s}{3r - s} - \frac{3}{(3r - s)(r - 2s)} \cdot \frac{2r + s}{2r + s} + \frac{4}{(2r + s)(3r - s)} \cdot \frac{r - 2s}{r - 2s}$

$= \frac{15r - 5s - 6r - 3s + 4r - 8s}{(2r + s)(3r - s)(r - 2s)}$

$= \frac{13r - 16s}{(2r + s)(3r - s)(r - 2s)}$

65. $\frac{m + k}{m - k} (\frac{3m}{m - k} - \frac{2k}{k - m})$

$= \frac{m + k}{m - k} (\frac{3m}{m - k} - \frac{(-1)2k}{(-1)k - m})$

$= \frac{m + k}{m - k} (\frac{3m + 2k}{m - k})$

$= \frac{(m + k)(3m + 2k)}{(m - k)^2}$

69. $\frac{9r}{6r + 2} - \frac{1}{6r^2 + 2r} + \frac{5}{r}$

$= \frac{9r}{6r + 2} - \frac{1}{r(6r + 2)} + \frac{5}{r}$

$= \frac{9r}{6r + 2} \cdot \frac{r}{r} - \frac{1}{r(6r + 2)} + \frac{5}{r} \cdot \frac{6r + 2}{6r + 2}$

$= \frac{9r^2 - 1 + 30r + 10}{r(6r + 2)} = \frac{9r^2 + 30r + 9}{r(6r + 2)}$

$= \frac{3(3r + 1)(r + 3)}{2r(3r + 1)} = \frac{3(r + 3)}{2r}$

73. $(\frac{2k}{k - 3} + \frac{5k}{2k + 1}) \div \frac{k}{2k + 1}$

$= (\frac{2k}{k - 3} \cdot \frac{2k + 1}{2k + 1} + \frac{5k}{2k + 1} \cdot \frac{k - 3}{k - 3}) \div \frac{k}{2k + 1}$ *Work inside parentheses*

$= [\frac{4k^2 + 2k + 5k^2 - 15k}{(k - 3)(2k + 1)}] \div \frac{k}{2k + 1}$

$= \frac{9k^2 - 13k}{(k - 3)(2k + 1)} \cdot \frac{2k + 1}{k}$ *Multiply*

$= \frac{k(9k - 13)}{(k - 3)(2k + 1)} \cdot \frac{2k + 1}{k}$ *Factor*

$= \frac{9k - 13}{k - 3}$

77. $(a - b)^{-1} + (a + b)^{-1}$

$= \frac{1}{a - b} + \frac{1}{a + b}$

$= \frac{1}{a - b} \cdot \frac{a + b}{a + b} + \frac{1}{a + b} \cdot \frac{a - b}{a - b}$

$= \frac{a + b + a - b}{(a - b)(a + b)}$

$= \frac{2a}{(a - b)(a + b)}$

81. $\frac{\frac{2}{3} + \frac{1}{6}}{\frac{5}{9} - \frac{1}{3}} = \frac{18(\frac{2}{3} + \frac{1}{6})}{18(\frac{5}{9} - \frac{1}{3})} = \frac{12 + 3}{10 - 6} = \frac{15}{4}$

**Section 4.4 (page 173)**

1. $\frac{\frac{3}{x}}{\frac{6}{x - 1}} = \frac{3}{x} \div \frac{6}{x - 1}$ *Write using division sign*

$= \frac{3}{x} \cdot \frac{x - 1}{6}$ *Reciprocal of* $\frac{6}{x - 1}$

$= \frac{3(x - 1)}{6x}$ *Multiply*

$= \frac{(x - 1)}{2x}$

5. $\frac{\frac{m - 1}{4m}}{\frac{m + 1}{m}} = \frac{m - 1}{4m} \div \frac{m + 1}{m}$ *Write using a division sign*

$= \frac{m - 1}{4m} \cdot \frac{m}{m + 1}$ *Reciprocal of* $\frac{m + 1}{m}$

$= \frac{m(m - 1)}{4m(m + 1)}$ *Multiply*

$= \frac{m - 1}{4(m + 1)}$

9. $$\frac{\frac{22m^4n^5}{3m^5}}{\frac{11m^3n^2}{9m^8}} = \frac{22m^4n^5}{3m^5} \div \frac{11m^3n^2}{9m^8}$$

*Write using a division sign*

$$= \frac{22m^4n^5}{3m^5} \cdot \frac{9m^8}{11m^3n^2}$$

$$= \frac{198m^{12}n^5}{33m^8n^2} \quad \textit{Multiply}$$

$$= 6m^4n^3 \quad \textit{Write in lowest terms}$$

13. $$\frac{\frac{p + 2q}{q^2}}{\frac{p^2 - 4q^2}{2q}} = \frac{p + 2q}{q^2} \div \frac{p^2 - 4q^2}{2q}$$

*Write using a division sign*

$$= \frac{p + 2q}{q^2} \cdot \frac{2q}{p^2 - 4q^2}$$

*Reciprocal of* $\frac{p^2 - 4q^2}{2q}$

$$= \frac{p + 2q}{q^2} \cdot \frac{2q}{(p + 2q)(p - 2q)}$$

*Factor*

$$= \frac{(p + 2q)2q}{q^2(p + 2q)(p - 2q)}$$

*Multiply*

$$= \frac{2}{q(p - 2q)}$$

17. $$\frac{\frac{1}{k} + \frac{1}{r}}{\frac{1}{k} - \frac{1}{r}} = \frac{\frac{1}{k}\cdot\frac{r}{r} + \frac{1}{r}\cdot\frac{k}{k}}{\frac{1}{k}\cdot\frac{r}{r} - \frac{1}{r}\cdot\frac{k}{k}}$$

*Write numerator and denominator with least common denominator*

$$= \frac{\frac{r}{kr} + \frac{k}{kr}}{\frac{r}{kr} - \frac{k}{kr}}$$

$$= \frac{\frac{r + k}{kr}}{\frac{r - k}{kr}} \quad \textit{Add or subtract as indicated}$$

$$= \frac{r + k}{kr} \div \frac{r - k}{kr} \quad \textit{Write using a division sign}$$

$$= \frac{r + k}{kr} \cdot \frac{kr}{r - k} \quad \textit{Change to multiplication}$$

$$= \frac{kr(r + k)}{kr(r - k)} \quad \textit{Multiply}$$

$$= \frac{r + k}{r - k}$$

21. $$\frac{\frac{1}{r} - \frac{4}{s}}{\frac{s^2 - 16r^2}{rs}} = \frac{\frac{1}{r}\cdot\frac{s}{s} - \frac{4}{s}\cdot\frac{r}{r}}{\frac{s^2 - 16r^2}{rs}}$$

*Write each fraction with least common denominator*

$$= \frac{\frac{s}{rs} - \frac{4r}{rs}}{\frac{s^2 - 16r^2}{rs}} = \frac{\frac{s - 4r}{rs}}{\frac{s^2 - 16r^2}{rs}}$$

*Simplify*

*Subtract on top*

$$= \frac{s - 4r}{rs} \cdot \frac{rs}{s^2 - 16r^2}$$

*Change to multiplication*

$$= \frac{s - 4r}{rs} \cdot \frac{rs}{(s - 4r)(s + 4r)}$$

*Factor where possible*

$$= \frac{rs(s - 4r)}{rs(s - 4r)(s + 4r)}$$

*Multiply*

$$= \frac{1}{s + 4r} \quad \textit{Write in lowest terms}$$

25. $$\frac{y - \frac{y - 3}{3}}{\frac{4}{9} + \frac{2}{3y}} = \frac{y\cdot\frac{9y}{9y} - \frac{y - 3}{3}\cdot\frac{3y}{3y}}{\frac{4}{9}\cdot\frac{y}{y} + \frac{2}{3y}\cdot\frac{3}{3}}$$

*Write with a common denominator*

$$= \frac{\frac{9y^2}{9y} - \frac{3y(y - 3)}{9y}}{\frac{4y}{9y} + \frac{6}{9y}}$$

$$= \frac{\frac{9y^2 - 3y(y - 3)}{9y}}{\frac{4y + 6}{9y}} \quad \textit{Combine}$$

$$= \frac{9y^2 - 3y(y - 3)}{9y} \cdot \frac{9y}{4y + 6}$$

*Multiply by reciprocal of* $\frac{4y + 6}{9y}$

$$= \frac{9y^2 - 3y^2 + 9y}{1} \cdot \frac{1}{4y + 6}$$

$$= \frac{6y^2 + 9y}{4y + 6} \quad \textit{Simplify}$$

$$= \frac{3y(2y + 3)}{2(2y + 3)} \quad \textit{Factor}$$

$$= \frac{3y}{2} \quad \textit{Write in lowest terms}$$

29. $$\frac{\frac{a+3}{a} - \frac{4}{a-1}}{\frac{a}{a-1} + \frac{1}{a}}$$

$$= \frac{\frac{a+3}{a}\cdot\frac{a-1}{a-1} - \frac{4}{a-1}\cdot\frac{a}{a}}{\frac{a}{a-1}\cdot\frac{a}{a} + \frac{1}{a}\cdot\frac{a-1}{a-1}}$$ *Write each fraction with least common denominator*

$$= \frac{\frac{(a+3)(a-1) - 4a}{a(a-1)}}{\frac{a^2 + a - 1}{a(a-1)}}$$

$$= \frac{\frac{a^2 + 2a - 3 - 4a}{a(a-1)}}{\frac{a^2 + a - 1}{a(a-1)}}$$ *Simplify*

$$= \frac{a^2 - 2a - 3}{a^2 + a - 1}$$

33. $$\frac{\frac{2}{a-1} + \frac{3}{a}}{4 - \frac{2}{1-a}} = \frac{\frac{2}{a-1}\cdot\frac{a}{a} + \frac{3}{a}\cdot\frac{a-1}{a-1}}{4\cdot\frac{a-1}{a-1} - \frac{2}{-(a-1)}}$$

*Write each fraction with least common denominator*

$$= \frac{\frac{2a + 3(a-1)}{a(a-1)}}{\frac{4(a-1)+2}{a-1}}$$

$$= \frac{\frac{5a-3}{a(a-1)}}{\frac{4a-2}{a-1}}$$ *Simplify*

$$= \frac{5a-3}{a(a-1)}\cdot\frac{a-1}{4a-2}$$ *Multiply by reciprocal of denominator*

$$= \frac{(a-1)(5a-3)}{a(a-1)(4a-2)}$$

$$= \frac{5a-3}{a(4a-2)}$$

$$= \frac{5a-3}{2a(2a-1)}$$

37. $$m^{-1} - 1 = \frac{1}{m} - 1 = \frac{1}{m} - \frac{m}{m}$$

$$= \frac{1-m}{m}$$

41. $$\frac{x^{-2}}{x^{-2} - y^{-2}} = \frac{\frac{1}{x^2}}{\frac{1}{x^2} - \frac{1}{y^2}}$$

$$= \frac{\frac{1}{x^2}}{\frac{1}{x^2}\cdot\frac{y^2}{y^2} - \frac{1}{y^2}\cdot\frac{x^2}{x^2}}$$ *Simplify denominator*

$$= \frac{\frac{1}{x^2}}{\frac{y^2 - x^2}{x^2y^2}}$$

$$= \frac{1}{x^2}\cdot\frac{x^2y^2}{y^2 - x^2}$$ *Multiply by reciprocal of denominator*

$$= \frac{y^2}{y^2 - x^2}$$

45. $$(a^{-1} + b^{-1})^{-1} = (\frac{1}{a} + \frac{1}{b})^{-1}$$

$$= (\frac{1}{a}\cdot\frac{b}{b} + \frac{1}{b}\cdot\frac{a}{a})^{-1}$$

*Simplify inside parentheses*

$$= (\frac{b+a}{ab})^{-1}$$

$$= \frac{ab}{b+a}$$

49. $$2 - \frac{1}{2 + \frac{1}{x}} = 2 - \frac{1}{2 + \frac{1}{x}}\cdot\frac{x}{x}$$

$$= 2 - \frac{x}{2x+1}$$

$$= \frac{(2x+1)2}{2x+1} - \frac{x}{2x+1}$$

$$= \frac{4x + 2 - x}{2x+1}$$

$$= \frac{3x+2}{2x+1}$$

53. $$\frac{1}{p+\frac{1}{p+\frac{1}{1+p}}} = \frac{1}{p+\frac{1}{\frac{p+p^2+1}{1+p}}}$$

$$= \frac{1}{\frac{p(p+p^2+1)+p+1}{p+p^2+1}}$$

$$= \frac{p^2+p+1}{p(p^2+p+1)+p+1}$$

$$= \frac{p^2+p+1}{p^3+p^2+2p+1}$$

57. $$\frac{2(m+1)^{-1}-(m-1)^{-1}}{(m+1)^{-1}+5(m-1)^{-1}}$$

$$= \frac{\frac{2}{m+1}-\frac{1}{m-1}}{\frac{1}{m+1}+\frac{5}{m-1}}$$

$$= \frac{\frac{2}{m+1}\cdot\frac{m-1}{m-1}-\frac{1}{m-1}\cdot\frac{m+1}{m+1}}{\frac{1}{m+1}\cdot\frac{m-1}{m-1}+\frac{5}{m-1}\cdot\frac{m+1}{m+1}}$$

$$= \frac{\frac{2m-2-m-1}{(m-1)(m+1)}}{\frac{m-1+5m+5}{(m-1)(m+1)}}$$

$$= \frac{m-3}{6m+4}$$

61. $$-\frac{8a^3b^7}{6a^5b} = -\frac{4b^6}{3a^2}$$

65. $$10x^3 - 8x^2 + 4x - (12x^3 + 5x^2 - 7x)$$

$$= 10x^3 - 8x^2 + 4x - 12x^3 - 5x^2 + 7x$$

$$= -2x^3 - 13x^2 + 11x$$

**Section 4.5 (page 178)**

1. $$\frac{5a^2-10a+5}{5} = \frac{5a^2}{5} - \frac{10a}{5} + \frac{5}{5}$$

*Divide each term by 5*

$$= a^2 - 2a + 1$$

*Write in lowest terms*

5. $$\frac{12a^3+8a^2-10a}{4a^3} = \frac{12a^3}{4a^3} + \frac{8a^2}{4a^3} - \frac{10a}{4a^3}$$

*Divide each term by $4a^3$*

$$= 3 + \frac{2}{a} - \frac{5}{2a^2}$$

9. $$\frac{x^2-2x-15}{x-5} = \frac{(x-5)(x+3)}{x-5}$$

*Factor*

$$= x + 3$$

13. $$\frac{3m^3+5m^2-5m+1}{3m-1}$$

Write as a long division:

$$\begin{array}{r|l} & m^2 + 2m - 1 \\ 3m - 1 & 3m^3 + 5m^2 - 5m + 1 \\ & \underline{3m^3 - m^2} \quad \textit{Subtract } 3m^3 - m^2 \textit{ Change signs} \\ & 6m^2 - 5m \\ & \underline{6m^2 - 2m} \quad \textit{Subtract} \\ & -3m + 1 \\ & \underline{-3m + 1} \\ & 0 \end{array}$$

Answer: $m^2 + 2m - 1$

17. $$\begin{array}{r|l} & a^2 - 5a - 9 \\ a - 3 & a^3 - 8a^2 + 6a - 3 \\ & \underline{a^3 - 3a^2} \\ & -5a^2 + 6a \\ & \underline{-5a^2 + 15a} \\ & -9a - 3 \\ & \underline{-9a + 27} \\ & -30 \quad \textit{Remainder} \end{array}$$

Answer: $a^2 - 5a - 9 - \dfrac{30}{a-3}$

21.

$$\begin{array}{r} 9t^2 - 4t + 1 \\ t^2 - t + 2\overline{)9t^4 - 13t^3 + 23t^2 - 9t + 2} \\ \underline{9t^4 - 9t^3 + 18t^2} \\ - 4t^3 + 5t^2 - 9t \\ \underline{- 4t^3 + 4t^2 - 8t} \\ t^2 - t + 2 \\ \underline{t^2 - t + 2} \\ 0 \end{array}$$

Answer: $9t^2 - 4t + 1$

25.

$$\begin{array}{r} 2y^2 + 3y - 1 \\ 2y^2 + 1\overline{)4y^4 + 6y^3 + 0y^2 + 3y - 1} \end{array}$$

*Write $+0y^2$ as placeholder*

$$\begin{array}{r} \underline{4y^4 + 2y^2} \\ 6y^3 - 2y^2 + 3y \\ \underline{6y^3 + 3y} \\ - 2y^2 - 1 \\ \underline{- 2y^2 - 1} \\ 0 \end{array}$$

Answer: $2y^2 + 3y - 1$

29. $\frac{16r^4 + 1}{2r - 1}$

$$\begin{array}{r} 8r^3 + 4r^2 + 2r + 1 \\ 2r - 1\overline{)16r^4 + 0r^3 + 0r^2 + 0r + 1} \\ \underline{16r^4 - 8r^3} \\ 8r^3 + 0r^2 \\ \underline{8r^3 - 4r^2} \\ 4r^2 + 0r \\ \underline{4r^2 - 2r} \\ 2r + 1 \\ \underline{2r - 1} \\ 2 \end{array}$$

Answer: $8r^3 + 4r^2 + 2r + 1 + \frac{2}{2r - 1}$

33.

$$\begin{array}{r} \frac{2}{3}x - 1 \\ 3x + 1\overline{)2x^2 - \frac{7}{3}x - 1} \\ \underline{2x^2 + \frac{2}{3}x} \\ - 3x - 1 \\ \underline{- 3x - 1} \\ 0 \end{array}$$

Answer: $\frac{2}{3}x - 1$

37.

$$\begin{array}{r} \frac{2}{3}a^3 - \frac{5}{3}a + \frac{5}{3} \\ 3a^2 + 3\overline{)2a^5 + 0a^4 - 3a^3 + 5a^2 + 0a - 8} \end{array}$$

*Write placeholders*

$$\begin{array}{r} \underline{2a^5 + 2a^3} \\ - 5a^3 + 5a^2 + 0a \\ \underline{- 5a^3 - 5a} \\ 5a^2 + 5a - 8 \\ \underline{5a^2 + 5} \\ 5a - 13 \end{array}$$

*Remainder*

Answer: $\frac{2}{3}a^3 - \frac{5}{3}a + \frac{5}{3} + \frac{5a - 13}{3a^2 + 3}$

41. Since $d = rt$, $r = \frac{d}{t}$.

$d = 2m^3 + 15m^2 + 13m - 63$ and $t = 2m + 9$

and

$$r = \frac{2m^3 + 15m^2 + 13m - 63}{2m + 9}.$$

$$\begin{array}{r} m^2 + 3m - 7 \\ 2m + 9\overline{)2m^3 + 15m^2 + 13m - 63} \\ \underline{2m^3 + 9m^2} \\ 6m^2 + 13m \\ \underline{6m^2 + 27m} \\ - 14m - 63 \\ \underline{- 14m - 63} \\ 0 \end{array}$$

The car's rate is $m^2 + 3m - 7$ kilometers per hour.

45. $x^2 + 2x - 9$

$= 1^2 + 2(1) - 9$ *Let x = 1*

$= 1 + 2 - 9$

$= -6$

49. $2x^3 + x^2 - 5x - 5$

$= 2(-2)^3 + (-2)^2 - 5(-2) - 4$

*Let x = -2*

$= 2(-8) + 4 + 10 - 4$

$= -16 + 4 + 10 - 4$

$= -6$

**Section 4.6 (page 183)**

1. $\dfrac{x^2 + 6x - 7}{x - 1}$

| 1) | 1 | 6 | -7 |
|---|---|---|---|
| | | 1 | 7 |
| | 1 | 7 | 0 |

Answer $X + 7$

5. $(p^2 - 3p + 5) \div (p + 1)$

| -1) | 1 | -3 | 5 |
|---|---|---|---|
| | | -1 | 4 |
| | 1 | -4 | 9 |

Answer: $p - 4 + \dfrac{9}{p + 1}$ *Write the remainder as a fraction*

9. $(6x^5 - 2x^3 - 4x^2 + 3x - 2) \div (x - 2)$

| 2) | 6 | 0 | - 2 | - 4 | 3 | - 2 |
|---|---|---|---|---|---|---|
| | | 12 | 24 | 44 | 80 | 166 |
| | 6 | 12 | 22 | 40 | 83 | 164 |

*Write zero for the missing $x^4$ term*

$= 6x^4 + 12x^3 + 22x^2 + 40x + 83 + \dfrac{164}{x - 2}$

13. $\dfrac{y^3 - 1}{y + 1}$

| -1) | 1 | 0 | 0 | -1 |
|---|---|---|---|---|
| | | -1 | 1 | -1 |
| | 1 | -1 | 1 | -2 |

*Write zero for the missing $y^2$ and y terms*

Answer: $y^2 - y + 1 + \dfrac{-2}{y + 1}$.

17. $\dfrac{.2m^3 - .28m^2 + .49m - .36}{m - .9}$

| .9) | .2 | -.28 | .49 | -.36 |
|---|---|---|---|---|
| | | .18 | -.09 | .36 |
| | .2 | -.10 | .40 | 0 |

Answer: $.2m^2 - .1m + .4$

21. $k = -4;\ P(r) = -r^3 + 8r^2 - 3r - 2$

| -4) | -1 | 8 | - 3 | - 2 |
|---|---|---|---|---|
| | | 4 | -48 | 204 |
| | -1 | 12 | -51 | 202 |

$P(-4) = 202$

25. $k = -2;\ P(x) = x^4 - 3x^3 + 5x^2 - 2x + 5$

| -2) | 1 | -3 | 5 | - 2 | 5 |
|---|---|---|---|---|---|
| | | -2 | 10 | -30 | 64 |
| | 1 | -5 | 15 | -32 | 69 |

$P(-2) = 69$

29. -2 is a solution to

$3a^3 + 2a^2 - 2a + 11 = 0$ if $P(-2) = 0$.

| -2) | 3 | 2 | -2 | 11 |
|---|---|---|---|---|
| | | -6 | 8 | -12 |
| | 3 | -4 | 6 | - 1 |

The remainder is not zero, so -2 is not a solution.

33. $-2$ is a solution to
$-7w^6 + 15w^5 - w^4 + 16w^3 - 4w + 4 = 0$
if $P(-2) = 0$.

$$\begin{array}{r|rrrrrr} -2) & -7 & 15 & -1 & 16 & -4 & 4 \\ & & 14 & -58 & 118 & -268 & 544 \\ \hline & -7 & 29 & -59 & 134 & -272 & 548 \end{array}$$

The remainder is not zero, so $-2$ is not a solution.

37. $2x^3 - x^2 - 13x + 24 = 0$; $x = -3$

$$\begin{array}{r|rrrr} -3) & 2 & -1 & -13 & 24 \\ & & -6 & 21 & -24 \\ \hline & 2 & -7 & 8 & 0 \end{array}$$

The remainder is zero, so $-3$ is a solution.

$2x^3 - x^2 - 13x + 24$

$= (x + 3)(2x^2 - 7x + 8)$

41. $2y^4 - 5y^3 + 11y^2 - 14y + 3 = 0$; $y = \frac{3}{2}$

$$\begin{array}{r|rrrrr} \frac{3}{2}) & 2 & -5 & 11 & -14 & 3 \\ & & 3 & -3 & 12 & -3 \\ \hline & 2 & -2 & 8 & -2 & 0 \end{array}$$

$\frac{3}{2}$ is a solution since the remainder is zero.

$2y^4 - 5y^3 + 11y^2 - 14y + 3$

$= (y - \frac{3}{2})(2y^3 - 2y^2 - 8y - 2)$

45. $k = -2$; $x^3 + x^2 - 3x - 2 = 0$

$$\begin{array}{r|rrrr} -2) & 1 & 1 & -3 & -2 \\ & & -2 & 2 & 2 \\ \hline & 1 & -1 & -1 & 0 \end{array}$$

*$-2$ is a solution*

$x^3 + x^2 - 3x - 2$

$= (x + 2)(x^2 - x - 1)$

The polynomial $x^2 - x - 1$ cannot be factored so no other solutions can be given; $-2$ is the only rational solution.

49. $7z - 2 = 5z + 4$

$7z = 5z + 6$

$2z = 6$

$z = 3$

Solution set: $\{3\}$

53. $q^2 - 4q + 3 = 0$

$(q - 3)(q - 1) = 0$

$q = 1$ or $q = 3$

Solution set: $\{1, 3\}$

**Section 4.7 (page 187)**

1. $\frac{p}{4} - \frac{p}{8} = 2$

$8(\frac{p}{4} - \frac{p}{8}) = 8 \cdot 2$ *Multiply each side by least common denominator 8*

$2p - p = 16$

$p = 16$

Solution set: $\{16\}$

5. $\frac{3y - 6}{2} = \frac{5y + 1}{6}$

$6(\frac{3y - 6}{2}) = 6(\frac{5y + 1}{6})$ *Multiply each side by 6*

$3(3y - 6) = 5y + 1$ *Simplify each side*

$9y - 18 = 5y + 1$

$4y = 19$ *Add -5 + 18 to each side*

$y = \frac{19}{4}$ *Divide each side by 4*

Solution set: $\{\frac{19}{4}\}$

Note: Students should always check that their solutions make true statements when substituted in the original equations. This is especially important in the following exercises where fractions with variables in the denominator are undefined for some values of the variables.

9. $$\frac{1}{a} + \frac{2}{3a} = \frac{1}{3}$$

$$3a(\frac{1}{a}) + 3a(\frac{2}{3a}) = 3a(\frac{1}{3})$$ *Multiply each term by 3a*

$$3 + 2 = a$$ *Simplify and solve*

$$5 = a$$

Solution set: {5}.

13. $$\frac{3}{4a - 8} - \frac{2}{3a - 6} = \frac{1}{36}$$

$$\frac{3}{4(a - 2)} - \frac{2}{3(a - 2)} = \frac{1}{36}$$ *Factor*

$$36(a - 2)[\frac{3}{4(a - 2)} - \frac{2}{3(a - 2)}] =$$

$$36(a - 2)\frac{1}{36}$$ *Least common denominator is 36(a - 2)*

$$9(3) - 12(2) = a - 2$$

*Distributive property*

$$27 - 24 = a - 2$$

$$3 = a - 2$$

$$5 = a$$

Solution set: {5}

17. $$\frac{5}{6q + 14} - \frac{2}{3q + 7} = \frac{1}{56}$$

$$\frac{5}{2(3q + 7)} - \frac{2}{3q + 7} = \frac{1}{56}$$

$$[56(3q + 7)][\frac{5}{2(3q + 7)} - \frac{2}{3q + 7}] =$$

$$[56(3q + 7)]\frac{1}{56}$$ *Least common denominator: 56(3q + 7)*

$$28(5) - 56(2) = 3q + 7$$

*Distributive property*

$$140 - 112 = 3q + 7$$

$$21 = 3q$$

$$7 = q$$

Solution set: {7}

21. $$\frac{2}{y - 5} \quad \frac{1}{y + 5} = \frac{11}{y^2 - 25}$$

$$\frac{2}{y - 5} \quad \frac{1}{y + 5} = \frac{11}{(y + 5)(y - 5)}$$

$$(y + 5)(y - 5)(\frac{2}{y - 5} + \frac{1}{y + 5})$$

$$= (y + 5)(y - 5)(\frac{11}{(y + 5)(y - 5)})$$

$$2(y + 5) + (y - 5) = 11$$

$$2y + 10 + y - 5 = 11$$

$$3y + 5 = 11$$

$$3y = 6$$

$$y = 2$$

Solution set: {2}

25. $$2 + \frac{7}{a} = \frac{4}{a^2}$$

$$a^2(2 + \frac{7}{a}) = a^2 \cdot \frac{4}{a^2}$$

$$2a^2 + 7a = 4$$

$$2a^2 + 7a - 4 = 0$$

$$(2a - 1)(a + 4) = 0$$

$$a = \frac{1}{2} \quad \text{or} \quad a = -4$$

Solution set: $\{-4, \frac{1}{2}\}$

29. $$\frac{2}{p} - \frac{3}{p - 2} = \frac{-3}{p^2 - 2p}$$

$$p[p - 2)[\frac{2}{9} - \frac{3}{p - 2}] = p(p - 2)[\frac{-3}{p(p - 2)}]$$

$$2(p - 2) - 3p = -3$$

$$2p - 4 - 3p = -3$$

$$-p - 4 = -3$$

$$-p = 1$$

$$p = -1$$

Solution set: {-1}.

33. $$\frac{7}{p-5} = \frac{p^2-10}{p^2-p-20} - 1$$

$$\frac{7}{p-5}(p-5)(p+4) =$$

$$\frac{p^2-10}{(p-5)(p+4)}(p-5)(p+4) - 1(p-5)(p+4)$$

$$7p + 28 = p^2 - 10 - p^2 + p + 20$$

$$6p = -18$$

$$p = -3$$

Solution set: $\{-3\}$

37. $$\frac{5m-22}{m^2-6m+9} - \frac{11}{m^2-3m} = \frac{5}{m}$$

$$\frac{5m-22}{(m-3)(m-3)} - \frac{11}{m(m-3)} = \frac{5}{m}$$

$$\left[\frac{5m-22}{(m-3)^2} - \frac{11}{m(m-3)}\right](m-3)^2m$$

$$= \frac{5}{m}(m-3)^2m$$

$$5m^2 - 22m - 11m + 33 = 5m^2 - 30m + 45$$

$$-3m = 12$$

$$m = -4$$

Solution set: $\{-4\}$

41. $$\frac{x}{.9} + \frac{3x}{.4} = 31$$

$$(.36)\left(\frac{x}{.9}\right) + (.36)\left(\frac{3x}{.4}\right) = (.36)(31)$$

$$.4x + .9(3x) = 11.16$$

$$.4x + 2.7x = 11.16$$

$$3.1x = 11.16$$

$$x = 3.6$$

Solution set: $\{3.6\}$

45. $$7p^{-1} + 2p^{-2} - 4 = 0$$

$$\frac{7}{p} + \frac{2}{p^2} - 4 = 0$$

$$[p^2]\left[\frac{7}{p} + \frac{2}{p^2} - 4\right] = 0[p^2]$$

$$7p + 2 - 4p^2 = 0$$

$$4p^2 - 7p - 2 = 0 \quad \textit{Multiply by -1}$$

$$(4p + 1)(p - 2) = 0$$

$$4p + 1 = 0 \quad \text{or} \quad p - 2 = 0$$

$$4p = -1$$

$$p = -\frac{1}{4} \quad \text{or} \quad p = 2$$

Solution set: $\{-\frac{1}{4}, 2\}$

49. $$\left(\frac{x}{x-1}\right)^2 + 2\left(\frac{x}{x-1}\right) - 3 = 0$$

Let $y = \frac{x}{x-1}$,

$$y^2 + 2y - 3 = 0$$

$$(y + 3)(y - 1) = 0 \quad \textit{Factor}$$

$$y + 3 = 0 \quad \text{or} \quad y = 1$$

$$y = -3 \quad \text{or} \quad y = 1$$

Substitute 1 for y:

$$\frac{x}{x-1} = 1$$

$$(x - 1)\left(\frac{x}{x-1}\right) = 1(x - 1)$$

$$x = x - 1$$

$$0 = -1 \quad \textit{False}$$

Solution set: $\{\frac{3}{4}\}$

53. Let x be the number.

| Half a number | added to | two-thirds of the number | gives | 10. |
|---|---|---|---|---|
| ↓ | ↓ | ↓ | ↓ | ↓ |
| $\frac{1}{2}x$ | $+$ | $\frac{2}{3}x$ | $=$ | 10 |

Solve this equation.

$$6\left(\frac{1}{2}x\right) + 6\left(\frac{2}{3}x\right) = 6(10)$$

$$3x + 4x = 60$$

$$7x = 60$$

$$x = \frac{60}{7}$$

The number is $\frac{60}{7}$.

57. Let x be the number.

$$\frac{1}{2} + \frac{1}{3} = \frac{1}{x}$$

$$6x\left(\frac{1}{2} + \frac{1}{3}\right) = 6x\left(\frac{1}{x}\right)$$

$$3x + 2x = 6$$

$$5x = 6$$

$$x = \frac{6}{5}$$

The number is $\frac{6}{5}$.

**Supplementary Exercises on Rational Expressions and Equations (page 189)**

1. $$\frac{a}{2} - \frac{a}{5} = 3$$

$$10\left(\frac{a}{2} - \frac{a}{5}\right) = 10 \cdot 3$$

$$5a - 2a = 30$$

$$3a = 30$$

$$a = 10$$

Solution set: {10}

5. $$\frac{5}{m} - \frac{10}{3m} = \frac{3}{3} \cdot \frac{5}{m} - \frac{10}{3m} = \frac{15}{3m} - \frac{10}{3m} = \frac{5}{3m}$$

9. $$\frac{r - 5}{3} - \frac{r - 2}{5} = -\frac{1}{3}$$

$$15\left(\frac{r - 5}{3} - \frac{r - 2}{5}\right) = 15\left(-\frac{1}{3}\right)$$

$$5(r - 5) - 3(r - 2) = -5$$

$$5r - 25 - 3r + 6 = -5$$

$$2r - 19 = -5$$

$$2r = 14$$

$$r = 7$$

Solution set: {7}

13. $$\frac{\frac{5}{m} - \frac{2}{m + 1}}{\frac{3}{m + 1} + \frac{1}{m}} = \frac{\frac{m + 1}{m + 1} \cdot \frac{5}{m} - \frac{2}{m + 1} \cdot \frac{m}{m}}{\frac{m}{m} \cdot \frac{3}{m + 1} + \frac{1}{m} \cdot \frac{m + 1}{m + 1}}$$

$$= \frac{\frac{5(m + 1)}{m(m + 1)} - \frac{2m}{m(m + 1)}}{\frac{3m}{m(m + 1)} + \frac{m + 1}{m(m + 1)}}$$

$$= \frac{\frac{5(m + 1) - 2m}{m(m + 1)}}{\frac{3m + m + 1}{m(m + 1)}}$$

$$= \frac{\frac{3m + 5}{m(m + 1)}}{\frac{4m + 1}{m(m + 1)}}$$

$$= \frac{3m + 5}{4m + 1}$$

17. $$\frac{9}{2 - x} \cdot \frac{4x - 8}{5} = \frac{9}{2 - x} \cdot \frac{4(x - 2)}{5}$$

$$= \frac{-9}{x - 2} \cdot \frac{4(x - 2)}{5}$$

$$= \frac{-36(x - 2)}{5(x - 2)}$$

$$= \frac{-36}{5}$$

21. $$\frac{5}{a^2 - 2a} + \frac{3}{a^2 - 4}$$

$$= \frac{5}{a(a - 2)} + \frac{3}{(a - 2)(a + 2)}$$

$$= \frac{a + 2}{a + 2} \cdot \frac{5}{a(a - 2)} + \frac{a}{a} \cdot \frac{3}{(a - 2)(a + 2)}$$

$$= \frac{5(a + 2)}{a(a^2 - 4)} + \frac{3a}{a(a^2 - 4)}$$

$$= \frac{5a + 10 + 3a}{a(a^2 - 4)}$$

$$= \frac{8a + 10}{a(a^2 - 4)}$$

$$= \frac{8a + 10}{a(a - 1)(a + 2)}$$

25. $$\frac{\frac{3}{k} - \frac{5}{m}}{\frac{9m^2 - 25k^2}{km^2}} = \frac{\frac{m^2}{m^2} \cdot \frac{3}{k} - \frac{mk}{mk} \cdot \frac{5}{m}}{\frac{9m^2 - 25k^2}{km^2}}$$

$$= \frac{\frac{3m^2}{m^2k} - \frac{5mk}{m^2k}}{\frac{9m^2 - 25k^2}{km^2}}$$

$$= \frac{\frac{3m^2 - 5mk}{m^2k}}{\frac{9m^2 - 25k^2}{m^2k}}$$

$$= \frac{3m^2 - 5mk}{9m^2 - 25k^2}$$

$$= \frac{m(3m - 5k)}{(3m + 5k)(3m - 5k)}$$

$$= \frac{m}{3m + 5k}$$

29. $\frac{8}{3p + 9} - \frac{2}{5p + 15} = \frac{8}{15}$

$$\frac{8}{3(p + 3)} - \frac{2}{5(p + 3)} = \frac{8}{15}$$

$$15(p + 3)[\frac{8}{3(p + 3)} - \frac{2}{5(p + 3)}]$$

$$= 15(p + 3)\cdot\frac{8}{15}$$

$$40 - 6 = 8(p + 3)$$

$$34 = 8p + 24$$

$$10 = 8p$$

$$p = \frac{10}{8} = \frac{5}{4}$$

Solution set: $\{\frac{5}{4}\}$.

33. $\frac{2}{q + 1} - \frac{3}{q^2 - q - 2} = \frac{3}{q - 2}$

$$\frac{2}{q + 1} - \frac{3}{(q + 1)(q - 2)} = \frac{3}{q - 2}$$

$$(q + 1)(q - 2)[\frac{2}{q + 1} - \frac{3}{(q + 1)(q - 2)}]$$

$$= (q + 1)(q - 2)\cdot\frac{3}{q - 2}$$

$$2(q - 2) - 3 = 3(q + 1)$$

$$2q - 4 - 3 = 3q + 3$$

$$2q - 7 = 3q + 3$$

$$-10 = q$$

Solution set: $\{-10\}$

37. $\frac{2x^2 - 11xy + 15y^2}{2x^2 + 5xy - 3y^2}\cdot\frac{8x^2 - 2xy - y^2}{8x^2 - 18xy - 5y^2}$

$$= \frac{(2x - 5y)(x - 3y)\cdot(4x + y)(2x - y)}{(2x - y)(x + 3y)\cdot(4x + y)(2x - 5y)}$$

$$= \frac{(2x - 5y)(4x + y)(2x - y)(x - 3y)}{(2x - 5y)(4x + y)(2x - y)(x + 3y)}$$

$$= \frac{x - 3y}{x + 3y}$$

**Section 4.8 (page 195)**

1. $I = \frac{E}{R}$

$R = \frac{E}{I}$ *Solve for R*

$r = \frac{8}{20} = \frac{2}{5}$ *Substitute 8 for E and 20 for I*

5. $\frac{1}{R} = \frac{1}{R_1} + \frac{1}{R_2}$

Substitute 8 for R and 12 for $R_2$ and solve for $R_1$:

$$\frac{1}{8} = \frac{1}{R_1} + \frac{1}{12}$$

$$24R_1(\frac{1}{8}) = 24R_1(\frac{1}{R_1}) + 24R_1(\frac{1}{12})$$

$$3R_1 = 24 + 2R_1$$

$$R_1 = 24$$

9. $F = \frac{Gm_1m_2}{d^2}$

$Fd^2 = Gm_1m_2$ *Multiply each side*

$\frac{Fd^2}{Gm_2} = m_1$ *Divide each side by $Gm_2$*

13. $\frac{V_1P_1}{T_1} = \frac{V_2P_2}{T_2}$

$V_1P_1 = \frac{V_2P_2}{T_2}\cdot T_1$ *Multiply each side by $T_1$*

$\frac{T_2V_1P_1}{V_2P_2} = T_1$ *Multiply each side by $\frac{T_2}{V_2P_2}$*

17. $S_n = \frac{n}{2}(a_1 + a_n)$

$2S_n = 2[\frac{n}{2}(a_1 + a_n)]$ *Multiply each side by 2*

$2S_n = n(a_1 + a_n)$ *Simplify*

$2S_n = na_1 + na_n$ *Remove parentheses*

$2S_n - na_n = na_1$ *Add $-na_n$ to each side*

$\frac{2S_n - na_n}{n} = a_1$ *Divide each side by n*

21. Let x represent the number.

$$\frac{15 + x}{17 + x} = \frac{5}{6}$$

$$6(17 + x)\frac{15 + x}{17 + x} = 6(17 + x)\frac{5}{6}$$

$$90 + 6x = 85 + 5x$$

$$x = -5$$

25. Let the three consecutive integers be

x, x + 1, x + 2

The third dividend by the second is one more than quotient of first and second integers.

$$\frac{x+2}{x+1} = \quad + \quad \frac{x}{x+1}$$

$$(x + 1)\frac{x + 2}{x - 1} = (x+1)1 + (x+1)\frac{x}{x+1}$$

$$x + 2 = x + 1 + x$$

$$1 = x$$

The integers are 1, 2, and 3.

29. Let x be the number of hours it will take Fred and John working together. The part done by Fred in one hour is $\frac{1}{5}$. The part done by John in one hour is $\frac{1}{6}$. The part done working together in one hour is $\frac{1}{x}$.

$$\frac{1}{5} + \frac{1}{6} = \frac{1}{x}$$

$$(6\cdot5\cdot x)\left(\frac{1}{5} + \frac{1}{6}\right) = (6\cdot5\cdot x)\left(\frac{1}{x}\right)$$

$$6x + 5x = 30$$

$$11x = 30$$

$$x = \frac{30}{11}$$

It will take them 30/11 hours.

33. Let x be the distance each way.

| | d | r | t |
|---|---|---|---|
| Bike | x | 12 | $\frac{x}{12}$ |
| Car | x | 36 | $\frac{x}{36}$ |

$d = rt$

$\frac{d}{r} = t$ *Solve for t*

Time of bike less time of car is difference in time.

$$\frac{x}{12} - \frac{x}{36} = \frac{1}{2}$$

$$36\left(\frac{x}{12}\right) - 36\left(\frac{x}{36}\right) = 36\left(\frac{1}{2}\right)$$

$$3x - x = 18$$

$$2x = 18$$

$$x = 9$$

The distance to school is 9 miles.

37. $y = \frac{x + z}{a - x}$ for x

$y(a - x) = \frac{x + z}{a - x}(a - x)$ *Multiply by (a - x)*

$ay - xy = x + z$

$ay - z = x + xy$

$ay - z = x(1 + y)$ *Factor*

$\frac{ay - z}{1 + y} = x$ *Divide both sides by (1 + y)*

41. $\frac{E}{e} = \frac{R + r}{r}$

$er\left(\frac{E}{e}\right) = er\left(\frac{R + r}{r}\right)$ *Multiply each side by er*

$Er = eR + er$ *Simplify*

$Er - er = eR$ *Isolate terms with r*

$r(E - e) = eR$ *Factor out r*

$r = \frac{eR}{E - e}$ *Divide each side by E - e*

45. Let x represent the number of hours required to fill the barrel with inlet and outlet pipes open. If the inlet pipe can fill it in 6 hours, then the rate of filling per hour is 1/6 of the barrel.

If the outlet pipe can empty it in 8 hours, then the rate of emptying per hour is 1/8 of the barrel. Since emptying the barrel works against filling the barrel, the rates are subtracted to get the rate with both pipes open.

| Rate of filling | less | rate of emptying | is | rate with both pipes open. |
|---|---|---|---|---|
| ↓ | ↓ | ↓ | ↓ | ↓ |
| $\frac{1}{6}$ | $-$ | $\frac{1}{8}$ | $=$ | $\frac{1}{x}$ |

Solve this equation.

$$24x(\frac{1}{6}) - 24x(\frac{1}{8}) = 24x(\frac{1}{x})$$

$$4x - 3x = 24$$

$$= 24$$

The barrel takes 24 hours to fill when both pipes are open.

49. Let x represent the distance for the first part of the trip. When x + 10 is the distance for the second part and x + x + 10 is the total distance.

| | d | r | t |
|---|---|---|---|
| First part | $x$ | 60 | $\frac{x}{60}$ |
| Second part | $x + 10$ | 50 | $\frac{x + 10}{50}$ |

$$d = rt$$

$$\frac{d}{r} = t$$

| Time for the second part | less | time for the first part | is | difference in times. |
|---|---|---|---|---|
| ↓ | ↓ | ↓ | ↓ | ↓ |
| $\frac{x + 10}{50}$ | $-$ | $\frac{x}{60}$ | $=$ | $\frac{1}{2}$ |

$$300(\frac{x + 10}{50} - \frac{x}{60}) = 300(\frac{1}{2})$$

$$300(\frac{x + 10}{50}) - 300(\frac{x}{10}) = 300(\frac{1}{2})$$

$$6(x + 10) - 5(x) = 150$$

$$6x + 60 - 5x = 150$$

$$x = 90$$

$$x + 10 = 90 + 10 = 100$$

$$x + x + 10 = 90 + 100 = 190$$

The distance to Carmel is 190 miles.

53. $\sqrt[3]{8^4} = 8^{4/3} = (8^{1/3})^4 = 2^4 = 16$

**Chapter 4 Review Exercises (page 200)**

1. $\frac{1}{p + 11}$

$p + 11 = 0$ *Set denominator equal to zero*

$p = -11$

When $p = -11$, this expression is undefined.

5. $\frac{25m^4p^3}{10m^5p} = \frac{5p^2}{2m}$

9. $\frac{25m^2 - k^2}{25m^2 - 10mk + k^2} = \frac{(5m - k)(5m + k)}{(5m - k)(5m - k)}$

$$= \frac{5m + k}{5m - k}$$

13. $\frac{y}{2y + 1}$; $2y^2 - 5y - 3$

$$\frac{2y^2 - 5y - 3}{2y + 1} = \frac{(2y + 1)(y - 3)}{2y + 1}$$

$$= y - 3,$$

Multiply numerator and denominator by $y - 3$.

$$\frac{y}{2y + 1} = \frac{y(y - 3)}{(2y + 1)(y - 3)}$$

$$= \frac{y(y - 3)}{2y^2 - 5y - 3}$$

17. $\dfrac{25p^3q^2}{8p^4q} \div \dfrac{15pq^2}{16p^5}$

$= \dfrac{25p^3q^2}{8p^4q} \cdot \dfrac{16p^5}{15pq^2}$

$= \dfrac{(25)(16)p^8q^2}{(15)(8)p^5q^3}$

$= \dfrac{10p^3}{3q}$

21. $\dfrac{z^2 - z - 6}{z - 6} \cdot \dfrac{z^2 - 6z}{z^2 + 2z - 15}$

$= \dfrac{(z - 3)(z + 2)}{z - 6} \cdot \dfrac{z(z - 6)}{(z + 5)(z - 3)}$

$= \dfrac{z(z - 6)(z - 3)(z + 2)}{(z - 6)(z - 3)(z + 5)}$

$= \dfrac{z(z + 2)}{z + 5}$

25. 12, 5x

Least common denominator:

$12(5x) = 60x$

29. $6k + 3,\ 10k + 5,\ 18k + 9$

$6k + 3 = 3(21 + 1)$,

$10k + 5 = 5(2k + 1)$, *Factor*

$18k + 9 = 9(2k + 1)$.

Least common denominator:

$(2k + 1)(9 \cdot 5) = (2k + 1)(45)$

33. $\dfrac{7}{3k} - \dfrac{16}{3k} = \dfrac{7 - 16}{3k} = \dfrac{-9}{3k} = \dfrac{-3}{k}$

37. $\dfrac{2m}{m^2 - 25} - \dfrac{3}{m + 5} = \dfrac{2m}{(m - 5)(m + 5)} - \dfrac{3}{m + 5}$

$= \dfrac{2m}{(m - 5)(m + 5)} - \dfrac{(m - 5)}{(m - 5)} \cdot \dfrac{3}{(m + 5)}$

$= \dfrac{2m - 3(m - 5)}{(m - 5)(m + 5)}$

$= \dfrac{-m + 15}{m^2 - 25}$

$= \dfrac{15 - m}{(m - 5)(m + 5)}$

41. $\dfrac{\dfrac{4m^5n^6}{mn}}{\dfrac{8m^7n^3}{m^4n^2}} = \dfrac{4m^4n^5}{8m^3n} = \dfrac{mn^4}{2}$

45. $\dfrac{\dfrac{2}{m - 3n}}{\dfrac{1}{3n - m}} = \dfrac{2}{m - 3n} \cdot \dfrac{3n - m}{1}$

$= \dfrac{-2}{3n - m} \cdot \dfrac{3n - m}{1} = -2$

49. $\dfrac{18r - 32}{12} = \dfrac{18r}{12} - \dfrac{32}{12}$

$= \dfrac{3r}{2} - \dfrac{8}{3}$

53. $\dfrac{8x^2 - 23x + 2}{x - 3}$

$$\begin{array}{r|l} & \quad\ \ 8x + 1 \\ x - 3 & 8x^2 - 23x + 2 \\ & \underline{8x^2 - 24x} \\ & \qquad\quad x + 2 \\ & \qquad\quad \underline{x - 3} \\ & \qquad\qquad\ 5 \end{array}$$ *Remainder*

Answer: $8x + 1 + \dfrac{5}{x - 3}$

57. $\dfrac{3p^2 - p - 2}{p - 1}$

$$\begin{array}{r|rrr} 1) & 3 & -1 & -2 \\ & & 3 & 2 \\ \hline & 3 & 2 & 0 \end{array}$$

Answer: $3p + 2$

61. $(3z^4 + 14z^3 - 22z^2 + 14z + 18) \div (z + 6)$

$$\begin{array}{r|rrrrr} -6) & 3 & 14 & -22 & 14 & 18 \\ & & -18 & 24 & -12 & -12 \\ \hline & 3 & -4 & 2 & 2 & 6 \end{array}$$

Answer $3z^3 - 4z^2 + 2z + 2 + \dfrac{6}{z + 6}$

65. $k = -1$; $P(x) = 3x^3 - 5x^2 + 4x - 1$

$$\begin{array}{r|rrrr} -1) & 3 & -5 & 4 & -1 \\ & & -3 & +8 & -12 \\ \hline & 3 & -8 & 12 & -13 \end{array}$$ *Remainder*

So $P(-1) = -13$

69. $\frac{y}{3} - \frac{y}{7} = 1$

$21(\frac{y}{3} - \frac{y}{7}) = 21(1)$

$7y - 3y = 21$

$4y = 21$

$y = \frac{21}{4}$

Solution set: $\{\frac{21}{4}\}$

73. $\frac{2}{m-1} + \frac{1}{m+1} = \frac{4m+1}{m^2-1}$

$\frac{2}{m-1} + \frac{1}{m+1} = \frac{4m+1}{(m-1)(m+1)}$

$(m-1)(m+1)(\frac{2}{m-1} + \frac{1}{m+1})$

$= (m-1)(m+1)(\frac{4m+1}{m-1)(m+1)})$

$2(m+1) + (m-1)$

$= 4m+1$

$2m+2+m-1 = 4m+1$

$3m+1 = 4m+1$

$m = 0$

Solution set: $\{0\}$

77. $\frac{1}{R} = \frac{1}{R_1} + \frac{1}{R_2}$

$\frac{1}{R} = \frac{1}{30} + \frac{1}{10}$ *Let $R_1 = 30$, $R_2 = 10$*

$30R(\frac{1}{R}) = 30R(\frac{1}{30} + \frac{1}{10})$

$30 = R + 3R$

$30 = 4R$

$R = \frac{30}{4} = \frac{15}{2}$

81. $\frac{V_1P_1}{T_1} = \frac{V_2P_2}{T_2}$

$\frac{V_1P_1}{T_1} \cdot T_1T_2 = \frac{V_2P_2}{T_2} \cdot T_1T_2$ *Multiply by least common denominator*

$V_1P_1T_2 = V_2P_2T_1$

$T_2 = \frac{V_2P_2T_1}{V_1P_1}$ *Divide by $V_1P_1$*

85. $S_n = \frac{n}{2}[2a_1 + (n-1)d]$; Solve for $a_1$

$\frac{2S_n}{n} = 2a_1 + (n-1)d$

*Multiply each side by $\frac{2}{n}$*

$\frac{2S_n}{n} - (n-1)d = 2a_1$

*Isolate $a_1$ term on right*

$\frac{2S_n}{n} - \frac{n}{n}(n-1)d = 2a_1$

$\frac{2S_n - n(n-1)d}{n} = 2a_1$

$\frac{2S_n - n(n-1)d}{2n} = a_1$

$\frac{S_n}{n} - \frac{(n-1)d}{2} = a_1$

89. Let t be the time to fill the tub with both taps open. In t minutes the hot water does $\frac{t}{12}$ the job and the cold water does $\frac{t}{8}$ the job.

$\frac{t}{12} + \frac{t}{8} = 1$

$24(\frac{t}{12} + \frac{t}{8}) - 24(1)$

$2t + 3t = 24$

$5t = 24$

$t = \frac{24}{5}$

It takes 24/5 mintues to fill the tub.

93.
```
                5p +  7
3p - 2 | 15p^2 + 11p - 17
         15p^2 - 10p
                 21p - 17
                 21p - 14
                       -3   Remainder
```

Answer: $5p + 7 + \frac{-3}{3p-2}$

97. $\frac{3m + 12}{8} \div \frac{5m + 20}{4}$

$= \frac{3m + 12}{8} \cdot \frac{4}{5m + 20}$ *Multiply by reciprocal*

$= \frac{3(m + 4)}{8} \cdot \frac{4}{5(m + 4)}$ *Factor*

$= \frac{3}{10}$

101. $A = \frac{Rr}{R + r}$; for r

$(R + r)A = (R + r)\frac{Rr}{R + r}$

$AR + Ar = Rr$

$Ar - Rr = -AR$ *Isolate terms with r on left*

$r(A - R) = -AR$ *Factor*

$r = \frac{-AR}{A - R} = \frac{AR}{R - A}$

105. $-\frac{1}{3h} - \frac{1}{h} = -\frac{4}{3}$

$3h(-\frac{1}{3h} - \frac{1}{h}) = 3h(-\frac{4}{3})$

$-1 - 3 = -4h$

$-4 = -4h$

$h = 1$

Solution set: $\{1\}$

**Chapter 4 Test (page 203)**

1. $\frac{m + 6}{2m^2 + 5m - 3}$

$2m^2 + 5m - 3 = 0$ *Set denominator equal 0*

$(2m - 1)(m + 3) = 0$ *Factor*

$2m - 1 = 0$ or $m + 3 = 0$

$2m = 1$

$m = \frac{1}{2}$ or $m = -3$

When $m = \frac{1}{2}$ or $-3$, the expression is undefined.

2. $\frac{3p^2 + 5p}{3p^2 - 7p - 20} = \frac{p(3p + 5)}{(p - 4)(3p + 5)}$

*Factor top and bottom*

$= \frac{p}{p - 4}$ *Write in lowest terms*

3. $\frac{\frac{5k^5}{9k^2}}{\frac{10k^6}{8k}} = \frac{5k^5}{9k^2} \cdot \frac{8k}{10k^6}$ *Multiply by reciprocal of $\frac{10k^6}{8k}$*

$= \frac{40k^6}{90k^6}$ *Multiply*

$= \frac{4}{9k^2}$ *Write in lowest terms*

4. $\frac{x^2 - 7x + 12}{x^2 + 2x - 15} \cdot \frac{x^2 - 25}{x^2 - 16}$

$= \frac{(x - 3)(x - 4)}{(x + 5)(x - 3)} \cdot \frac{(x + 5)(x - 5)}{(x + 4)(x - 4)}$

*Factor top and bottom*

$= \frac{x - 5}{x + 4}$ *Multiply and write in lowest terms*

5. $\frac{y^2 + y - 12}{y^3 + 9y^2 + 20y} \div \frac{y^2 - 9}{y^3 + 3y^2}$

$= \frac{y^2 + y - 12}{y^3 + 9y^2 + 20y} \cdot \frac{y^3 + 3y^2}{y^2 - 9}$

*Multiply by reciprocal*

$= \frac{(y + 4)(y - 3)}{y(y + 4)(y + 5)} \cdot \frac{y^2(y + 3)}{(y + 3)(y - 3)}$

*Factor*

$= \frac{y}{y + 5}$

6. $9z, 12z^2, 6z^3$

$9z = 3 \cdot 3 \cdot z$

$12z^2 = 2 \cdot 2 \cdot 3 \cdot z \cdot z$

$6z^3 = 2 \cdot 3 \cdot z \cdot z \cdot z$

Least common denominator:

$2 \cdot 2 \cdot 3 \cdot 3 \cdot z \cdot z \cdot z = 36z^2$

7. $m^2 + m - 6,\ m^2 + 3m$

$m^2 + m - 6 = (m + 3)(m - 2)$

$m^2 + 3m = m(m + 3)$

Least common denominator:

$m(m + 3)(m - 2)$

8. $\frac{4}{k} + \frac{1}{5k} = \frac{4}{k}\cdot\frac{5}{5} + \frac{1}{5k}$ *Write each fraction with the least common denominator*

$= \frac{20 + 1}{5k}$ *Add*

$= \frac{21}{5k}$ *Simplify*

9. $\frac{1}{a - b} + \frac{5}{b - a} = \frac{1}{a - b} - \frac{5}{a - b}$

$= \frac{-4}{a - b}$ or $\frac{4}{b - a}$

10. $\frac{4}{t^2 - 16} - \frac{5}{2t + 8}$

$= \frac{4}{(t + 4)(t - 4)} - \frac{5}{2(t + 4)}$

*Factor each denominator*

$= \frac{4}{(t + 4)(t - 4)}\cdot\frac{2}{2} \quad \frac{5}{2(t + 4)}\cdot\frac{(t - 4)}{(t - 4)}$

*Write each fraction with least common denominator*

$= \frac{8 - 5(t - 4)}{2(t + 4)(t - 4)}$ *Simplify the numerator*

$= \frac{8 - 5t + 20}{t(t + 4)(t - 4)}$

$= \frac{28 - 5t}{2(t + 4)(t - 4)}$

11. $\frac{\frac{6}{m - 2}}{\frac{8}{3m - 6}} = \frac{6}{m - 2}\cdot\frac{3m - 6}{8}$ *Multiply by reciprocal of denominator*

$= \frac{6}{m - 2}\cdot\frac{3(m - 2)}{8}$ *Factor where possible*

$= \frac{9}{4}$

12. $\frac{\frac{3}{k - 1} - \frac{2}{k}}{\frac{4}{k} + \frac{2}{k - 1}} = \frac{k(k - 1)(\frac{3}{k - 1} - \frac{2}{k})}{k(k - 1)(\frac{4}{k} + \frac{2}{k - 1})}$

$= \frac{3k - 2(k - 1)}{4(k - 1) + 2k}$

$= \frac{3k - 2k + 2}{4k - 4 + 2k}$

$= \frac{k + 2}{6k - 4}$

13. $\frac{8z^3 - 16z^2 + 24z}{8z^2}$

$= \frac{8z^3}{8z^2} - \frac{16z^2}{8z^2} + \frac{24z}{8z^2}$

$= z - 2 + \frac{3}{z}$

14. $2q + 3 \,\overline{)\, 2q^2 - 13q - 16}$, quotient $q - 8$

$2q^2 + 3q$

$-16q - 16$

$-16q - 24$

$8$ *Remainder*

Answer: $q - 8 + \frac{8}{2q + 3}$

15. $2y + 1 \,\overline{)\, 6y^4 - 3y^3 + 5y^2 + 6y - 9}$, quotient $3y^3 - 3y^2 + 4y + 1$

$6y^4 + 3y^3$

$-6y^3 + 5y^2$

$-6y^3 - 3y^2$

$8y^2 + 6y$

$8y^2 + 4y$

$2y - 9$

$2y + 1$

$-10$

*Remainder*

Answer: $3y^3 - 3y^2 + 4y + 1 + \frac{-10}{2y + 1}$

16. $9x^5 + 40x^4 - 23x^3 + 8x^2 - 6x + 22 \div x + 5$

| -5) | 9 | 40 | -23 | 8 | -6 | 22 |
|---|---|---|---|---|---|---|
| | | -45 | 25 | -10 | 10 | -20 |
| | 9 | -5 | 2 | -2 | 4 | 2 |

*Remainder*

Answer:

$$9x^4 - 5x^3 + 2x^2 - 2x + 4 + \frac{2}{x+5}$$

17. $x^4 - 8x^3 + 21x^2 - 14x - 24 = 0$; is 4 a solution?

| 4) | 1 | -8 | 21 | -14 | -24 |
|---|---|---|---|---|---|
| | | 4 | -16 | 20 | 24 |
| | 1 | -4 | 5 | 6 | 0 *Remainder* |

Since the remainder is 0, 4 is a solution.

18. $1 - \frac{3}{x} = \frac{1}{2}$

$2x(1) - 2x(\frac{3}{x}) = 2x(\frac{1}{2})$ *Multiply each term by 2x*

$2x - 6 = x$ *Simplify*

$x = 6$ *Solve for x*

Solution set: {6}.

19. $\frac{1}{a} - \frac{1}{a-2} = \frac{4}{3a}$

$3a(a - 2)(\frac{1}{a}) - 3a(a - 2)(\frac{1}{a-2})$

$= 3a(a - 2)(\frac{4}{3a})$

*Multiply each term by 3a(a - 2)*

$3(a - 2) - 3a = 4(a - 2)$ *Simplify*

$3a - 6 - 3a = 4a - 8$

$-6 = 4a - 8$

$2 = 4a$

$\frac{1}{2} = a$

Solution set: $\{\frac{1}{2}\}$

20. $\frac{2}{m-3} - \frac{3}{m+3} = \frac{12}{(m+3)(m-3)}$

*Factor the denominator*

$(m - 3)(m + 3)[\frac{2}{m-3}]$

$- (m - 3)(m + 3)[\frac{3}{m+3}]$

$= (m - 3)(m + 3)[\frac{12}{(m+3)(m-3)}]$

*Multiply each term by the least common denominator*

$2(m + 3) - 3(m - 3) = 12$ *Simplify*

$2m + 6 - 3m + 9 = 12$

$-m + 15 = 12$

$-m = -3$

$m = 3$

Notice that 3 causes the denominator of the first and last term to become 0. Therefore, there is no solution. Solution set: ∅

21. $\frac{2}{z} - \frac{5}{m} = \frac{1}{y}$ Solve for m.

$myz(\frac{2}{z}) - myz(\frac{5}{m}) = myz(\frac{1}{y})$

*Multiply each term by the least common denominator*

$2my - 5yz = mz$ *Simplify*

$2my - mz = 5yz$

*Isolate the m terms alone on one side*

$m(2y - z) = 5yz$ *Factor out m*

$m = \frac{5yz}{2y - z}$

*Divide each side by 2y - z*

22. $\frac{k}{y} = \frac{r+2}{r-2}$; Solve for r

$y(r - 2)(\frac{k}{y}) = y(r - 2)(\frac{r+2}{r-2})$

*Multiply each term by the least common denominator*

$k(r - 2) = y(r + 2)$ *Simplify*

$kr - 2k = ry + 2y$

$kr - ry = 2k + 2y$

*Isolate the r terms alone on one side*

$r(k - y) = 2k + 2y$ *Factor out r*

$r = \frac{2k + 2y}{k - y}$ *Divide each side by k - y*

23. Let x be the smaller of the consecutive integer. So x + 1 is the larger.

| Reciprocal of smaller | less | twice | reciprocal of larger |
|---|---|---|---|
| ↓ | ↓ | ↓ | ↓ |
| $\frac{1}{x}$ | $-$ | $2\cdot$ | $\frac{1}{x+1}$ |

| is | -5 | times | reciprocal of their product |
|---|---|---|---|
| ↓ | ↓ | ↓ | ↓ |
| $=$ | $-5$ | $\cdot$ | $\frac{1}{x(x+1)}$ |

Solve this equation.

$$x(x+1)(\tfrac{1}{x}) - x(x+1)(\tfrac{2}{x+1}) = x(x+1)(\tfrac{-5}{x(x+1)})$$

$$x + 1 - 2x = -5$$
$$1 - x = -5$$
$$-x = -6$$
$$x = 6$$
$$x + 1 = 7$$

The numbers are 6 and 7.

24. If Wayne can do a job in 9 hours, then his rate per hour is 1/9 of the job. If Susan takes 5 hours to do the same job, then her rate is 1/5 of the job. If together, they take x hours, then their rate per hour is 1/x of the job.

| Wayne's rate | and | Susan's rate | is | their rate together. |
|---|---|---|---|---|
| ↓ | ↓ | ↓ | ↓ | ↓ |
| $\frac{1}{9}$ | $+$ | $\frac{1}{5}$ | $=$ | $\frac{1}{x}$ |

Solve this equation.

$$45x(\tfrac{1}{9}) + 45x(\tfrac{1}{5}) = 45x(\tfrac{1}{x})$$
$$5x + 9x = 45$$
$$14x = 45$$
$$x = \frac{45}{14}$$

It takes then $\frac{45}{14}$ hours working together.

25. Let x be the speed of the boat in still water. Then, with a current of 3 miles per hour, the boat's speed downstream (with the current) is x + 3 and the speed upstream (against the current) is x - 3.

| | d | r | t |
|---|---|---|---|
| Downstream | 36 | x + 3 | $\frac{36}{x+3}$ |
| Upstream | 24 | x - 3 | $\frac{24}{x-3}$ |

The two times are equal.

$$\frac{36}{x+3} = \frac{24}{x-3}$$

Solve this equation.

$$(x+3)(x-3)(\tfrac{36}{x+3}) = (x+3)(x-3)(\tfrac{24}{x-3})$$
$$36(x-3) = 24(x+3)$$
$$36x - 108 = 24x + 72$$
$$12x = 180$$
$$x = 15$$

The speed of the boat in still water is 15 miles per hour.

**CHAPTER 5 RATIONAL EXPONENTS AND RADICALS**

**Section 5.1 (page 209)**

1. $(121)^{1/2} = 11$ *Since* $121 > 0$ *and* $11^2 = 121$

5. $(512)^{1/3} = 8$ *Since* $8^3 = 512$

9. $-6561^{1/4} = -(6561^{1/4})$
$= -(9)$ *Since* $9^4 = 6561$
$= -9$

13. $(\frac{16}{25})^{1/2} = \frac{16^{1/2}}{25^{1/2}} = \frac{4}{5}$ *Since* $(\frac{4}{5})^2 = \frac{16}{25}$ *and* $\frac{16}{25} > 0$

17. $(100)^{3/2} = (100^{1/2})^3$
$= (10)^3$
$= 1000$

21. $-144^{3/2} = -(144^{1/2})^3$
$= -12^3$
$= -1728$

25. $(\frac{16}{81})^{3/4} = [(\frac{16}{81})^{1/4}]^3$
$= (\frac{2}{3})^3$
$= \frac{8}{27}$

29. $(\frac{16}{625})^{-1/4} = \frac{(16)^{-1/4}}{(625)^{-1/4}}$
$= \frac{\frac{1}{16^{1/4}}}{\frac{1}{625^{1/4}}}$
$= \frac{\frac{1}{2}}{\frac{1}{5}}$
$= \frac{1}{2}\cdot\frac{5}{1} = \frac{5}{2}$

33. $9^{3/5}\,9^{7/5} = 9^{3/5 + 7/5}$
$= 9^{10/5} = 9^2 = 81$

37. $x^{5/3}\,x^{-2/3} = x^{5/3 + (-2/3)}$
$= x^{3/3} = x^1 = x$

41. $(8p^9q^6)^{2/3} = 8^{2/3}p^{9\cdot 2/3}q^{6\cdot 2/3}$
$= (8^{1/3})^2p^{18/3}q^{12/3}$
$= 2^2p^6q^4$
$= 4p^6q^4$

45. $(\frac{z^{10}}{x^{12}})^{1/4} = \frac{z^{10\cdot 1/4}}{x^{12\cdot 1/4}}$
$= \frac{z^{10/4}}{x^{12/4}}$
$= \frac{z^{5/2}}{x^3}$

49. $\frac{p^{1/5}p^{7/10}p^{1/2}}{(p^3)^{-1/5}} = \frac{p^{1/5+7/10+1/2}}{p^{-3/5}}$
$= \frac{p^{2/10+7/10+5/10}}{p^{-6/10}}$
$= \frac{p^{14/10}}{p^{-6/10}}$
$= p^{14/10-(-6/10)}$
$= p^{20/10}$
$= p^2$

53. $(\frac{p^{-1/4}q^{-3/2}}{3^{-1}p^{-2}q^{-2/3}})^{-2}$
$= \frac{p^{-1/4(-2)}q^{-3/2(-2)}}{3^{-1(-2)}p^{-2(-2)}q^{-2/3(-2)}}$
$= \frac{p^{1/2}q^3}{3^2p^4q^{4/3}}$
$= \frac{q^{3-4/3}}{3^2p^{4-1/2}}$
$= \frac{q^{9/3-4/3}}{3^2p^{7/2}} = \frac{q^{5/3}}{3^2p^{7/2}} = \frac{q^{5/3}}{9p^{7/2}}$

57. $k^{1/4}(k^{3/2} - k^{1/2})$
$= k^{1/4+3/2} - k^{1/4+1/2}$
$= k^{1/4+6/4} - k^{1/4+2/4}$
$= k^{7/4} - k^{3/4}$

61. $x^{-3/5}(x^{11/5} - 4x)$
$= x^{-3/5+11/5} - 4x^{-3/5+1}$
$= x^{8/5} - 4x^{-3/5+5/5}$
$= x^{8/5} - 4x^{2/5}$

65. $m^3 - 3m^{5/2} = m^{5/2+1/2} - 3m^{5/2+0}$
$= m^{5/2} \cdot m^{1/2} - 3m^{5/2} \cdot m^0$
$= m^{5/2}(m^{1/2} - 3m^0)$
$= m^{5/2}(m^{1/2} - 3)$

69. $2m^{1/8} - m^{5/8} = m^{1/8} \cdot 2 - m^{1/8} \cdot m^{4/8}$
$= m^{1/8}(2 - m^{1/2})$

73. $L = (\frac{S}{a})^{1/2}$
$= (\frac{1000}{\frac{2}{5}})^{1/2}$ *Substitute given numbers*
$= (\frac{1000 \cdot 5}{2})^{1/2}$ *Multiply by reciprocal of denominator*
$= (2500)^{1/2}$
$= 50$ centimeters

77. $\frac{x^{5a/3}}{x^{a/6}} = x^{5a/3 - a/6}$
$= x^{10a/6 - a/6}$
$= x^{9a/6}$
$= x^{3a/2}$

81. $\frac{x^{-1/b}y^{-1/a}}{x^{2/b}y^{-2/b}} = \frac{y^{-1/a-(-2/b)}}{x^{2/b-(-1/b)}}$
$= \frac{y^{-1/a+2/b}}{x^{3/b}}$
$= \frac{y^{-b/ab+2a/ab}}{x^{3/b}}$ *Common denominator is ab*
$= \frac{y^{(2a-b)/ab}}{x^{3/b}}$
$= x^{-3/b}y^{(2a-b)/ab}$

85. $(x^{1/2} + y^{1/2})^2 \neq x + y$
Let x = 16 and y = 25.
$(x^{1/2} + y^{1/2})^2 = [(16)^{1/2} + (25)^{1/2}]^2$
$= (4 + 5)^2$
$= 9^2$
$= 81$
$x + y = 16 + 25 = 41$
Since $81 \neq 41$,
$(x^{1/2} + y^{1/2})^2 \neq x + y.$

89. $\sqrt[3]{64} = 4$ *Since* $4^3 = 64$

93. $\sqrt[7]{-128} = -2$ *Since* $(-2)^7 = 128$

**Section 5.2 (page 214)**

1. $\sqrt{81} = 9$ *Since* $9^2 = 81$

5. $\sqrt[3]{-216} = 6$ *Since* $(-6)^3 = -216$

9. $-\sqrt[4]{625} = -5$ *Since* $-(5^4) = -625$

13. $-\sqrt[4]{6561} = -9$ *Since* $-(9^4) = -6561$

17. $-\sqrt[5]{-1024} = 4$ *Since* $-(-4^5) = -(-1024)$ *or* $4^5 = 1024$

21. $\sqrt[6]{4096} = 4$ *Since $4^6 = 4096$*

25. $\sqrt[3]{m^{15}} = (m^{1/3})^{15} = m^{15/3} = m^5$

29. $\sqrt{(-6)^2} = |-6| = 6$

33. $\sqrt{7} = 2.646$

37. $-\sqrt{82} = -9.055$

41. $-\sqrt{510} = 22.583$

45. $s = \frac{1}{2}(a + b + c)$

$= \frac{1}{2}(4 + 3 + 5)$ *Substitute*

$= 6$

$k = \sqrt{s(s - a)(s - b)(s - c)}$

$= \sqrt{6(6 - 4)(6 - 3)(6 - 5)}$ *Substitute*

$= \sqrt{6(2)(3(1)} = \sqrt{36}$

$= 6$

The area is 6 square meters.

49. $9^{1/5} = \sqrt[5]{9}$

52. $(3py^2)^{1/4} = \sqrt[4]{3py^2}$

57. $(2y + x)^{2/3} = [(2y + x)^{1/3}]^2$

$= (\sqrt[3]{2y + x})^2$

61. $(3m^4 + 2k^2)^{-2/3}$

$= [(3m^4 + 2k^2)^{1/3}]^{-2}$

$= (\sqrt[3]{3m^4 + 2k^2})^{-2}$

$= 1/(\sqrt[3]{3m^4 + 2k^2})^2$

65. $\sqrt[3]{7} = 7^{1/3}$

69. $\sqrt[3]{6^4} = (6^4)^{1/3} = 6^{4/3}$

73. $\sqrt[4]{x^3} = (x^3)^{1/4} = x^{3 \cdot 1/4} = x^{3/4}$

77. $\dfrac{\sqrt{x^5}}{x^4} = \dfrac{(x^5)^{1/2}}{x^4} = \dfrac{x^{5 \cdot 1/2}}{x^4} = \dfrac{x^{5/2}}{x^{8/2}}$

$= x^{5/2-8/2} = x^{-3/2}$

$= \dfrac{1}{x^{3/2}}$

81. $\sqrt{m + r} = (m + r)^{1/2}$

*Note that $(m + r)^{1/2} \neq m^{1/2}r^{1/2}$. However, $(m \cdot r)^{1/2} = m^{1/2}r^{1/2}$.*

85. $\sqrt[3]{m - 5n} = (m - 5n)^{1/3}$

*Note that $(m - 5n)^{1/3} \neq m^{1/3} - (5n)^{1/3}$.*

89. $2\sqrt{m} - 5\sqrt[3]{m^2} = 2m^{1/2} - 5m^{2/3}$

93. $\sqrt{16z^2} = |4| \cdot |z| = 4|z|$

97. $\sqrt{(r - 2q)^2} = |r - 2q|$

101. $\dfrac{-5}{\sqrt{p^4 - 2p^2q^2 + q^4}}$

$= \dfrac{-5}{\sqrt{(p^2 - q^2)^2}}$ *Factor denominator*

$= \dfrac{-5}{|p^2 - q^2|}$

105. $\sqrt{\sqrt[3]{m}} = \sqrt{(m)^{1/3}} = (m^{1/3})^{1/2}$
$= m^{1/6}$

109. $3x^2 + 2 = 6x$

*Substitute* $\frac{(3 + \sqrt{3})}{3}$ *for x*

$3(\frac{3 + \sqrt{3}}{3})^2 + 2 = 6(\frac{3 + \sqrt{3}}{3})$

$3(\frac{9 + 6\sqrt{3} + 3}{9}) + 2 = 2(3 + \sqrt{3})$

$\frac{12 + 6\sqrt{3}}{3} + 2 = 6 + 2\sqrt{3}$

*Simplify*

$12 + 6\sqrt{3} + 6 = 18 + 6\sqrt{3}$

*Multiply each side by 3*

$18 + 6\sqrt{3} = 18 + 6\sqrt{3}$ *True*

$\frac{3 + \sqrt{3}}{3}$ is a solution.

113. $(6x^2)^{-1}(2x^3)^{-2} = 6^{-1}x^{-2}2^{-2}x^{-6}$
$= \frac{x^{-2+(-6)}}{6 \cdot 2^2}$
$= \frac{x^{-8}}{24}$
$= \frac{1}{24x^8}$

**Section 5.3 (page 222)**

1. $\sqrt{18} = \sqrt{9} \cdot \sqrt{2} = 3\sqrt{2}$

5. $\sqrt{76} = \sqrt{4} \cdot \sqrt{19} = 2\sqrt{19}$

9. $\sqrt[3]{128} = \sqrt[3]{64} \cdot \sqrt[3]{2} = 4\sqrt[3]{2}$

13. $-\sqrt[4]{1250} = -(\sqrt[4]{625} \cdot \sqrt[4]{2}) = -5\sqrt[4]{2}$

17. $\sqrt{\frac{72}{25}} = \frac{\sqrt{72}}{\sqrt{25}} = \frac{\sqrt{36}\sqrt{12}}{5} = \frac{6\sqrt{2}}{5}$

21. $\sqrt{100y^{10}} = \sqrt{10y^5 \cdot 10y^5} = 10y^5$

25. $-\sqrt{144m^{10}z^2} = -(\sqrt{12m^5z\ 12m^5z})$
$= -12m^5z$

29. $\sqrt[4]{\frac{1}{16}m^{12}x^{16}} = \frac{\sqrt[4]{m^{12}x^{16}}}{\sqrt[4]{16}} = \frac{\sqrt[4]{(m^3)^4(x^4)^4}}{2}$
$= \frac{m^3x^4}{2}$

33. $\sqrt{7x^5y^6} = \sqrt{x^4y^6} \cdot \sqrt{7x}$
$= x^2y^3\sqrt{7x}$ *Since all the variables are positive we omit absolute value signs*

37. $\sqrt[3]{24z^5x^9} = \sqrt[3]{8z^3x^9} \cdot \sqrt[3]{3z^2} = 2zx^3 \cdot \sqrt[3]{3z^2}$

41. $\sqrt[4]{32k^5m^{10}} = \sqrt[4]{16k^4m^8} \cdot \sqrt[4]{2km^2}$
$= 2km^2\sqrt[4]{2km^2}$ *Omit absolute value signs*

45. $\sqrt[3]{\frac{y^{10}}{27}} = \frac{\sqrt[3]{y^{10}}}{\sqrt[3]{27}}$
$= \frac{\sqrt[3]{y^9} \cdot \sqrt[3]{y}}{3}$
$= \frac{y^3\sqrt[3]{y}}{3}$

49. $\frac{15}{\sqrt{5}} = \frac{15 \cdot \sqrt{5}}{\sqrt{5} \cdot \sqrt{5}} = \frac{15\sqrt{5}}{5} = 3\sqrt{5}$

53. $\frac{\sqrt{7}}{\sqrt{5}} = \frac{\sqrt{7} \cdot \sqrt{5}}{\sqrt{5} \cdot \sqrt{5}} = \frac{\sqrt{35}}{5}$

57. $\frac{9}{\sqrt{20}} = \frac{9 \cdot \sqrt{5}}{\sqrt{20} \cdot \sqrt{5}} = \frac{9\sqrt{5}}{\sqrt{100}} = \frac{9\sqrt{5}}{10}$

61. $\sqrt{\frac{5}{18}} + \frac{\sqrt{5}}{\sqrt{18}} = \frac{\sqrt{5}}{3\sqrt{2}}$ *Since* $\sqrt{18} = \sqrt{9}\sqrt{2} = 3\sqrt{2}$

$= \frac{\sqrt{5}\cdot\sqrt{2}}{3\sqrt{2}\cdot\sqrt{2}} = \frac{\sqrt{10}}{3\cdot 2} = \frac{\sqrt{10}}{6}$

65. $\sqrt{\frac{5m}{k}} = \frac{\sqrt{5m}}{\sqrt{k}} = \frac{\sqrt{5m}\sqrt{k}}{\sqrt{k}\sqrt{k}} = \frac{\sqrt{5mk}}{k}$

69. $-\sqrt{\frac{48k^2}{z}} = \frac{-\sqrt{48k^2}}{\sqrt{z}} = \frac{-\sqrt{16k^2}\sqrt{3}\cdot\sqrt{z}}{\sqrt{z}\cdot\sqrt{z}}$

$= \frac{-4k\sqrt{3z}}{z}$

73. $\sqrt[3]{\frac{9}{32}} = \frac{\sqrt[3]{9}}{\sqrt[3]{32}} = \frac{\sqrt[3]{9}\cdot\sqrt[3]{2}}{\sqrt[3]{32}\ \sqrt[3]{2}} = \frac{\sqrt[3]{18}}{\sqrt[3]{64}} = \frac{\sqrt[3]{18}}{4}$

77. $\sqrt[3]{\frac{r^{15}}{s^8}} = \frac{\sqrt[3]{r^{15}}}{\sqrt[3]{s^8}} = \frac{\sqrt[3]{(r^5)^3}}{\sqrt[3]{s^6}\sqrt[3]{s^2}}$

$= \frac{r^5}{s^2\sqrt[3]{s^2}}$ *Since* $\sqrt[3]{s^6} = s^2$

$= \frac{r^5\cdot\sqrt[3]{s}}{s^2\sqrt[3]{s^2}\cdot\sqrt[3]{s}} = \frac{r^5\sqrt[3]{s}}{s^2\sqrt[3]{s^3}}$

$= \frac{r^5\sqrt[3]{s}}{s^2\cdot s} = \frac{r^5\sqrt[3]{s}}{s^3}$

81. $\frac{\sqrt[5]{a^2b^3}}{\sqrt[5]{a^3}\sqrt[5]{ab^7}}$

$= \frac{\sqrt[5]{a^2b^3}}{\sqrt[5]{a^4b^7}}$

$= \frac{\sqrt[5]{a^2b^3}}{\sqrt[5]{a^4b^2}\sqrt[5]{b^5}} = \frac{\sqrt[5]{a^2b^3}}{b\sqrt[5]{a^4b^2}}$

$= \frac{\sqrt[5]{a^2b^3}}{b\sqrt[5]{a^4b^2}}\cdot\frac{\sqrt[5]{ab^3}}{\sqrt[5]{ab^3}} = \frac{\sqrt[5]{a^3b^6}}{b\sqrt[5]{a^5b^5}} = \frac{\sqrt[5]{a^3b^5b}}{b(ab)}$

$= \frac{b\sqrt[5]{a^3b}}{b\cdot ab} = \frac{\sqrt[5]{a^3b}}{ab}$

85. $\sqrt[10]{m^{15}} = m^{15/10} = m^{3/2} = \sqrt{m^3}$

$= m\sqrt{m}$

89. $\sqrt[6]{256} = \sqrt[6]{2^8} = 2^{8/6} = 2^{4/3}$

$= 2^{3/3}\cdot 2^{1/3} = 2^1 2^{1/3}$

$= 2\sqrt[3]{2}$

93. $\sqrt[3]{2}\cdot\sqrt[5]{3}$

The common index of 3 and 5 is 15. Thus,

$\sqrt[3]{2} = 2^{1/3} = 2^{5/15} = \sqrt[15]{2^5}$

$\sqrt[5]{3} = 3^{1/5} = 3^{3/15} = \sqrt[15]{3^3}$.

$\sqrt[3]{2}\cdot\sqrt[5]{3} = \sqrt[15]{2^5}\ \sqrt[15]{3^3}$

$= \sqrt[15]{2^5 3^3}$

$= \sqrt[15]{32\cdot 27}$

$= \sqrt[15]{864}$

97. $d = \sqrt{\frac{k}{I}}$

$d = \sqrt{\frac{700}{12}} = \frac{\sqrt{700}}{\sqrt{12}} = \frac{\sqrt{100.7}}{\sqrt{4.3}}$

$= \frac{10\sqrt{7}\cdot\sqrt{3}}{2\sqrt{3}\cdot\sqrt{3}} = \frac{10\sqrt{21}}{6} = \frac{5\sqrt{21}}{3}$

The distance is $(5\sqrt{21}/3$ or $5(4.583)/3 = 7.638$ feet.

101. $9q + 2q - 5q^2 - q^2 = -6q^2 + 11q$

**Section 5.4 (page 226)**

1. $\sqrt{36} + \sqrt{100} = 6 + 10 = 16$

*Simplify each radical and add*

5. $6\sqrt{18} + \sqrt{32} - 2\sqrt{50}$

$= 6\sqrt{9\cdot 2} + \sqrt{16\cdot 2} - 2\sqrt{25\cdot 2}$

$= 6\cdot 3\sqrt{2} + 4\sqrt{2} - 2\cdot 5\sqrt{2}$

$= 18\sqrt{2} + 4\sqrt{2} - 10\sqrt{2}$

$= 12\sqrt{2}$

9. $2\sqrt{5} - 3\sqrt{20} - 4\sqrt{45}$

$= 2\sqrt{5} - 3\sqrt{4\cdot 5} - 4\sqrt{9\cdot 5}$

$= 2\sqrt{5} - 6\sqrt{5} - 12\sqrt{5}$

$= -16\sqrt{5}$

13. $3\sqrt{2x} - \sqrt{8x} - \sqrt{72x}$

$= 3\sqrt{2x} - \sqrt{4\cdot 2x} - \sqrt{36\cdot 2x}$

$= 3\sqrt{2x} - 2\sqrt{2x} - 6\sqrt{2x}$

$= -5\sqrt{2x}$

17. $7q\sqrt{10} - 2q\sqrt{40} + 8q\sqrt{90}$

$= 7q\sqrt{10} - 2q\sqrt{4\cdot 10} + 8q\sqrt{9\cdot 10}$

$= 7q\sqrt{10} - 4q\sqrt{10} + 24q\sqrt{10}$

$= 27q\sqrt{10}$

21. $\sqrt[3]{54} - 2\sqrt[3]{16}$

$= \sqrt[3]{27\cdot 2} - 2\sqrt[3]{8\cdot 2}$

$= 3\sqrt[3]{2} - 4\sqrt[3]{2}$

$= -\sqrt[3]{2}$

25. $\sqrt[3]{x^2y} - \sqrt[3]{8x^2y} = \sqrt[3]{x^2y} - \sqrt[3]{8}\sqrt[3]{x^2y}$

$= \sqrt[3]{x^2y} - 2\sqrt[3]{x^2y}$

$= -\sqrt[3]{x^2y}$

29. $5\sqrt[4]{32} + 3\sqrt[4]{162} = 5\sqrt[4]{16}\cdot\sqrt[4]{2} + 3\sqrt[4]{81}\cdot\sqrt[4]{2}$

$= 5\cdot 2\sqrt[4]{2} + 3\cdot 3\sqrt[4]{2}$

$= 10\sqrt[4]{2} + 9\sqrt[4]{2}$

$= 19\sqrt[4]{2}$

33. $3\sqrt[4]{x^5y} - 2x\sqrt[4]{xy}$

$= 3\sqrt[4]{x^4}\sqrt[4]{xy} - 2x\sqrt[4]{xy}$

$= 3x\sqrt[4]{xy} - 2x\sqrt[4]{xy}$

$= x\sqrt[4]{xy}$

37. $\dfrac{2\sqrt{5}}{3} + \dfrac{1}{\sqrt{5}} = \dfrac{2\sqrt{5}}{3} + \dfrac{1\cdot\sqrt{5}}{\sqrt{5}\sqrt{5}}$ *Simplify second term*

$= \dfrac{2\sqrt{5}}{3} + \dfrac{\sqrt{5}}{5}$

$= \dfrac{10\sqrt{5}}{15} + \dfrac{3\sqrt{5}}{15}$ *Find common denominator*

$= \dfrac{13\sqrt{5}}{15}$

41. $\dfrac{\sqrt{32}}{3} + \dfrac{2\sqrt{2}}{3} - \dfrac{1}{\sqrt{8}}$

$= \dfrac{\sqrt{16}\sqrt{2}}{3} + \dfrac{2\sqrt{2}}{3} - \dfrac{1}{\sqrt{4}\sqrt{2}}$ *Simplify each radical*

$= \dfrac{4\sqrt{2}}{3} + \dfrac{2\sqrt{2}}{3} - \dfrac{1}{2\sqrt{2}}$

$= \dfrac{6\sqrt{2}}{3} - \dfrac{1\cdot\sqrt{2}}{2\sqrt{2}\ \sqrt{2}}$

$= 2\sqrt{2} - \dfrac{\sqrt{2}}{4}$

$= \dfrac{8\sqrt{2}}{4} - \dfrac{\sqrt{2}}{4}$

$= \dfrac{7\sqrt{2}}{4}$

45. $3\sqrt{x} + \frac{2}{\sqrt{x}} + \frac{1}{x} = 3\sqrt{x} + \frac{2\sqrt{x}}{\sqrt{x}\sqrt{x}} + \frac{\sqrt{1}}{\sqrt{x}}$

$= 3\sqrt{x} + \frac{2\sqrt{x}}{x} + \frac{1\sqrt{x}}{\sqrt{x}\sqrt{x}}$

$= 3\sqrt{x} + \frac{2\sqrt{x}}{x} + \frac{\sqrt{x}}{x}$

$= 3\sqrt{x} + \frac{3\sqrt{x}}{x}$

$= \frac{3x\sqrt{x}}{x} + \frac{3\sqrt{x}}{x}$

$= \frac{3x\sqrt{x} + 3\sqrt{x}}{x}$

49. $2\sqrt[3]{3} - \frac{5}{\sqrt[3]{9}} = 2\sqrt[3]{3} - \frac{5\sqrt[3]{3}}{\sqrt[3]{9}\sqrt[3]{3}}$

$= 2\sqrt[3]{3} - \frac{5\sqrt[3]{3}}{\sqrt[3]{27}}$

$= 2\sqrt[3]{3} - \frac{5\sqrt[3]{3}}{3}$

$= \frac{6\sqrt[3]{3}}{3} - \frac{5\sqrt[3]{3}}{3}$

$= \frac{\sqrt[3]{3}}{3}$

53. $P = a + b + c$

$P = 2\sqrt{45} + \sqrt{75} + 3\sqrt{20}$

$= 2\sqrt{5\cdot 9} + \sqrt{25\cdot 3} + 3\sqrt{4\cdot 5}$

$= 6\sqrt{5} + 5\sqrt{3} + 6\sqrt{5}$

$= 12\sqrt{5} + 5\sqrt{3}$ or 35.493 centimeters.

57. $\sqrt{3} = 1.732$

$\sqrt{12} = 3.464$ suggests $\sqrt{12} = 2\sqrt{3}$

61. $(3a - 2b)(5a + 7b)$

$= (3a)(5a) + (3a)(7b) + (-2b)(5a) + (-2b)(7b)$

$= 15a^2 + 11ab - 14b^2$

65. $(7r - 2s)(7r + 2s) = (7r)^2 - (2s)^2$

$= 49r^2 - 4s^2$

**Section 5.5 (page 230)**

1. $3(5 - \sqrt{6}) = 15 - 3\sqrt{6}$

5. $5(\sqrt{72} - \sqrt{8}) = 5(\sqrt{36}\sqrt{2} - \sqrt{4}\sqrt{2}$

$= 5(6\sqrt{2} - 2\sqrt{2})$

$= 5(4\sqrt{2})$

$= 20\sqrt{2}$

9. $(\sqrt{7} + 3)(\sqrt{7} - 3)$

$= \sqrt{7}\sqrt{7} - 3\sqrt{7} + 3\sqrt{7} - 9$ *Use FOIL method*

$= 7 - 9$

$= -2$

13. $(\sqrt{8} - \sqrt{2})(\sqrt{8} + \sqrt{2}) = (\sqrt{8})^2 - (\sqrt{2})^2$

*Product of the sum and difference of two terms*

$= 8 - 2$

$= 6$

17. $(\sqrt{11} - \sqrt{7})(\sqrt{2} + \sqrt{5})$

$= \sqrt{11}\sqrt{2} + \sqrt{11}\sqrt{5} + -\sqrt{7}\sqrt{2} - \sqrt{7}\sqrt{5}$

$= \sqrt{22} + \sqrt{55} - \sqrt{14} - \sqrt{35}$

21. $(2\sqrt{3} + \sqrt{5})(3\sqrt{3} - 2\sqrt{5})$

$= 2\sqrt{3}\cdot 3\sqrt{3} + 2\sqrt{3}\cdot(-2\sqrt{5}) + \sqrt{5}\cdot 3\sqrt{3} + \sqrt{5}\cdot(-2\sqrt{5})$

$= 6\sqrt{9} - 4\sqrt{15} + 3\sqrt{15} - 2\sqrt{25}$

$= 6\cdot 3 - 4\sqrt{15} + 3\sqrt{15} - 2\cdot 5$

$= 18 - 4\sqrt{15} + 3\sqrt{15} - 10$

$= 8 - \sqrt{15}$

25. $(\sqrt{21} - \sqrt{5})^2$

$= (\sqrt{21})^2 - 2(\sqrt{21})(\sqrt{5}) + (\sqrt{5})^2$

$= 21 - 2\sqrt{105} + 5$

$= 26 - 2\sqrt{105}$

29. $(2 + \sqrt[3]{6})(2 - \sqrt[3]{6}) = (2)^2 - (\sqrt[3]{6})^2$

$= 4 - \sqrt[3]{6^2}$

$= 4 - \sqrt[3]{36}$

33. $(3\sqrt{x} - \sqrt{5})(2\sqrt{x} + 1)$

$= 3\sqrt{x}\cdot 2\sqrt{x} + 3\sqrt{x}\cdot 1 - \sqrt{5}\cdot 2\sqrt{x} - \sqrt{5}\cdot 1$

$= 6\sqrt{x^2} + 3\sqrt{x} - 2\sqrt{5x} - \sqrt{5}$

$= 6x + 3\sqrt{x} - 2\sqrt{5x} - \sqrt{5}$

37. $(5\sqrt{z} + 1)^2$

$= (5\sqrt{z})^2 + 2(5\sqrt{z})(1) + (1)^2$

$= 5\sqrt{z}\cdot 5\sqrt{z} + 10\sqrt{z} + 1$

$= 25\sqrt{z^2} + 10\sqrt{z} + 1$

$= 25z + 10\sqrt{z} + 1$

41. $2\sqrt{3} + 2 = 2(\sqrt{3} + 1)$ *Factor out 2*

45. $2\sqrt{20} - 4\sqrt{7} = 2\cdot\sqrt{4}\sqrt{5} - 4\sqrt{7}$ *Simplify*

$= 2\cdot 2\sqrt{5} - 4\sqrt{7}$

$= 4\sqrt{5} - 4\sqrt{7}$

$= 4(\sqrt{5} - \sqrt{7})$ *Factor out 4*

49. $\dfrac{30 - 20\sqrt{6}}{10}$

$= \dfrac{30}{10} - \dfrac{20\sqrt{6}}{10}$

$= 3 - 2\sqrt{6}$

53. $\dfrac{16 - 4\sqrt{8}}{12} = \dfrac{16 - 4\cdot 2\sqrt{2}}{12}$

$= \dfrac{16 - 8\sqrt{2}}{12}$

$= \dfrac{4(4 - 2\sqrt{2})}{12}$

$= \dfrac{4 - 2\sqrt{2}}{3}$

57. $\dfrac{3}{4 + \sqrt{5}} = \dfrac{3(4 - \sqrt{5})}{(4 + \sqrt{5})(4 - \sqrt{5})}$

$= \dfrac{3(4 - \sqrt{5})}{(4)^2 - (\sqrt{5})^2}$

$= \dfrac{3(4 - \sqrt{5})}{16 - 5}$

$= \dfrac{3(4 - \sqrt{5})}{11}$

61. $\dfrac{2}{\sqrt{2} + \sqrt{5}} = \dfrac{2(\sqrt{2} - \sqrt{5})}{(\sqrt{2} + \sqrt{5})(\sqrt{2} - \sqrt{5})}$

$= \dfrac{2(\sqrt{2} - \sqrt{5})}{(\sqrt{2})^2 - (\sqrt{5})^2}$

$= \dfrac{2(\sqrt{2} - \sqrt{5})}{2 - 5}$

$= \dfrac{2(\sqrt{2} - \sqrt{5})}{-3}$

$= \dfrac{-2(\sqrt{2} - \sqrt{5})}{3}$

$= \dfrac{-2\sqrt{2} + 2\sqrt{5}}{3}$

65. $\dfrac{1}{2 - 3\sqrt{2}} = \dfrac{2 + 3\sqrt{2}}{(2 - 3\sqrt{2})(2 + 3\sqrt{2})}$

$= \dfrac{2 + 3\sqrt{2}}{(2)^2 - (3\sqrt{2})^2}$

$= \dfrac{2 + 3\sqrt{2}}{4 - 9\cdot 2}$

$= \dfrac{2 + 3\sqrt{2}}{4 - 18}$

$= \dfrac{2 + 3\sqrt{2}}{-14}$

$= \dfrac{-(2 + 3\sqrt{2})}{14}$

69. $$\frac{\sqrt{2}-1}{\sqrt{2}+\sqrt{3}} = \frac{(\sqrt{2}-1)(\sqrt{2}-\sqrt{3})}{(\sqrt{2}+\sqrt{3})(\sqrt{2}-\sqrt{3})}$$
$$= \frac{2\cdot\sqrt{2}+\sqrt{2}\cdot(-\sqrt{3})-1\cdot\sqrt{2}-1\cdot(-\sqrt{3})}{(\sqrt{2})^2-(\sqrt{3})^2}$$
$$= \frac{2-\sqrt{6}-\sqrt{2}+\sqrt{3}}{2-3}$$
$$= \frac{2-\sqrt{6}-\sqrt{2}+\sqrt{3}}{-1}$$
$$= -(2-\sqrt{6}-\sqrt{2}+\sqrt{3})$$
$$= -2+\sqrt{6}+\sqrt{2}-\sqrt{3}$$

73. $$\frac{m-4}{\sqrt{m}+2} = \frac{(m-4)(\sqrt{m}-2)}{(\sqrt{m}+2)(\sqrt{m}-2)}$$
$$= \frac{(m-4)(\sqrt{m}-2)}{(\sqrt{m})^2-(2)^2}$$
$$= \frac{(m-4)(\sqrt{m}-2)}{m-4}$$
$$= \sqrt{m}-2$$

77. $$\frac{\sqrt{m}-\sqrt{3r}}{\sqrt{m}+\sqrt{3r}} = \frac{(\sqrt{m}-\sqrt{3r})(\sqrt{m}-\sqrt{3r})}{(\sqrt{m}+\sqrt{3r})(\sqrt{m}-\sqrt{3r})}$$
$$= \frac{(\sqrt{m}-\sqrt{3r})^2}{(\sqrt{m})^2-(\sqrt{3r})^2}$$
$$= \frac{(\sqrt{m})^2-2(\sqrt{m})(\sqrt{3r})+(\sqrt{3r})^2}{m-3r}$$
$$= \frac{m-2\sqrt{3mr}+3r}{m-3r}$$

81. $$\frac{6-\sqrt{2}}{4} = \frac{(6-\sqrt{2})(6+\sqrt{2})}{4(6+\sqrt{2})}$$
$$= \frac{6^2-(\sqrt{2})^2}{4(6+\sqrt{2})}$$
$$= \frac{36-2}{4(6+\sqrt{2})}$$
$$= \frac{34}{4(6+\sqrt{2})}$$
$$= \frac{17}{2(6+\sqrt{2})}$$

85. $$\frac{3\sqrt{a}+\sqrt{b}}{b} = \frac{(3\sqrt{a}+\sqrt{b})(3\sqrt{a}-\sqrt{b})}{b(3\sqrt{a}-\sqrt{b})}$$
$$= \frac{(3\sqrt{a})^2-(\sqrt{b})^2}{b(3\sqrt{a}-\sqrt{b})}$$
$$= \frac{9a-b}{b(3\sqrt{a}-\sqrt{b})}$$

89. $$\frac{2}{\sqrt[3]{5}-\sqrt[3]{3}} = \frac{2(\sqrt[3]{25}+\sqrt[3]{15}+\sqrt[3]{9})}{(\sqrt[3]{5}-\sqrt[3]{3})(\sqrt[3]{25}+\sqrt[3]{15}+\sqrt[3]{9})}$$
$$= \frac{2(\sqrt[3]{25}+\sqrt[3]{15}+\sqrt[3]{9})}{2}$$ *See Exercise 87*
$$= \sqrt[3]{25}+\sqrt[3]{15}+\sqrt[3]{9}$$

93. $p^2 - 5p + 4 = 0$
$(p-4)(p-1) = 0$ *Factor*
$p - 4 = 0$ or $p - 1 = 0$ *Place each factor equal to 0*
$p = 4$ or $p = 1$
Solution set: $\{4, 1\}$

## Section 5.6 (page 236)

You always should check possible solutions in the original equation. In the following exercises only values that produce false statements are shown.

1. $\sqrt{r-2} = 3$
$(\sqrt{r-2})^2 = 3^2$
$r - 2 = 9$
$r = 11$
Solution set: $\{11\}$

5. $\sqrt{a} + 5 = 0$
$\sqrt{a} = -5$ *Make sure radical is alone*
$(\sqrt{a})^2 = (-5)^2$
$a = 25$
Check in original equation:
$\sqrt{25} + 5 = 0$
$5 + 5 = 0$ *False*
Solution set: $\emptyset$

9. $4 - \sqrt{x - 2} = 0$

$4 = \sqrt{x - 2}$ *Get the radical alone on one side*

$(4)^2 = (\sqrt{x - 2})^2$ *Square both sides*

$16 = x - 2$

$18 = x$

Solution set: {18}

13. $2\sqrt{x} = \sqrt{3x + 4}$

$(2\sqrt{x})^2 = (\sqrt{3x + 4})^2$

$4x = 3x + 4$

$x = 4$

Solution set: {4}

17. $k = \sqrt{k^2 + 4k - 20}$

$(k)^2 = (\sqrt{k^2 + 4k - 20})^2$

$k^2 = k^2 + 4k - 20$

$-4k = -20$

$k = 5$

Solution set: {5}

21. $\sqrt{x^2 + 3x - 3} = x + 1$

$(\sqrt{x^2 + 3x - 3})^2 = (x + 1)^2$

$x^2 + 3x - 3 = x^2 + 2x + 1$

$x = 4$

Solution set: {4}

25. $\sqrt{r^2 + 9r + 3} = -r$

$(\sqrt{r^2 + 9r + 3})^2 = (-r)^2$

$r^2 + 9r + 3 = r^2$

$9r = -3$

$r = -\frac{1}{3}$

Solution set: $\{-\frac{1}{3}\}$

29. $\sqrt[3]{2x + 5} = \sqrt[3]{6x + 1}$

$(\sqrt[3]{2x + 5})^3 = (\sqrt[3]{6x + 1})^3$

$2x + 5 = 6x + 1$

$-4x = -4$

$x = 1$

Solution set: {1}

33. $\sqrt[3]{2m - 1} = \sqrt[3]{m + 13}$

$(\sqrt[3]{2m - 1})^3 = (\sqrt[3]{m + 13})^3$

$2m - 1 = m + 13$

$m = 14$

Solution set: {14}

37. $\sqrt[3]{x - 8} + 2 = 0$

$\sqrt[3]{x - 8} = -2$

$(\sqrt[3]{x - 8})^3 = (-2)^3$

$x - 8 = -8$

$x = 0$

Solution set: {0}

41. $\sqrt{k + 2} - \sqrt{k - 3} = 1$

$\sqrt{k + 2} = 1 + \sqrt{k - 3}$

$(\sqrt{k + 2})^2 = (1 + \sqrt{k - 3})^2$

$k + 2 = 1^2 + 2\sqrt{k - 3} + (\sqrt{k - 3})^2$

$k + 2 = 1 + 2\sqrt{k - 3} + k - 3$

$(4)^2 = (2\sqrt{k - 3})^2$

$16 = 4(k - 3)$

$16 = 4k - 12$

$28 = 4k$

$7 = k$

Solution set: {7}

45. $(3p + 4)^{1/2} - (2p - 4)^{1/2} = 2$

$\sqrt{3p + 4} = \sqrt{2p - 4} + 2$

$(\sqrt{3p + 4})^2 = (\sqrt{2p - 4} + 2)^2$

$3p + 4 = 2p - 4 + 2(2)\sqrt{2p - 4} + 4$

$(p + 4)^2 = (4\sqrt{2p - 4})^2$

$p^2 + 8p + 16 = 16(2p - 4)$

$p^2 + 8p + 16 = 32p - 64$

$p^2 - 24p + 80 = 0$

$(p - 20)(p - 4) = 0$

$p = 20$ or $p = 4$ *Check both solutions*

Solution set: $\{4, 20\}$

49. $\sqrt{2\sqrt{x + 11}} = \sqrt{4x + 2}$

$(\sqrt{2\sqrt{x + 11}})^2 = (\sqrt{4x + 2})^2$

$2\sqrt{x + 11} = 4x + 2$

$(2\sqrt{x + 11})^2 = (4x + 2)^2$

$4(x + 11) = 16x^2 + 2(2)(4x) + 4$

$4x + 44 = 16x^2 + 16x + 4$

$0 = 16x^2 + 12x - 40$

$0 = 4(4x^2 + 3x - 10)$

$0 = 4(4x - 5)(x + 2)$

$4x - 5 = 0$ or $x + 2 = 0$

$x = \frac{5}{4}$ or $x = -2$

Check $x = -2$:

$\sqrt{2\sqrt{-2 + 11}} = \sqrt{4(-2) + 2}$

$\sqrt{2\sqrt{10}} = \sqrt{-8 + 2}$

$\sqrt{2\sqrt{10}} = \sqrt{-6}$ *False*

Only $\frac{5}{4}$ works so the solution set is $\{\frac{5}{4}\}$.

53. $r^2 - 4r + 3 = 0$

$(r - 3)(r - 1) = 0$

$r - 3 = 0$ or $r - 1 = 0$

$r = 3$ or $r = 1$

The solution set is $\{1, 3\}$.
Check by substituting back in the original equation.

57. $6z^2 = 7z + 3$

$6z^2 - 7z - 3 = 0$

$(3z + 1)(2z - 3) = 0$

$3z + 1 = 0$ or $2z - 3 = 0$

$3z = -1$ or $2z = 3$

$z = \frac{-1}{3}$ or $z = \frac{3}{2}$

Solution set: $\{\frac{-1}{3}, \frac{3}{2}\}$

**Section 5.7 (page 242)**

1. $\sqrt{-144} = \sqrt{144}\cdot\sqrt{-1} = i\sqrt{144}$
$= 12i$

5. $\sqrt{-3} = \sqrt{3}\cdot\sqrt{-1}$
$= i\sqrt{3}$

9. $4\sqrt{-12} - 3\sqrt{-3}$
$= 4i\sqrt{12} - 3i\sqrt{3}$
$= 4i\sqrt{4\cdot 3} - 3i\sqrt{3}$
$= 8i\sqrt{3} - 3i\sqrt{3}$
$= 5i\sqrt{3}$

13. $\sqrt{-5}\cdot\sqrt{-5}$
$= i\sqrt{5}\cdot i\sqrt{5}$
$= 5i^2$
$= 5(-1)$
$= -5$

17. $\sqrt{-16}\cdot\sqrt{-100} = i\sqrt{16}\cdot i\sqrt{100}$
$= 4i\cdot 10i$
$= 40i^2$
$= -40$

21. $\frac{\sqrt{-200}}{\sqrt{-100}} = \frac{\sqrt{-1}\sqrt{200}}{\sqrt{-1}\sqrt{100}}$
$= \frac{i\sqrt{200}}{i\sqrt{100}}$
$= \frac{\sqrt{200}}{\sqrt{100}}$
$= \sqrt{\frac{200}{100}}$
$= \sqrt{2}$

25. $\frac{\sqrt{-288}}{\sqrt{-8}} = \frac{\sqrt{-1}\sqrt{288}}{\sqrt{-1}\sqrt{8}}$
$= \frac{i\sqrt{144 \cdot 2}}{i\sqrt{4 \cdot 2}}$
$= \frac{12i\sqrt{2}}{2i\sqrt{2}}$
$= 6$

29. $(1+i) + (2-i) = (1+2) + i(1-1)$
$= 3$
$= 3 + 0i$
*In standard form a + bi*

33. $(-3-2i) + (3+2i) = (-3+3) + (2i-2i)$
$= 0 + 0i$

37. $(5-8i) - (1+7i) = (5-1) + (-8i-7i)$
$= 4 - 15i$

41. $(-2-3i) - (5+5i) = (-2-5) + (-3i-5i)$
$= -7 - 8i$

45. $(4 + i) + (1 + 5i) + (3 + 2i)$
$= (4 + 1 + 3) + (i + 5i + 2i)$
$= 8 + 8i$

49. $(2i)(5i) = 10i^2$
$= -10 + 0i$

53. $2i(4 - 3i) = 8i - 6i^2$
$= 6 + 8i$

57. $(3 + 2i)(-1 + 2i)$
$= -3 + 6i - 2i + 4i^2$
$= -3 + 4i - 4$
$= -7 + 4i$

61. $(3 + 5i)(-1 + 9i)$
$= -3 + 27i - 5i + 45i^2$
$= -3 + 22i - 45$
$= -48 + 22i$

65. $(-5+2i)^2 = (-5+2i)(-5+2i)$
$= 25 - 10i - 10i + 4i^2$
$= 25 - 20i - 4$
$= 21 - 20i$

69. $(4 + 2i)(4 - 2i) = 16 - 4i^2$
$= 16 - 4(-1)$
$= 20 + 0i$

73. $(8 - 9i)(8 + 9i) = 64 - 81i^2$
$= 64 - 81(-1)$
$= 145 + 0i$

77. $\frac{i}{2+i} = \frac{i}{2+i}\,\frac{2-i}{2-i}$
$= \frac{i(2-i)}{4-i^2}$
$= \frac{i(2-i)}{4-(-1)}$
$= \frac{i(2-i)}{5}$
$= \frac{1}{5} + \frac{2}{5}i$

81. $\frac{2-3i}{2+3i} = \frac{2-3i}{2+3i} \cdot \frac{2-3i}{2-3i}$
$= \frac{4-6i-6i+9i^2}{4-9i^2}$
$= \frac{4-12i+9(-1)}{4-9(-1)}$
$= \frac{-5-12i}{13}$
$= \frac{-5}{13} - \frac{12}{13}i$

85. $\frac{2+6i}{2+i} = \frac{2+6i}{2+i} \cdot \frac{2-i}{2-i}$
$= \frac{4-2i+12i-6i^2}{4-i^2}$
$= \frac{4+10i+6}{4-(-1)}$
$= \frac{10+10i}{5}$
$= 2 + 2i$

89. $\frac{-6}{5i} = \frac{-6}{5i}\cdot\frac{i}{i}$

$= \frac{-6i}{5i^2}$

$= \frac{-6i}{5(-1)} = \frac{6i}{5}$

$= 0 + \frac{6}{5}i$

93. $i^{17} = i^{16}\cdot i = (i^2)^8\cdot i = (-1)^8\cdot i = i$

97. $i^{-10} = \frac{1}{i^{10}}$

$= \frac{1}{(i^2)^5}$

$= \frac{1}{(-1)^5}$

$= \frac{1}{-1}$

$= -1$

101. $\frac{5i}{2 - 3i} - \frac{3i}{4 + i}$

$= \frac{5i}{2 - 3i}\cdot\frac{2 + 3i}{2 + 3i} - \frac{3i}{4 + i}\cdot\frac{4 - i}{4 - i}$

*Multiply by conjugates*

$= \frac{10i + 15i^2}{4 - 9i^2} - \frac{12i - 3i^2}{16 - 1^2}$

$= \frac{10i - 15}{13} - \frac{12i + 3}{17}$

$= \frac{17(10i - 15) - 13(12i + 3)}{221}$

*Get each fraction to the least common denominator*

$= \frac{170i - 255 - 156i - 39}{221}$

$= -\frac{294}{221} + \frac{14}{221}i$

105. $I = \frac{E}{R + (X_L - X_C)i}$

$I = \frac{2 + 3i}{5 + (4 - 3)i} = \frac{2 + 3i}{5 + i}$

$= \frac{2 + 3i}{5 + i}\cdot\frac{5 - i}{5 - i} = \frac{10 - 2i + 15i - 3i^2}{25 - i^2}$

$= \frac{13 + 13i}{26} = \frac{13}{26} + \frac{13i}{26}$

$= \frac{1}{2} + \frac{1}{2}i$

109. The square roots of 16 are 4 and -4.

113. The square roots of 12 are $2\sqrt{3}$ and $-2\sqrt{3}$.

**Chapter 5 Review Exercises (page 245)**

1. $100^{3/2} = (100^{1/2})^3 = 10^3 = 1000$

5. $(\frac{81}{10,000})^{-3/4} = (\frac{10,000}{81})^{3/4}$

$= \frac{(10,000^{1/4})^3}{(81^{1/4})^3}$

$= \frac{(10)^3}{(3)^3} = \frac{1000}{27}$

9. $\frac{y^{-1/3}\cdot y^{5/6}}{y} = y^{-1/3+5/6-1}$

$= y^{-2/6+5/6-6/6}$

$= y^{-3/6}$

$= \frac{1}{y^{1/2}}$

13. $\sqrt{1764} = 42$ *Since* $42^2 = 1764$

17. $\sqrt[5]{-32} = -2$ *Since* $(-2)^5 = -32$

21. $\sqrt{3^{17}} = 3^{17/2}$
$= 3^{16/2}3^{1/2}$
$= 3^8 3^{1/2}$
$= 3^8\sqrt{3}$

25. $\sqrt{6}\cdot\sqrt{13} = \sqrt{78}$

29. $\sqrt{125} = \sqrt{25}\cdot\sqrt{5} = 5\sqrt{5}$

33. $\sqrt{\frac{49}{81}} = \frac{\sqrt{49}}{\sqrt{81}} = \frac{7}{9}$

37. $\sqrt[8]{s^4} = {}^{4/8} = s^{1/2} = \sqrt{s}$

41. $\frac{\sqrt{6}}{\sqrt{5}} = \frac{\sqrt{6}\sqrt{5}}{\sqrt{5}\sqrt{5}} = \frac{\sqrt{30}}{5}$

45. $-\sqrt[3]{\frac{9}{25}} = -\frac{\sqrt[3]{9}\ \sqrt[3]{5}}{\sqrt[3]{25}\cdot\sqrt[3]{5}}$
$= -\frac{\sqrt[3]{45}}{5}$

49. $2\sqrt{8} - 3\sqrt{50}$
$= 2\sqrt{4\cdot 2} - 3\sqrt{25\cdot 2}$
$= 2\cdot 2\sqrt{2} - 3\cdot 5\sqrt{2}$
$= 4\sqrt{2} - 3\cdot 5\sqrt{2}$
$= -11\sqrt{2}$

53. $-6\sqrt[4]{32} + \sqrt[4]{512}$
$= -6\sqrt[4]{16}\sqrt[4]{2} + \sqrt[4]{256}\sqrt[4]{2}$
$= -6\cdot 2\sqrt[4]{2} + 4\sqrt[4]{2}$
$= -12\sqrt[4]{2} + 4\sqrt[4]{2}$
$= -8\sqrt[4]{2}$

57. $(3\sqrt{2} + 1)(2\sqrt{2} - 3)$
$= (3\sqrt{2})(2\sqrt{2}) + (3\sqrt{2})(-3) +$
$(1)(2\sqrt{2}) + (1)(-3)$
$= 6\cdot\sqrt{2}\cdot\sqrt{2} - 9\sqrt{2} + 2\sqrt{2} - 3$
$= 6\cdot\overline{2} - 7\sqrt{2} - 3$
$= 12 - 7\sqrt{2} - 3$
$= 9 - 7\sqrt{2}$

61. $\frac{-5}{\sqrt{6} - \sqrt{3}} = \frac{-5(\sqrt{6} + \sqrt{3})}{(\sqrt{6} - \sqrt{3})(\sqrt{6} + \sqrt{3})}$
$= \frac{-5(\sqrt{6} + \sqrt{3})}{(\sqrt{6})^2 - (\sqrt{3})^2} = \frac{-5(\sqrt{6} + \sqrt{3})}{6 - 3}$
$= \frac{-5(\sqrt{6} + \sqrt{3})}{3}$

65. $\sqrt{2z - 3} - 3 = 0$
$2z - 3 = 3$
$(\sqrt{2z - 3})^2 = 3^2$
$2z - 3 = 9$
$2z = 12$
$z = 6$
Solution set: {6}

69. $\sqrt{p^2 + 3p + 7} = p + 2$
$(\sqrt{p^2 + 3p + 7})^2 = (p + 2)^2$ *Square both sides*
$p^2 + 3p + 7 = p^2 + 4p + 4$
$3 = p$ *Simplify*
Solution set: {3}

73. $\sqrt{-25} = \sqrt{-1}\sqrt{25} = i\cdot 5 = 5i$

77. $(-2 + 5i) + (-8 - 7i)$
$= (-2 - 8) + (5i - 7i)$
$= -10 - 2i$

81. $\frac{\sqrt{-72}}{\sqrt{-8}} = \frac{i\sqrt{72}}{i\sqrt{8}}$
$= \frac{i\sqrt{36\cdot 2}}{i\sqrt{4\cdot 2}} = \frac{6i\sqrt{2}}{2i\sqrt{2}}$
$= 3$

85. $i^{11} = i^{10}\cdot i$
$= (i^2)^5\cdot i$
$= (-1)^5\cdot i = -i$

89. $-\sqrt{27y} + 2\sqrt{75y}$

$= -\sqrt{9 \cdot 3y} + 2\sqrt{25 \cdot 3y}$

$= -3\sqrt{3y} + 10\sqrt{3y}$

$= 7\sqrt{3y}$

93. $2\sqrt{54m^3} + 5\sqrt{96m^3}$

$= 2\sqrt{9 \cdot 6 \cdot m^2 \cdot m} + 5\sqrt{16 \cdot 6 \cdot m^2 \cdot m}$

$= 2 \cdot 3 \cdot m\sqrt{6m} + 5 \cdot 4 \cdot m\sqrt{6m}$

$= 6m\sqrt{6m} + 20m\sqrt{6m}$

$= 26m\sqrt{6m}$

97. $\sqrt[3]{216} = 6$ *Since* $6^3 = 216$

101. $\left(\frac{1}{32}\right)^{-7/5}$

$= \left(\frac{32}{1}\right)^{7/5} = (32^{1/5})^7 = 2^7 = 128$

105. $(\sqrt{11} + 3\sqrt{5})(\sqrt{11} + 5\sqrt{5})$

$= \sqrt{11} \cdot \sqrt{11} + \sqrt{11} \cdot 5\sqrt{5} + 3\sqrt{5} \cdot \sqrt{11} + 3\sqrt{5} \cdot 5\sqrt{5}$

$= 11 + 5\sqrt{55} + 3\sqrt{55} + 15\sqrt{25}$

$= 11 + 8\sqrt{55} + 75$

$= 86 + 8\sqrt{55}$

109. $\frac{\sqrt{50}}{\sqrt{-2}} = \frac{\sqrt{50}}{i\sqrt{2}}$

$= \frac{\sqrt{50} - i\sqrt{2}}{i\sqrt{2} - i\sqrt{2}}$

$= \frac{-\sqrt{100}}{-2i^2}$

$= \frac{-10i}{2} = -5i$

**Chapter 5 Test (page 247)**

1. $27^{4/3} = (\sqrt{27})^4$

$= (3)^4$

$= 81$

2. $3^{1/2} \cdot 3^{5/2} = 3^{1/2+5/2} = 3^{6/2} = 3^3 = 27$

3. $\frac{9^{-3/4}}{9^{-1/4}} = 9^{-3/4-(-1/4)}$

$= 9^{-1/2}$

$= \frac{1}{9^{1/2}}$

$= \frac{1}{3}$

4. $\left(\frac{a^{-2}a^{-1/2}}{a^{-3/4} \cdot a}\right)^{-1} = \frac{a^2a^{1/2}}{a^{3/4} \cdot a^{-1}}$

$= \frac{a^{5/2}}{a^{3/4 - 1}}$

$= \frac{a^{10/4}}{a^{-1/4}}$

$= a^{10/4-(-1/4)}$

$= a^{11/4}$

5. $r^{-2/3}(r^{5/3} + r^{-1/3})$

$= r^{-2/3+5/3} + r^{-2/3+(-1/3)}$

$= r^{3/3} + r^{-3/3}$

$= r^1 + r^{-1}$

$= r + \frac{1}{r}$

6. $\sqrt[3]{-1000} = -10$ *Since* $(-10)^3 = -1000$

7. $\sqrt{100y^{14}} = 10y^7$

8. $-\sqrt[4]{16m^4p^{12}} = -2mp^3$

9. $\sqrt[4]{-256}$ is not a real number since even roots of negative numbers are not real.

10. $\sqrt[3]{a^4}\cdot\sqrt[3]{a^7} = \sqrt[3]{a^{11}}$
$= \sqrt[3]{a^9}\sqrt[3]{a^2}$
$= a^3\sqrt[3]{a^2}$

11. $\sqrt{108x^5} = \sqrt{36x^4}\cdot\sqrt{3x}$
$= 6x^2\sqrt{3x}$

12. $\sqrt[4]{16a^5b^{11}} = \sqrt[4]{16a^4b^8}\cdot\sqrt[4]{ab^3}$
$= 2ab^2\sqrt[4]{ab^3}$

13. $\sqrt[3]{3}\cdot\sqrt[3]{18} = \sqrt[3]{54}$
$= \sqrt[3]{27}\cdot\sqrt[3]{2} = 3\sqrt[3]{2}$

14. $8\sqrt{20} + 3\sqrt{80} - 2\sqrt{500}$
$= 8\cdot\sqrt{4}\cdot\sqrt{5} + 3\sqrt{16}\cdot\sqrt{5} - 2\sqrt{100}\cdot\sqrt{5}$
$= 8\cdot 2\sqrt{5} + 3\cdot 4\sqrt{5} - 2\cdot 10\cdot\sqrt{5}$
$= 16\sqrt{5} + 12\sqrt{5} - 20\sqrt{5}$
$= 8\sqrt{5}$

15. $(\sqrt{5} - \sqrt{3})(3\sqrt{5} + \sqrt{3})$
$= 3\sqrt{5}\cdot\sqrt{5} + \sqrt{5}\cdot\sqrt{3} - \sqrt{3}\cdot 3\sqrt{5} - \sqrt{3}\cdot\sqrt{3}$
$= 15 - 2\sqrt{15} - 3$
$= 12 - 2\sqrt{15}$

16. $(2\sqrt{5} - 3)(\sqrt{5} + 6)$
$= 2\sqrt{5}\cdot\sqrt{5} + 2\cdot 6\sqrt{5} + -3\cdot\sqrt{5} + (-3\cdot 6)$
$= 2\cdot 5 + 12\sqrt{5} - 3\sqrt{5} - 18$
$= 10 - 18 + 9\sqrt{5}$
$= -8 + 9\sqrt{5}$

17. $\frac{-9}{\sqrt{40}} = \frac{-9}{2\sqrt{10}}$ $\quad\sqrt{40} = \sqrt{4}\cdot\sqrt{10} = 2\sqrt{10}$
$= \frac{-9\cdot\sqrt{10}}{2\sqrt{10}\cdot\sqrt{10}}$
$= \frac{-9\sqrt{10}}{20}$

18. $\frac{3}{\sqrt[3]{4}} = \frac{3\sqrt[3]{2}}{\sqrt[3]{4}\cdot\sqrt[3]{2}} = \frac{3\sqrt[3]{2}}{\sqrt[3]{8}} = \frac{3\sqrt[3]{2}}{2}$

19. $\frac{-8}{\sqrt{7} - \sqrt{5}} = \frac{-8(\sqrt{7} + \sqrt{5})}{(\sqrt{7} - \sqrt{5})(\sqrt{7} + \sqrt{5})}$
$= \frac{-8(\sqrt{7} + \sqrt{5})}{(\sqrt{7})^2 - (\sqrt{5})^2}$
$= \frac{-8(\sqrt{7} + \sqrt{5})}{7 - 5}$
$= \frac{-8(\sqrt{7} + \sqrt{5})}{2}$
$= -4(\sqrt{7} + \sqrt{5})$

20. $4\sqrt{p} = \sqrt{13p + 27}$
$(4\sqrt{p})^2 = (\sqrt{13p + 27})^2$
$16p = 13p + 27$
$3p = 27$
$p = 9$
Solution set: $\{9\}$

21. $\sqrt{y^2 - 5y + 3} = 4 + y$
$(\sqrt{y^2 - 5y + 3})^3 = (4 + y)^2$
$y^2 - 5y + 3 = y^2 + 8y + 16$
$-13 = 13y$
$-1 = y$
Solution set: $\{-1\}$

22. $\sqrt{y + 5} + \sqrt{2y + 9} = 2$
$\sqrt{y + 5} = 2 - \sqrt{2y + 9}$
$(\sqrt{y + 5})^2 = (2 - \sqrt{2y + 9})^2$
$y + 5 = 4 - 4\sqrt{2y + 9}$
$+ 2y + 9$
$y + 5 - 4 - 2y - 9 = -4\sqrt{2y + 9}$
$-y - 8 = -4\sqrt{2y + 9}$
$y + 8 = 4\sqrt{2y + 9}$
$(y + 8)^2 = (4\sqrt{2y + 9})^2$
$y^2 + 16y + 64 = 16(2y + 9)$
$y^2 + 16y + 64 = 32y + 144$
$y^2 - 16y - 80 = 0$
$(y + 4)(y - 20) = 0$
$y + 4 = 0$ or $y - 20 = 0$
$y = -4$ or $y = 20$

Check:

$\sqrt{20 + 5} + \sqrt{2 \cdot 20 + 9} = 2$ *Let y = 20*

$25 + \quad 49 = 2$

$5 + 7 = 2$

$12 = 2$ *False*

Reject 20 as a solution.

Solution set: $\{-4\}$

23. $(-6 + 4i) - (8 - 7i) + 10i$

$= (-6 - 8) + (4i + 7i + 10i)$

$= -14 + 21i$

24. $(1 + 5i)(3 + i)$

$= 3 \cdot 1 + 1 \cdot i + 5i \cdot 3 + 5i \cdot i$

$= 3 + 16i + 5(-1)$

$= -2 + 16i$

25. $\frac{1 + 2i}{3 - i} = \frac{1 + 2i}{3 - i} \cdot \frac{3 + i}{3 + i}$

$= \frac{3 + i + 6i + 2i^2}{9 - i^2}$

$= \frac{3 + 7i + 2(-1)}{9 - (-1)}$

$= \frac{1 + 7i}{10} = \frac{1}{10} + \frac{7}{10}i$

## CHAPTER 6 QUADRATIC EQUATIONS AND INEQUALITIES

### Section 6.1 (page 253)

1. $b^2 = 49$

$b = 7$ or $b = -7$ *Take square root of each side*

Solution set: $\{-7, 7\}$

5. $k^2 = 12$

$k = \sqrt{12}$ or $k = 1 - \sqrt{12}$

$k = 2\sqrt{3}$ or $k = 1 - 2\sqrt{3}$ *Take the square root of each side*

Solution set: $\{2\sqrt{3}, -2\sqrt{3}\}$

9. $(p + 6)^2 = 9$

$p + 6 = 3$ or $p + 6 = -3$ *Take the square root of each side*

$p = -3$ or $p = -9$ *Solve each equation*

Solution set: $\{-3, -9\}$

13. $(4p + 1)^2 = 12$

$4p + 1 = \sqrt{12}$ or $4p + 1 = -\sqrt{12}$ *Take the square root of each side*

$4p = -1 + 2\sqrt{3}$ $\quad$ $4p = -1 - 2\sqrt{3}$

*$12 = 2\sqrt{3}$*

$p = \frac{-1 + 2\sqrt{3}}{4}$ or $p = \frac{-1 - 2\sqrt{3}}{4}$

Solution set: $\{\frac{-1 + 2\sqrt{3}}{4}, \frac{-1 - 2\sqrt{3}}{4}\}$

17. $a^2 = -72$

$a = \sqrt{-72}$ or $a = -\sqrt{-72}$ *Take the square root of each side*

$a = 6i\sqrt{2}$ or $a = -6i\sqrt{2}$

*$\sqrt{-72} = 6i\sqrt{2}$*

Solution set: $\{-6i\sqrt{2}, 6i\sqrt{2}\}$

21. $(2m - 1)^2 = -3$

$2m - 1 = \sqrt{-3}$ or $2m - 1 = -\sqrt{-3}$ *Take the square root of each side*

$2m = 1 + i\sqrt{3}$ $\quad$ $2m = 1 - i\sqrt{3}$

*$\sqrt{-3} = i\sqrt{3}$*

$m = \frac{1 + i\sqrt{3}}{2}$ or $m = \frac{1 - i\sqrt{3}}{2}$

Solution set: $\frac{1 + i\sqrt{3}}{2}, \frac{1 - i\sqrt{3}}{2}$

25. $y^2 - 10y$

$\frac{1}{2}(-10) = -5$ *Take half of coefficient of first-degree term*

$(-5)^2 = 25$ *Square result*

Add 25 to make perfect square trinomial"

$y^2 - 10y + 25.$

29. $x^2 + 11x$

$\frac{1}{2}(11) = \frac{11}{2}$ *Half of coefficient of first-degree term*

$(\frac{11}{2})^2 = \frac{121}{4}$ *Square result*

Add $\frac{121}{4}$:

$x^2 + 11x + \frac{121}{4}.$

33. $x^2 - 2x - 15 = 0$

$x^2 - 2x = 15$ *Add 15 to each side*

$x^2 - 2x + 1 = 15 + 1$ *Half of -2; square result; add to each side*

$(x - 1)^2 = 16$ *Write left side as perfect square*

$x - 1 = 4$ or $x - 1 = -4$

$x = 5$ or $x = -3$

Solution set: $\{5, -3\}$

37. $2m^2 + 5m - 3 = 0$

$2m^2 + 5m = 3$ *Add 3 to both sides*

$m^2 + \frac{5}{2}m = \frac{3}{2}$ *Divide both sides by 2*

$m^2 + \frac{5}{2}m + \frac{25}{16} = \frac{3}{2} + \frac{25}{16}$

*Take half of 5 and square it; add to both sides*

$(m + \frac{5}{4})^2 = \frac{24}{16} + \frac{25}{16}$ *Write left side as perfect square*

$(m + \frac{5}{4})^2 = \frac{49}{16}$

$m + \frac{5}{4} = \sqrt{\frac{49}{16}}$ or $m + \frac{5}{4} = -\sqrt{\frac{49}{16}}$

$m = -\frac{5}{4} + \frac{7}{7}$ $\qquad m = -\frac{5}{4} - \frac{7}{4}$

$m = \frac{2}{4} = \frac{1}{2}$ or $m = -\frac{12}{4} = -3$

Solution set: {1/2, -3}

41. $m^2 = -6m - 7$

$m^2 + 6m = -7$

$m^2 + 6m + 9 = -7 + 9$

$(m + 3)^2 = 2$

$m + 3 = \sqrt{2}$

$m = -3 + \sqrt{2}$

or

$m + 3 = -\sqrt{2}$

$m = -3 - \sqrt{2}$

Solution set: $\{-3 + \sqrt{2}, -3 - \sqrt{2}\}$

45. $q^2 + \frac{7}{2}q = 2$

$q^2 + \frac{7}{2}q + \frac{49}{16} = 2 + \frac{49}{16}$

$(q + \frac{7}{4})^2 = \frac{81}{16}$

$q + \frac{7}{4} = \frac{9}{4}$ or $q + \frac{7}{4} = -\frac{9}{4}$

$q = \frac{2}{4} = \frac{1}{2}$ or $q = -\frac{16}{4} = -4$

Solution set: {-4, 1/2}

49. $9a^2 - 24a = -13$

$a^2 - \frac{24}{9}a = -\frac{13}{9}$

$a^2 - \frac{8}{3}a = -\frac{13}{9}$

$a^2 - \frac{8}{3}a + \frac{16}{9} = -\frac{13}{9} + \frac{16}{9}$

$(a - \frac{4}{3})^2 = \frac{1}{3}$

$a = \frac{4}{3} = \sqrt{\frac{1}{3}}$ or $a - \frac{4}{3} = -\sqrt{\frac{1}{3}}$

Since $\sqrt{\frac{1}{3}} = \frac{1}{\sqrt{3}} = \frac{1 \cdot \sqrt{3}}{\sqrt{3} \cdot \sqrt{3}} = \frac{\sqrt{3}}{3}$.

$a = \frac{4}{3} + \frac{\sqrt{3}}{3}$ or $a = \frac{4}{3} - \frac{\sqrt{3}}{3}$

Solution set: $\{\frac{4 + \sqrt{3}}{3}, \frac{4 - \sqrt{3}}{3}\}$

53. $m^2 + 6m + 10 = 0$

$m^2 + 6m = -10$

$m^2 + 6m + 9 = -10 + 9$

$(m + 3)^2 = -1$

$m + 3 = \sqrt{-1}$ or $m + 3 = -\sqrt{-1}$

$m = -3 + i$ or $m = -3 - i$

Solution set: $\{-3 + i, -3 - i\}$

57. $25y^2 + 46 = 70y$

$25y^2 - 70y = -46$

$y^2 - \frac{70}{25}y = -\frac{46}{25}$

$y^2 - \frac{14}{5}y = -\frac{46}{25}$

$y^2 - \frac{14}{5}y + \frac{49}{25} = -\frac{46}{25} + \frac{49}{25}$

$(y - \frac{7}{5})^2 = \frac{-46 + 49}{25}$

$(y - \frac{7}{5})^2 = \frac{3}{25}$

$y - \frac{7}{5} = \sqrt{\frac{3}{25}}$ or $y - \frac{7}{5} = -\sqrt{\frac{3}{25}}$

$y = \frac{7}{5} + \frac{\sqrt{3}}{5}$ or $y = \frac{7}{5} - \frac{\sqrt{3}}{5}$

Solution set: $\{\frac{7 + \sqrt{3}}{5}, \frac{7 - \sqrt{3}}{5}\}$

61. $9x^2 - 30x + 29 = 0$

$$x^2 - \frac{30}{9}x = -\frac{29}{9}$$

$$x^2 - \frac{10}{3}x = -\frac{29}{9}$$

$$x^2 - \frac{10}{3}x + \frac{25}{9} = -\frac{29}{9} + \frac{25}{9}$$

$$(x - \frac{5}{3})^2 = -\frac{4}{9}$$

$x - \frac{5}{3} = \frac{2}{3}i$ or $x - \frac{5}{3} = -\frac{2}{3}i$

$x = \frac{5}{3} + \frac{2}{3}i$ or $x = \frac{5}{3} - \frac{2}{3}i$

Solution set: $\{\frac{5}{3} + \frac{2}{3}i, \frac{5}{3} - \frac{2}{3}i\}$

65. $4x^2 = b^2 + 16$

$x^2 = \frac{b^2 + 16}{4}$ *Multiply each side by 1/4*

$x = \sqrt{\frac{b^2 + 16}{4}}$ or $x = -\sqrt{\frac{b^2 + 16}{4}}$ *Take the square root of each side*

$x = \frac{\sqrt{b^2 + 16}}{2}$ or $x = -\frac{\sqrt{b^2 + 16}}{4}$

Solution set: $\{\frac{\sqrt{b^2 + 16}}{2}, -\frac{\sqrt{b^2 + 16}}{2}\}$

69. $(3x + 7b)^2 = 25$

$3x + 7b = \sqrt{25}$

$3x + 7b = 5$

$3x = -7b + 5$

$x = \frac{-7b + 5}{3}$

or

$3x + 7b = -\sqrt{25}$

$3x + 7b = -5$

$3x = -7b - 5$

$x = \frac{-7b - 5}{3}$

Solution set: $\{\frac{-7b + 5}{3}, \frac{-7b - 5}{3}\}$

73. $\sqrt{b^2 - 4ac}$

$= \sqrt{(-25)^2 - 4(18)(-3)}$

$= \sqrt{625 + 216}$

$= \sqrt{841}$

$= 29$

**Section 6.2 (page 258)**

1. $m^2 + 8m + 15 = 0$ *Use the quadratic formula*

$a = 1, b = 8, c = 15.$

$$m = \frac{-b \pm \sqrt{b^2 - 4ac}}{2a}$$

$$= \frac{-8 \pm \sqrt{8^2 - 4(1)(15)}}{2(1)}$$

$$= \frac{-8 \pm \sqrt{4}}{2}$$

$$= \frac{-8 \pm 2}{2}$$

$m = \frac{-8 + 2}{2} = -3$ or $m = \frac{-8 - 2}{2} = -5$

Solution set: $\{-3, -5\}$

5. $2x^2 + 4x + 1 = 0$

$x = \frac{-b \pm \sqrt{b^2 - 4ac}}{2a}$ *$a = 2$, $b = 4$, $c = 1$*

$$= \frac{-(4) \pm \sqrt{(4)^2 - 4(2)(1)}}{2(2)}$$

$$= \frac{-4 \pm \sqrt{16 - 8}}{4}$$

$$= \frac{-4 \pm \sqrt{8}}{4}$$

$= \frac{-4 \pm 2\sqrt{2}}{4}$ *$\sqrt{8} = \sqrt{4} \cdot \sqrt{2} = 2\sqrt{2}$*

$= \frac{2(-2 \pm \sqrt{2})}{4}$ *Factor out 2*

$$= \frac{-2 \pm \sqrt{2}}{2}$$

Solution set: $\{\frac{-2 + \sqrt{2}}{2}, \frac{-2 - \sqrt{2}}{2}\}$

9. $2y^2 = 2y + 1$

$2y^2 - 2y - 1 = 0$ *Rewrite the equation, a = 2, b = -2, c = -1*

$$y = \frac{-b \pm \sqrt{b^2 - 4ac}}{2a}$$

$$= \frac{2 \pm \sqrt{(-2)^2 - 4(2)(-1)}}{2(2)}$$

$$= \frac{2 \pm \sqrt{12}}{4}$$

$$= \frac{2 \pm 2\sqrt{3}}{4}$$

$$y = \frac{1 \pm \sqrt{3}}{2}$$

Solution set: $\{\frac{1 + \sqrt{3}}{2}, \frac{1 - \sqrt{3}}{2}\}$

13. $5m^2 + 8m + 2 = 0$

$$m = \frac{-b \pm \sqrt{b^2 - 4ac}}{2a}$$ *a = 5, b = 8, c = 2*

$$= \frac{-8 \pm \sqrt{8^2 - 4(5)(2)}}{2(5)}$$

$$= \frac{-8 \pm \sqrt{24}}{10}$$

$$m = \frac{-8 \pm 2\sqrt{6}}{10}$$

$$= \frac{-4 \pm \sqrt{6}}{5}$$

Solution set: $\{\frac{-4 + \sqrt{6}}{5}, \frac{-4 - \sqrt{6}}{5}\}$

17. $4r^2 - 3r = 5$

$4r^2 - 3r - 5 = 0$ *Rewrite the equation; a = 4, b = -3, c = -5*

$$x = \frac{-b \pm \sqrt{b^2 - 4ac}}{2a}$$

$$= \frac{-(-3) \pm \sqrt{(-3)^2 - 4(4)(-5)}}{2(4)}$$

$$= \frac{3 \pm \sqrt{9 + 80}}{8}$$

$$= \frac{3 \pm \sqrt{89}}{8}$$

Solution set:

$\{\frac{3 + \sqrt{89}}{8}, \frac{3 - \sqrt{89}}{8}\}$

21. $4k(k + 1) = 1$

$4k^2 + 4k = 1$

$4k^2 + 4k - 1 = 0$ *Rewrite the equation; a = 4, b = 4, c = -1*

$$x = \frac{-b \pm \sqrt{b^2 - 4ac}}{2a}$$

$$= \frac{-4 \pm \sqrt{(4)^2 - 4(4)(-1)}}{2(4)}$$

$$= \frac{-4 \pm \sqrt{16 + 16}}{8}$$

$$= \frac{-4 \pm \sqrt{32}}{8}$$ $\sqrt{32} = \sqrt{16}\cdot\sqrt{2} = 4\sqrt{2}$

$$= \frac{-4 \pm 4\sqrt{2}}{8}$$

$$= \frac{4(-1 \pm \sqrt{2})}{8}$$ *Factor out 4*

$$= \frac{-1 \pm \sqrt{2}}{2}$$

Solution set: $\{\frac{-1 + \sqrt{2}}{2}, \frac{-1 - \sqrt{2}}{2}\}$

25. $3x^2 + 2x = 2$

$3x^2 + 2x - 2 = 0$

$$x = \frac{-b \pm \sqrt{b^2 - 4ac}}{2a}$$ *a = 3, b = 2, c = -2*

$$= \frac{-2 \pm \sqrt{2^2 - 4(3)(-2)}}{2(3)}$$

$$= \frac{-2 \pm \sqrt{4 + 24}}{6}$$

$$= \frac{-2 \pm \sqrt{28}}{6}$$

$$x = \frac{-2 \pm 2\sqrt{7}}{6}$$

$$= \frac{-1 \pm \sqrt{7}}{3}$$

Solution set: $\{\frac{-1 + \sqrt{7}}{3}, \frac{-1 - \sqrt{7}}{3}\}$

29. $3x^2 + 4x + 2 = 0$

$x = \frac{-b \pm \sqrt{b^2 - 4ac}}{2a}$ *a = 3, b = 4, c = 2*

$= \frac{-4 \pm \sqrt{4^2 - 4(3)(2)}}{2(3)}$

$= \frac{-4 \pm \sqrt{-8}}{6}$

$x = \frac{-4 \pm 2i\sqrt{2}}{6}$

$= \frac{-2 \pm i\sqrt{2}}{3}$

Solution set: $\{-\frac{2}{3} + \frac{\sqrt{2}}{3}i, -\frac{2}{3} - \frac{\sqrt{2}}{3}i\}$

33. $4z^2 = 4z - 7$

$4z^2 - 4z + 7 = 0$

$z = \frac{-b \pm \sqrt{b^2 - 4ac}}{2a}$ *a = 4, b = -4, c = 7*

$= \frac{-(-4) \pm \sqrt{(-4)^2 - 4(4)(7)}}{2(4)}$

$= \frac{4 \pm \sqrt{-96}}{8}$

$z = \frac{4 \pm \sqrt{-16 \cdot 6}}{8} = \frac{4 \pm 4i\sqrt{6}}{8}$

$= \frac{1 \pm i\sqrt{6}}{2}$

Solution set: $\{\frac{1}{2} + \frac{\sqrt{6}}{2}i, \frac{1}{2} - \frac{\sqrt{6}}{2}i\}$

37. $3w^2 - w + 4 = 0$

$w = \frac{-b \pm \sqrt{b^2 - 4ac}}{2a}$ *a = 3, b = -1, c = 4*

$= \frac{-(-1) \pm \sqrt{(-1)^2 - 4(3)(4)}}{2(3)}$

$= \frac{1 \pm \sqrt{-47}}{6}$

$= \frac{1 \pm i\sqrt{47}}{6}$

Solution set: $\{\frac{1}{6} + \frac{\sqrt{47}}{6}i, \frac{1}{6} - \frac{\sqrt{47}}{6}i\}$

41. $2iz^2 - 3z + 2i = 0$

$x = \frac{-b \pm \sqrt{b^2 - 4ac}}{2a}$ *a = 2i, b = -3, c = 2i*

$= \frac{-(-3) \pm \sqrt{(-3)^2 - 4(2i)(2i)}}{2(2i)}$

$= \frac{3 \pm \sqrt{9 - 16i^2}}{4i}$

$= \frac{3 \pm \sqrt{9 + 16}}{4i}$ *$-16i^2 = -16(-1) = 16$*

$= \frac{3 \pm \sqrt{25}}{4i}$

$= \frac{3 \pm 5}{4i}$

$\frac{3 + 5}{4i} = \frac{8}{4i} = \frac{2}{i} = \frac{2 \cdot i}{i \cdot i} = \frac{2i}{i^2} = -2i$

*Write solutions without i in denominator*

$\frac{3 - 5}{4i} = \frac{-2}{4i} = \frac{-1}{2i} = \frac{-1 \cdot i}{2i \cdot i} = \frac{-i}{2i^2} = \frac{i}{2} = \frac{1}{2}i$

Solution set: $\{-2i, \frac{1}{2}i\}$

45. $2z^2 - 5z = 9$

$2z^2 - 5z - 9 = 0$

$x = \frac{-b \pm \sqrt{b^2 - 4ac}}{2a}$ *a = 2, b = -5, c = -9*

$= \frac{-(-5) \pm \sqrt{(-5)^2 - 4(2)(-9)}}{2(2)}$

$= \frac{5 \pm \sqrt{25 + 72}}{4}$

$= \frac{5 \pm \sqrt{97}}{4}$ *$\sqrt{97} \approx 9.849$ to the nearest thousandth*

$\frac{5 + \sqrt{97}}{4} \approx \frac{5 + 9.849}{4} = 3.712$

$\frac{5 - \sqrt{97}}{4} \approx \frac{5 - 9.849}{4} = -1.212$

Solution set: $\{3.712, -1.212\}$

49. Let x be the time the job takes Michael; then Jonathan's time would be x - 2. Together they do the job in 7 hours, giving the equation

$$\frac{1}{x} + \frac{1}{x - 2} = \frac{1}{7}.$$

$7(x - 2) + 7x = x(x - 2)$ *Multiply by least common denominator, $7x(x - 2)$*

$x^2 - 16x + 14 = 0$

$$x = \frac{-(-16) \pm \sqrt{(-16)^2 - 4(1)(14)}}{2(1)}$$ *$a = 1$, $b = -16$, $c = 14$*

$$= \frac{16 \pm 10\sqrt{2}}{2}$$

$$= 8 \pm 5\sqrt{2} \approx 8 \pm 7.07$$

$8 + 7.07 = 15.07 = 15.1$

$8 - 7.07 = .93$

Only the solution 15.1 makes sense in the problem. Thus the job would take Michael 15.1 hours and Jonathan 13.1 hours working alone.

53. Let x represent the time it takes Dick and x + 2 the time for Tom.

$$\frac{1}{x} + \frac{1}{x + 2} = \frac{1}{12}$$ *Multiply by l.c.d, $12x(x + 2)$*

$12(x + 2) + 12x = x(x + 2)$

$x^2 - 22x - 24 = 0$ *$a = 1$, $b = -22$, $c = -24$*

$$x = \frac{-(-22) \pm \sqrt{(-22)^2 - 4(1)(-24)}}{2(1)}$$

$$= \frac{22 \pm \sqrt{580}}{2}$$

$$\approx \frac{22 \pm 24.083}{2}$$

$$\frac{22 + 24.083}{2} = 23.0$$

$$\frac{22 - 24.083}{2} = -2.083$$

The solution 23.0 applies in this problem. Dick takes 23 hours planting flowers alone while Tom takes 25 hours.

57. $2\sqrt{5}m^2 + 3m - 4\sqrt{5} = 0$

$$m = \frac{-b \pm \sqrt{b^2 - ac}}{2a}$$ *$a = 2\sqrt{5}$, $b = 3$, $c = -4\sqrt{5}$*

$$= \frac{-3 \pm \sqrt{(3)^2 - 4(2\sqrt{5})(-4\sqrt{5})}}{2(2\sqrt{5})}$$

$$= \frac{-3 \pm \sqrt{169}}{4\sqrt{5}} = \frac{-3 \pm 13}{4\sqrt{5}}$$

$$\frac{-3 + 13}{4\sqrt{5}} = \frac{5}{2\sqrt{5}} \cdot \frac{\sqrt{5}}{\sqrt{5}} = \frac{5\sqrt{5}}{10} = \frac{\sqrt{5}}{2}$$

$$\frac{-3 - 13}{4\sqrt{5}} = \frac{-16}{4\sqrt{5}} = \frac{-4}{\sqrt{5}} \cdot \frac{\sqrt{5}}{\sqrt{5}} = -\frac{4\sqrt{5}}{5}$$

Solution set: $\{\frac{\sqrt{5}}{2}, -\frac{4\sqrt{5}}{5}\}$

61. $a = 4$, $b = -2$, $c = -3$

$$b^2 - 4ac = (-2)^2 - 4(4)(-3) = 52$$

$$-\frac{b}{a} = -\frac{(-2)}{4} = \frac{1}{2}$$

$$\frac{c}{a} = -\frac{3}{4}$$

65. If x = 3, then x - 3 = 0, since 3 - 3 = 0.

**Section 6.3 (page 265)**

1. $x^2 + 4x + 1 = 0$ *$a = 1$, $b = 4$, $c = 1$*

$$b^2 - 4ac = 4^2 - 4(1)(1)$$
$$= 16 - 4$$
$$= 12$$

which is not a perfect square, so the solutions of the equation will be (c) two different irrational numbers.

5. $25a^2 + 20a = 2$

$25a^2 + 20a - 2 = 0$ *a = 25, b = 20, c = -2*

$$b^2 - 4ac = (20)^2 - 4(25)(-2)$$
$$= 400 + 200$$
$$= 600$$

which is not a perfect square, so the solutions of the equations will be (c) two different irrational numbers.

9. $49p^2 - 42p + 5 = 0$ *a = 49, b = -42, c = 5*

$$b^2 - 4ac = (-42)^2 - 4(49)(5)$$
$$= 784$$

which is a perfect square, so the solutions of the equation will be (a) two different rational numbers.

13. $p^2 + bp + 25 = 0$ *a = 1, c = 25*

For there to be exactly one rational solution, we need

$$b^2 - 4ac = 0.$$
$$b^2 - 4(1)(25) = 0$$
$$b^2 = 100$$

$b = 10$ or $b = -10$

17. $9y^2 - 30y + c = 0$

For there to be only one rational solution, we need

$$(-30)^2 - 4(9)(c) = 0.$$
$$900 - 36c = 0$$
$$36c = 900$$
$$c = 25$$

21. $16m^2 - 40m + c = 0$

For there to be only one rational solution, we need

$$(-40)^2 - (4)(16)(c) = 0.$$
$$1600 - 64c = 0$$
$$64c = 1600$$
$$c = 25$$

25. $8k^2 + 38k - 33$

This polynomial can be factored if $8k^2 + 38k - 33 = 0$ has rational solutions.
Use the discriminant to check:

$$b^2 - 4ac = (38)^2 - 4(8)(-33) = 2500.$$

Since 2500 is a perfect square, the polynomial can be factored.

$$8k^2 + 38k - 33 = (4k - 3)(2k + 11)$$

29. $30w^2 + 41w - 6$

Use the discriminant to see if this polynomial can be factored.

$$b^2 - 4ac = (41)^2 - 4(30)(-6)$$
$$= 2401$$

Since 2401 is a perfect square, the polynomial can be factored.

$$30w^2 + 41w - 6 = (15w - 2)(2w + 3)$$

33. $20p^2 + 52p - 63$

The discriminant is

$$b^2 - 4ac = (52)^2 - 4(20)(-63)$$
$$= 7744$$

Since 7744 is a perfect square, the polynomial can be factored.

$$20p^2 + 52p - 63 = (10p - 9)(2p + 7)$$

37. $5x^2 = 2x + 1$

$5x^2 - 2x - 1 = 0$ *$a = 5$, $b = -2$, $c = -1$*

$\frac{-b}{a} = \frac{-(-2)}{5} = \frac{2}{5}$ *Sum of the solutions*

$\frac{c}{a} = \frac{-1}{5}$ *Product of the solutions*

41. $16x^2 = 49$

$16x^2 - 49 = 0$ *$a = 16$, $b = 0$, $c = -49$*

$-\frac{b}{a} = -\frac{0}{16}$ *Sum of the solutions*

$\frac{c}{a} = -\frac{49}{16}$ *Product of the solutions*

45. $\frac{3}{5}x^2 - \frac{2}{5}x - 1 = 0$

$3x^2 - 2x - 5 = 0$ *Multiply each term by 5*
*$a = 3$, $b = -2$, $c = -5$*

$-\frac{b}{a} = -\frac{(-2)}{3} = \frac{2}{3}$ *Sum of the solutions*

$\frac{c}{a} = -\frac{5}{3} = -\frac{5}{3}$ *Product of the solutions*

49. $x^2 = 6x + 3;\ x_1 = 3 + 2\sqrt{3},$
$x_2 = 2 + 3\sqrt{3},\ x^2 - 6x - 3 = 0$
*$a = 1$, $b = -6$, $c = -3$*

$\frac{-b}{a} = \frac{-(-6)}{1} = 6$ *Sum of solutions*

$3 + 2\sqrt{3} + 2 + 3\sqrt{3} = 5 + 5\sqrt{3}$ *Add $x_1 + x_2$*

$x_1 + x_2 \neq 6$ so the solutions are incorrect.

$\frac{c}{a} = \frac{-3}{1} = -3$ *Product of the solutions*

$(3 + 2\sqrt{3})(2 + 3\sqrt{3})$ *Multiply $x_1$ and $x_2$*
$= (3)(2) + 3(3\sqrt{3}) + 2\sqrt{3}(2) + 2\sqrt{3}(3\sqrt{3})$
$= 6 + 13\sqrt{3} + 18$
$= 24 + 13\sqrt{3}$

$(x_1)(x_2) \neq -3$ so this also shows the solution is incorrect.

53. Solution set: $\{\frac{1}{2}, \frac{2}{3}\}$

$x = \frac{1}{2}$ or $x = \frac{2}{3}$

$(x - \frac{1}{2})(x - \frac{2}{3}) = 0$ *Zero factor property*

$x^2 - \frac{2}{3}x - \frac{1}{2}x + \frac{2}{6} = 0$ *Use FOIL*

$6x^2 - 4x - 3x + 2 = 0$
*Multiply by common denominator*

$6x^2 - 7x + 2 = 0$

57. Solution set: $\{1 + 3i, 1 - 3i\}$
$x = 1 + 3i$ or $x = 1 - 3i$
$[x - (1 + 3i)][x - (1 - 3i)] = 0$
$(x - 1 - 3i)(x - 1 + 3i) = 0$
*Multiply the two factors*

$x^2 - 2x + 10 = 0$

61. $x_1 + x_2 = -3\sqrt{2},\ x_1x_2 = \frac{-\sqrt{2}}{2}$

$x_1 + x_2 = -3\sqrt{2}$
$x_1 = -3\sqrt{2} - x_2$
*Find $x_1$ using the first equation*

$x_1x_2 = \frac{-\sqrt{2}}{2}$

$(-3\sqrt{2} - x)x = \frac{-\sqrt{2}}{2}$ *Substitute into the second equation for $x_1$*

$-3x\sqrt{2} - x^2 = \frac{-\sqrt{2}}{2}$

$-6x\sqrt{2} - 2x^2 = -\sqrt{2}$ *Multiply each term by common denominator*

$2x^2 + 6x\sqrt{2} - \sqrt{2} = 0$

65. $2py^2 + 4y + 6p = 0$; sum of the roots is 4/3. Use the quadratic formula to find the factors of the equation.

$$y = \frac{-4 \pm \sqrt{(4)^2 - 4(2p)(6p)}}{2(2p)} \quad a = 2p,\ b = 4,\ c = 6p$$

$$= \frac{-4 \pm \sqrt{16 - 48p^2}}{4p}$$

$$= \frac{-4 \pm \sqrt{4(4 - 12p^2)}}{4p}$$

$$= \frac{-4 \pm 2\sqrt{4 - 12p^2}}{4p}$$

$$= \frac{-2 \pm \sqrt{4 - 12p^2}}{2p}$$

$$\left(\frac{-2 + \sqrt{4 - 12p^2}}{2p}\right) + \left(\frac{-2 - \sqrt{4 - 12p^2}}{2p}\right) = \frac{4}{3}$$

*Sum property*

$$(6p)\left(\frac{-2 + \sqrt{4 - 12p^2}}{2p}\right) + (6p)\left(\frac{-2 - \sqrt{4 - 12p^2}}{2p}\right) = \frac{4}{3}(6p)$$

*Multiply by l.c.d.*

$$(-6 + 3\sqrt{4 - 12p^2}) + (-6 - 3\sqrt{4 - 12p^2}) = 8p$$

*Solve for p*

$$-12 = 8p$$

$$\frac{-3}{2} = p$$

Solution set: {-3/2}

69. If $x^2 = b$ in $4x^4 - 5x^2 + 2$, then we have the equation $4b^2 - 5b + 2$.

73.

$$\sqrt{2x + 3} = 9$$

$$(\sqrt{2x + 3})^2 = (9)^2$$

$$2x + 3 = 81$$

$$2x = 78$$

$$x = 39$$

Check answer in original equation.

Solution set: {39}

## Section 6.4 (page 272)

1.

$$1 + \frac{3}{x} - \frac{28}{x^2} = 0$$

$$1(x^2) + \frac{3}{x}(x^2) - \frac{28}{x^2}(x^2) = 0(x^2)$$

*Multiply by l.c.d.*

$$x^2 + 3x - 28 = 0$$

$$(x + 7)(x - 4) = 0$$

$$x + 7 = 0 \quad \text{or} \quad x - 4 = 0$$

$$x = -7 \quad \text{or} \quad x = 4$$

*Check solutions in equation*

Solution set: {-7, 4}

5.

$$3 = \frac{1}{y + 2} + \frac{2}{(y + 2)^2}$$

$$3(y + 2)^2 = \frac{1}{y + 2}(y + 2)^2 + \frac{2}{(y + 2)^2}(y + 2)^2$$

$$3y^2 + 12y + 12 = y + 2 + 2$$

$$3y^2 + 11y + 8 = 0$$

$$(3y + 8)(y + 1) = 0$$

$$3y + 8 = 0 \quad \text{or} \quad y + 1 = 0$$

$$3y = -8$$

*Check solutions in original equation*

$$y = \frac{-8}{3} \quad \text{or} \quad y = -1$$

Solution set: $\{\frac{-8}{3}, -1\}$

9.

$$\frac{2r}{r - 3} + \frac{4}{r} - 6 = 0$$

$$\frac{2r}{r-3}[r(r-3)] + \frac{4}{r}[r(r-3)] - 6[r(r-3)] = 0$$

$$2r^2 + 4r - 12 - 6r^2 + 18r = 0$$

$$-4r^2 + 22r - 12 = 0$$

$$-2(2r^2 - 11r + 6) = 0$$

$$2r^2 - 11r + 6 = 0 \quad \textit{Divide by 2}$$

$$r = \frac{-(-11) \pm \sqrt{(-11)^2 - 4(2)(6)}}{2(2)}$$

*Use quadratic formula: a = 2, b = -11, c = 6*

$$= \frac{11 \pm \sqrt{73}}{4}$$

Solution set: $\{\frac{11 + \sqrt{73}}{4}, \frac{11 - \sqrt{73}}{4}\}$

13. $x\sqrt{2} = \sqrt{5x - 2}$

$(x\sqrt{2})^2 = (\sqrt{5x - 2})^2$ *Square both sides*

$2x^2 = 5x - 2$

$2x^2 - 5x + 2 = 0$ *Put into $ax^2 + bx + c = 0$ form*

$x = \dfrac{-(-5) \pm \sqrt{(-5)^2 - 4(2)(2)}}{2(2)}$ *$a = 2$, $b = -5$, $c = 2$*

$= \dfrac{5 \pm \sqrt{25 - 16}}{4}$

$= \dfrac{5 \pm \sqrt{9}}{4}$

$= \dfrac{5 \pm 3}{4}$

$x = \dfrac{5 + 3}{4} = 2$ or $x = \dfrac{5 - 3}{4} = \dfrac{1}{2}$ *Check each solution back in the original equation*

Solution set: {2, 1/2}

17. $\sqrt{2x + 3} = 2 + \sqrt{x - 2}$

$2x + 3 = 4 + 4\sqrt{x - 2} + x - 2$ *Square both sides*

$2x + 3 = 4\sqrt{x - 2} + x + 2$

$x + 1 = 4\sqrt{x - 2}$

$x^2 + 2x + 1 = 16(x - 2)$ *Square both sides*

$x^2 + 2x + 1 = 16x - 32$

$x^2 - 14x + 33 = 0$

$x = \dfrac{-b \pm \sqrt{b^2 - 4ac}}{2a}$ *$a = 1$, $b = -14$, $c = 33$*

$= \dfrac{14 \pm \sqrt{(14)^2 - 4(1)(33)}}{2(1)}$

$= \dfrac{14 \pm \sqrt{64}}{2}$

$= \dfrac{14 \pm 8}{2}$

$x = 11$ or $x = 3$ *Check each solution back in the original equation*

Solution set: {3, 11}

21. $\sqrt{2r + 12} - \sqrt{r + 4} = 2$

$\sqrt{2r + 12} = 2 + \sqrt{r + 4}$

$2r + 12 = 4 + 4\sqrt{r + 4} + r + 4$ *Square each side*

$2r + 12 = r + 8 + 4\sqrt{r + 4}$

$r + 4 = 4\sqrt{r + 4}$

$r^2 + 8r + 16 = 16(r + 4)$ *Square each side*

$r^2 + 8r + 16 = 16r + 64$

$r^2 - 8r - 48 = 0$ *$a = 1$, $b = -8$, $c = -48$*

$r = \dfrac{8 \pm \sqrt{(-8)^2 - 4(1)(-48)}}{2(1)}$

$r = \dfrac{8 \pm \sqrt{256}}{2}$

$r = \dfrac{8 \pm 16}{2}$

$r = 12$ or $r = -4$

Solution set: {12, -4}

25. $\sqrt{-z} - \sqrt{5 + z} = 1$

$\sqrt{-z} = 1 + \sqrt{5 + z}$

$-z = 1 + 2\sqrt{5 + z} + 5 + z$ *Square each side*

$-2z - 6 = 2\sqrt{5 + z}$

$-z - 3 = \sqrt{5 + z}$ *Divide by 2*

$z^2 + 6z + 9 = 5 + z$ *Square each side again*

$z^2 + 5z + 4 = 0$ *$a = 1$, $b = 5$, $c = 4$*

$z = \dfrac{-5 \pm \sqrt{5^2 - 4(1)(4)}}{2(1)}$

$z = \dfrac{-5 \pm \sqrt{25 - 16}}{2}$

$z = \dfrac{-5 \pm 3}{2}$

$z = -1$ or $z = -4$

Check: $\sqrt{-(-1)} - \sqrt{5 + (-1)} = 1$

$1 - 2 = 1$

$-1 = 1$ *False*

Solution set: $\{-4\}$

29. $3(2x + 5)^2 + 2(2x + 5) = 2$ *Let $a = 2x + 5$*

$3a^2 + 2a - 2 = 2$ *$a = 3$, $b = 2$, $c = -2$*

$$a = \frac{-(2) \pm \sqrt{(2)^2 - 4(3)(-2)}}{2(3)}$$

$$= \frac{-2 \pm \sqrt{28}}{6}$$

$$= \frac{-2 \pm 2\sqrt{7}}{6}$$

$$a = \frac{-1 + \sqrt{7}}{3}$$

$2x + 5 = \frac{-1 + \sqrt{7}}{3}$ *Substitute $2x + 5$ for $a$*

$6x + 15 = -1 + \sqrt{7}$

$6x = -16 + \sqrt{7}$

$x = \frac{-16 + \sqrt{7}}{6}$

or

$$a = \frac{-1 - \sqrt{7}}{3}$$

$2x + 5 = \frac{-1 - \sqrt{7}}{3}$ *Substitute $2x + 5$ for $a$*

$6x + 15 = -1 - \sqrt{7}$

$6x = -16 - \sqrt{7}$

$x = \frac{-16 - \sqrt{7}}{6}$

Solution set: $\{\frac{-16 + \sqrt{7}}{6}, \frac{-16 - \sqrt{7}}{6}\}$

33. Let $x = z^2$, then $x^2 = z^4$. So,

$z^4 - 37z^2 + 36 = 0$ becomes

$x^2 - 37x + 36 = 0$

$(x - 36)(x - 1) = 0$ *Factor*

Set each factor equal to 0.

$x - 36 = 0$ or $x - 1 = 0$

$x = 36$ or $x = 1$

Since $z^2 = 36$, $z = \pm 6$, also, $z^2 = 1$ gives $z = \pm 1$.

Solution set: $\{-1, 1, -6, 6\}$.

37. $2x^4 + x^2 - 3 = 0$

$2m^2 + m - 3 = 0$ *Let $x^2 = m$*

$(2m + 3)(m - 1) = 0$ *Factor*

$2m + 3 = 0$ or $m - 1 = 0$

$2m = -3$

$m = \frac{-3}{2}$ or $m = 1$

$x^2 = \frac{-3}{2}$ or $x^2 = 1$ *Substitute $x^2$ for $m$*

$x = \pm\sqrt{\frac{-3}{2}}$ or $x = \pm 1$

$x = \pm i\sqrt{\frac{3}{2}} = \pm i\frac{\sqrt{3} \cdot \sqrt{2}}{\sqrt{2} \cdot \sqrt{2}} = \pm\frac{i\sqrt{6}}{2}$

Solution set: $\{-1, 1, \frac{\sqrt{6}}{2}i, \frac{\sqrt{6}}{2}i\}$

41. Let k represent the speed of the plane in still air. Then, the speed of the plane with the wind is k + 15 miles per hour. In the same manner, the speed of the plane against the wind is k - 15 miles per hour.

Now, since $rt = d$, then $t = \frac{d}{r}$.

The distance of the plane with the wind is 810 miles per hour and the speed is k + 15, so $t = \frac{810}{k + 15}$.

In the same manner, for the plane against the wind, we have,

$$t = \frac{720}{k - 15}.$$

Since the total of the times with and against the wind is 6 hours, then

$$\frac{810}{k + 15} + \frac{720}{k - 15} = 6$$

$\frac{810}{k + 15}(k + 15)(k - 15) +$

$\frac{720}{k - 15}(k + 15)(k - 15)$ *Multiply by the l.c.d*

$= 6(k + 15)(k - 15)$ *Simplify*

$810(k - 15) + 720(k + 15)$

$= 6(k^2 - 225)$

$810k - 12,150 + 720k + 10,800$

$= 6k^2 - 1350$

$1530k - 1350$

$= 6k^2 - 1350$

$0 = 6k^2 - 1530k$ *Get 0 alone on one side*

$0 = 6k(k - 255)$ *Factor*

$k = 0$ or $k = 255$ *Solve*

Since $k = 0$ is nonsense for this problem, the plane's speed is 255 miles per hour.

45. Let x represent the number of hours required for one pipe to fill the tank alone. Then, the other pipe alone takes x + 3 hours to fill the tank.
The one pipe can fill 1/x of the tank per hour working alone, while the other fills 1/(x + 3) working alone. Since together they take 2 hours to fill the tank, then working together they can fill 1/2 of the tank per hour.
Adding together their individual rates per hour and setting this equal to their rate working together, we have,

$$\frac{1}{x} + \frac{1}{x + 3} = \frac{1}{2}.$$

Solve this equation.

$2x(x + 3)(\frac{1}{x}) + 2x(x + 3)(\frac{1}{x + 3})$

$= 2x(x + 3)(\frac{1}{2})$

*Multiply by the l.c.d.*

$2(x + 3) + 2x = x(x + 3)$ *Simplify*

$2x + 6 + 2x = x^2 + 3x$

$0 = x^2 - x - 6$ *Get 0 alone on one side*

$0 = (x - 3)(x + 2)$ *Factor*

$x = 3$ or $x = -2$ *Solve*

Since x represents hours, then $x = -2$ would not be considered a possible answer. So one pipe takes $x = 3$ hours, while the other pipe takes $x + 3 = 3 + 3 = 6$ hours.

49. $2m^6 + 11m^3 + 5 = 0$

$2x^2 + 11x + 5 = 0$ *Using substitution* $m^3 = x$

$(2x + 1)(x + 5) = 0$

$2x + 1 = 0$

$x = -\frac{1}{2}$

$m^3 = -\frac{1}{2}$

$m = \sqrt[3]{-\frac{1}{2}}$

$= -\frac{\sqrt[3]{1} \cdot \sqrt[3]{4}}{\sqrt[3]{3} \cdot \sqrt[3]{4}}$

$= -\frac{\sqrt[3]{4}}{2}$

or

$x + 5 = 0$

$x = -5$

$m^3 = -5$ *Use the 2 solutions for $m^3$*

$m = \sqrt[3]{-5}$

$= -\sqrt[3]{5}$

Solution set: $\{-\sqrt[3]{4}/2, -\sqrt[3]{5}\}$

53. $\left(\frac{p^2 + 5p}{6}\right)^2 - 7\left(\frac{p^2 + 5p}{6}\right) + 6 = 0$ *Substitute $x$ for $\frac{p^2 + 5p}{6}$*

$x^2 - 7x + 6 = 0$

$(x - 6)(x - 1) = 0$

$x - 6 = 0$ or $x - 1 = 0$

$x = 6$ or $x = 1$

To find p, use the two solutions for x

$\frac{p^2 + 5p}{6} = 6$

$p^2 + 5p - 36 = 0$

$(p + 9)(p - 4) = 0$

$p = -9$ or $p = 4$

$\frac{p^2 + 5p}{6} = 1$

$p^2 + 5p - 6 = 0$

$(p + 6)(p - 1) = 0$

$p = -6$ or $p = 1$

Solution set: $\{-9, 4, -6, 1\}$

57. $(x^2 + x)^2 = 8(x^2 + x) - 12$

$m^2 - 8m + 12 = 0$ *Substitute $m$ for $x^2 + x$*

$(m - 6)(m - 2) = 0$

$m = 6$ or $m = 2$

To find x, use the 2 solutions for m

$x^2 + x = 6$

$x^2 + x - 6 = 0$

$(x + 3)(x - 2) = 0$

$x = -3$ or $x = 2$

$x^2 + x = 2$

$x^2 + x - 2 = 0$

$(x + 2)(x - 1) = 0$

$x = -2$ or $x = 1$

Solution set: $\{-3, -2, 1, 2,\}$

61. $F = \frac{9}{5}C + 32$; for C

$F - 32 = \frac{9}{5}C$

$\frac{F - 32}{9/5} = C$

$(F - 32)\frac{5}{9} = C$ *Multiply by reciprocal of 9/5*

65. Let x be the first integer and x + 1 the second integer.

$\underbrace{\text{Sum of twice the first}}$ and $\underbrace{\text{three times the second}}$

$2x + 3(x + 1) = 113$

$2x + 3x + 3 = 113$

$5x = 110$

$x = 22$

The first integer is 22; the second is 23.

**Section 6.5 (page 278)**

1. $R = \frac{k}{d^2}$; for d

$d^2 \cdot R = k$ *Multiply both sides by $d^2$*

$d^2 = \frac{k}{R}$ *Divide both sides by R*

$d = \pm\sqrt{\frac{k}{R}}$ *Take the square root*

$= \pm\frac{\sqrt{k}}{\sqrt{R}}$

$= \pm\frac{\sqrt{k}\cdot\sqrt{R}}{\sqrt{R}\cdot\sqrt{R}}$ *Rationalize the denominator*

$= \pm\frac{\sqrt{kR}}{R}$

5. $V = \frac{1}{3}\pi r^2 h$; for r

$3V = \pi r^2 h$ *Multiply by 3*

$\frac{3V}{\pi h} = r^2$ *Divide each side by* $\pi h$

$\pm\sqrt{\frac{3V}{\pi h}} = r$ *Take the square root of both sides*

$r = \frac{\pm\sqrt{3V}\cdot\sqrt{\pi h}}{\sqrt{\pi h}\cdot\sqrt{\pi h}}$ *Rationalize the denominator*

$r = \frac{\pm\sqrt{3V\pi h}}{h}$

9. $M = \sqrt{\frac{2L}{p}}$, for p

$(M)^2 = (\sqrt{\frac{2L}{p}})^2$ *Square both sides*

$p\cdot M^2 = \frac{2L}{p}\cdot p$ *Multiply by p*

$M^2 p = 2L$

$p = \frac{2L}{M^2}$ *Divide by* $M^2$

13. Let a = the shorter leg = 7;
let b = the longer leg;
let c = the hypotneuse = b + 1.
Use the Pythagorean formula:

$$c^2 = a^2 + b^2$$
$$(b + 1)^2 = (7)^2 + b^2$$
$$b^2 + 2b + 1 = 49 + b^2$$
$$2b = 48$$
$$b = 24$$

The longer leg is 24 meters long.

17. Let h be the height of the tower. Then the distance to the top of the tower from a point 30 meters from the base is 2h + 2. This configuration forms a right triangle and so we have

$$h^2 + (30)^2 = (2h + 2)^2$$
$$h^2 + 900 = 4h^2 + 8h + 4$$
$$3h^2 + 8h - 896 = 0$$
$$(3h + 56)(h - 16) = 0$$
$$h = \frac{-56}{3} \quad \text{or} \quad h = 16$$

Discard the negative solution since height must be positive. The height is 16 meters.

21. Let x be the width of the sheet metal. Then the length is 2x - 4.

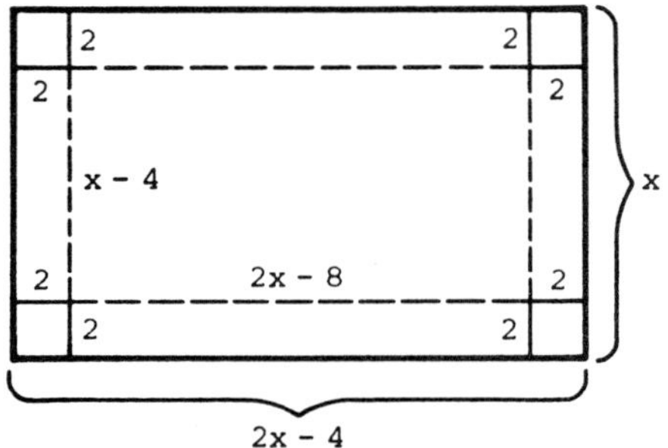


By cutting out 2 inch squares from each corner we get a rectangle with width x - 4 and length (2x - 4) - 4 = 2x - 8. The uncovered box then has height 2 inches, length 2x - 8 inches, and width x - 4 inches. So we have,

$$256 = V = 2(x - 4)(2x - 8)$$
$$256 = 4(x - 4)(x - 4)$$
$$64 = (x - 4)^2$$
$$x - 4 = 8 \quad \text{or} \quad x - 4 = -8$$
$$x = 12 \quad \text{or} \quad x = -4$$

Discard $x = -4$ as a solution for the problem since the width cannot be negative.
So, the width is 12 inches, the length is $2(12) - 4 = 20$ inches.

25. Use the Pythagorean formula,

$$x^2 + (x + 1)^2 = (x + 4)^2$$

$$x^2 + x^2 + 2x + 1 = x^2 + 8x + 16$$ *Simplify*

$$x^2 - 6x - 15 = 0$$

$$x = \frac{6 \pm \sqrt{36 + 60}}{2}$$ *Solve by the quadratic formula*

$$= \frac{6 \pm \sqrt{96}}{2}$$

$$\approx \frac{6 \pm 9.798}{2}$$

Reject the solution that gives a negative answer since length is positive.

$$x = \frac{6 + 9.798}{2}$$

$$= 7.899$$

The sides are about

7.899, $7.899 + 1 = 8.899$, and $7.899 + 4 = 11.899$.

29. $10p^2c^2 + 7pcr - 12r^2 = 0$; for r *Write in quadratic form*

$$12r^2 - 7pcr - 10p^2c^2 = 0$$ *Use the quadratic formula*

$$r = \frac{-(-7pc) \pm \sqrt{(-7pc)^2 - 4(12)(-10p^2c^2)}}{2(12)}$$ *$a = 12$, $b = -7pc$, $c = -10p^2c^2$*

$$= \frac{7pc \pm \sqrt{529p^2c^2}}{24}$$

$$= \frac{7pc \pm 23pc}{24}$$

$$r = \frac{7pc + 23pc}{24} = \frac{30pc}{24} = \frac{5}{4}pc$$

$$r = \frac{7pc - 23pc}{24} = \frac{-16}{24}pc = \frac{-2}{3}pc$$

Solution set: $\{r = \frac{5}{4}pc,\ r = \frac{-2}{3}pc\}$

33. (a) Use the formula $s = 20t^2 - 300t$ with $t = 10$.

$$s = 20(10)^2 - 300(10)$$
$$= -1000$$

The particle moved away from the starting point in a negative direction 1000 centimeters.

(b) At the starting point, $s = 0$.

$$0 = 20t^2 - 300t$$
$$0 = 10t(2t - 30)$$

$10t = 0$ or $2t - 30 = 0$

$2t = 30$

$t = 0$ or $t = 15$

At $t = 0$, the particle begins to move. At $t = 15$, or 15 seconds after the start, the particle returns to the starting point.

37. $$D = 13t^2 - 100t$$

$$200 = 13t^2 - 100t$$

$$13t^2 - 100t - 200 = 0$$ *Write in quadratic form*

$a = 13$, $b = -100$, $c = -200$ *Solve with quadratic formula*

$$t = \frac{-(-100) \pm \sqrt{(-100)^2 - 4(13)(-200)}}{2(13)}$$

$$= \frac{100 \pm \sqrt{20{,}400}}{26}$$

$$t = \frac{50 \pm 10\sqrt{51}}{13} = 9.3$$

$$t = \frac{50 - 10\sqrt{51}}{13} = -1.6$$ *Discard since time is positive*

The time would be 9.3 seconds.

41. $\frac{1}{a - 1} - \frac{2}{a + 1}$

$= \frac{a + 1}{a + 1} \cdot \frac{1}{a - 1} - \frac{2}{a + 1} \cdot \frac{a - 1}{a - 1}$

$= \frac{a + 1}{a^2 - 1} - \frac{2a - 2}{a^2 - 1}$

$= \frac{a + 1 - (2a - 2)}{a^2 - 1}$

$= \frac{a + 1 - 2a + 2}{a^2 - 1}$

$= \frac{-a + 3}{a^2 - 1}$

$= \frac{3 - a}{(a - 1)(a + 1)}$

**Section 6.6 (page 285)**

In Exercises 1-41, see the answer graphs in the textbook.

1. $(2x + 1)(x - 5) > 0$

The numbers -1/2 and 5 are solutions of the quadratic equation

$(2x + 1)(x - 5) = 0.$

The numbers -1/2 and 5 divide the number line into three regions, as shown in the figure.

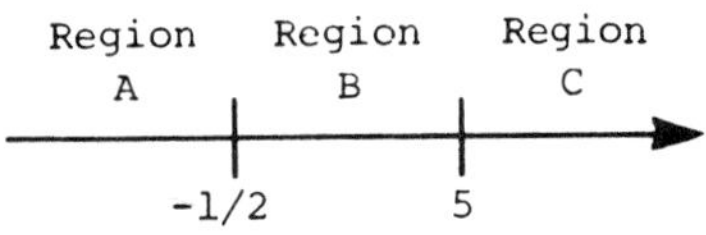


Check a point from each region as follows:

Region A: If $x = -1$, then

$(2x + 1)(x - 5) > 0.$ *True*

Region B: If $x = 0$, then

$(2x + 1)(x - 5) < 0.$ *False*

Region C: If $x = 6$, then

$(2x + 1)(x - 5) > 0.$ *True*

Thus, the solution set includes points in both Regions A and C.

Solution set: $(-\infty, -\frac{1}{2}) \cup (5, +\infty)$

5. $3r^2 + 10r \leq 8$

$3r^2 + 10r - 8 \leq 0$ *Subtract 8*

Solve the quadratic equation.

$3r^2 + 10r - 8 = 0$

$(r + 4)(3r - 2) = 0$

$r = -4$ or $r = 2/3$

The numbers -4 and 2/3 divide the number line into three regions.

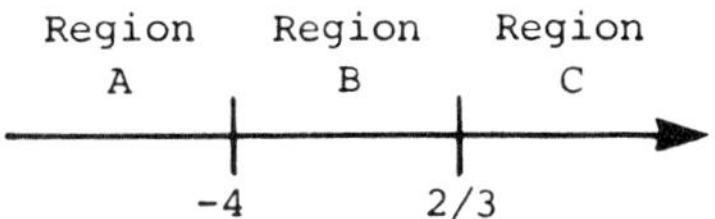


Check a point from each region as follows:

Region A: If $r = -5$, then

$(r + 4)(3r - 2) > 0.$ *False*

Region B: If $r = 0$, then

$(r + 4)(3r - 2) < 0.$ *True*

Region C: If $r = 1$, then

$(r + 4)(3r - 2) > 0.$ *False*

If $r = -4$ or $r = \frac{2}{3}$, then

$(r + 4)(3r - 2) = 0.$ *True*

Thus, only the points in Region B (together with the endpoints -4 and $\frac{2}{3}$) satisfy the inequality $(r + 4)(3r - 2) \leq 0.$

Solution set: $[-4, 2/3]$

9. $4y^2 + 7y + 3 < 0$

First, write the quadratic equation as

$$4y^2 + 7y + 3 = 0,$$
$$(y + 1)(4y + 3) = 0 \quad \textit{Solved by factoring}$$
$$y = -1 \quad \text{or} \quad y = -\frac{3}{4}$$

The numbers -1 and $-\frac{3}{4}$ divide the number line into three regions.

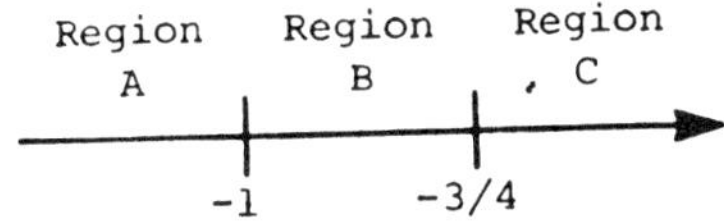


Check a point from each region as follows:
Region A: If y = -2, then

$$(y + 1)(4y + 3) > 0.$$

Region B: If $y = -\frac{7}{8}$, then

$$(y + 1)(4y + 3) < 0.$$

Region C: If y = 0, then

$$(y + 1)(4y + 3) > 0.$$

Based on these results only the points in Region B satisfy the inequality $(y + 1)(4y + 3) < 0$.
Solution set: (-1, -3/4)

13. $2x^2 + 4x + 1 > 0$

First, write the quadratic equation as

$$2x^2 + 4x + 1 = 0.$$

The numbers $\frac{-2 + \sqrt{2}}{2}$ and $\frac{-2 - \sqrt{2}}{2}$ are solutions to this quadratic equation, and divide the number line into three regions.

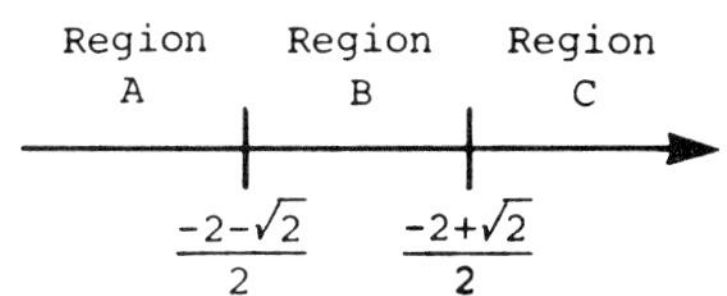


Check a point from each region as follows:
Region A: If x = -2, then

$$2x^2 + 4x + 1 > 0.$$

Region B: If x = -1, then

$$2x^2 + 4x + 1 < 0.$$

Region C: If x = 0, then

$$2x^2 + 4x + 1 > 0.$$

Based on these results, the solution set includes points in both Regions A and C.
Solution set:

$$(-\infty, \frac{-2 - \sqrt{2}}{2}) \cup (\frac{-2 + \sqrt{2}}{2}, +\infty)$$

17. $3k^2 - 5k \leq 0$

Solve the quadratic equation

$$3k^2 - 5k = 0.$$
$$k(3k - 5) = 0 \quad \textit{Factor out } k$$
$$k = 0 \quad \text{or} \quad 3k - 5 = 0$$
$$k = \frac{5}{3}$$

Divide the number line into three regions.

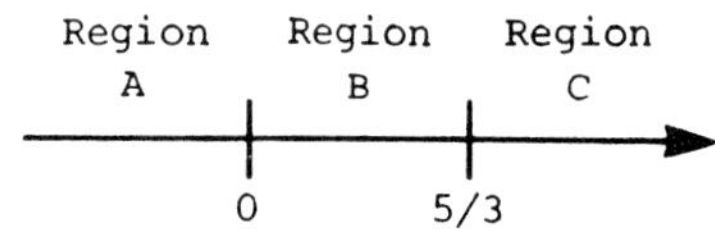


Check a point from each region:
Region A: If k = -1, then

$$3k^2 - 5k > 0.$$

Region B: If k = 1, then

$$3k^2 - 5k < 0.$$

Region C: If k = 2, then

$$3k^2 - 5k > 0.$$

If k = 0 or k = 5/3, then
$k(3k - 5) = 0$.
Thus, the solution set includes points in Region B (together with the endpoints 0 and 5/3) satisfy the inequality

$$3k^2 - 5k \le 0.$$

Solution set: [0, 5/3]

21. $(a - 4)(2a + 3)(3a - 1) \le 0$

Solve the equation

$$(a - 4)(2a + 3)(3a - 1) = 0.$$

$$a - 4 = 0 \quad \text{or} \quad 2a + 3 = 0 \quad \text{or} \quad 3a - 1 = 0$$

$$a = 4 \qquad a = \frac{-3}{2} \qquad a = \frac{1}{3}$$

Divide the number line into regions.

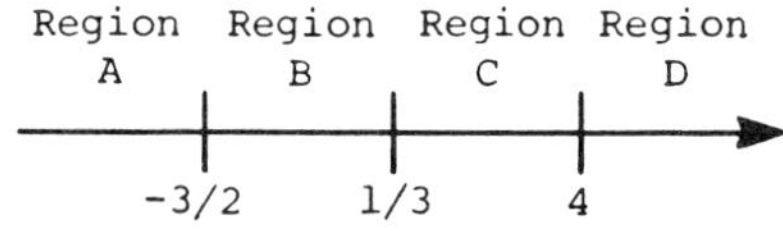


Test a point from each region:
Region A: If a = -2, then

$$(a - 4)(2a + 3)(3a - 1) < 0.$$

Region B: If a = 0, then

$$(a - 4)(2a + 3)(3a - 1) > 0.$$

Region C: If a = 1, the

inequality < 0.

Region D: If a = 5, the

inequality > 0.

If a = 4 or a = -3/2 or a = 1/3, then the inequality = 0.
Regions A and C (together with endpoints 4, -3/2, and 1/3) satisfy the inequality.

Solution set: $(-\infty, \frac{-3}{2}] \cup [\frac{1}{3}, 4]$

25. $$\frac{8}{x - 2} \ge 2$$

$$\frac{8}{x - 2} = 2 \quad \textit{Solve the equation}$$

$$\frac{8}{x - 2}(x - 2) = 2(x - 2)$$

$$8 = 2x - 4$$

$$6 = x$$

$$x - 2 = 0 \quad \textit{Set the denominator equal to 0}$$

$$x = 2$$

Divide the number line into regions.

Region A | Region B | Region C

2 | 6

Region A: If x = 1, then

$$\frac{8}{x - 2} - 2 < 0.$$

Region B: If x = 3, then

$$\frac{8}{x - 2} - 2 > 0.$$

Region C: If x = 7, then

$$\frac{8}{x - 2} - 2 < 0.$$

Only Region B satisfies the inequality. The number 6 can be used in the solution but not 2, since it makes the denominator equal 0.
Solution set: (2, 6]

29. $$\frac{-2}{k + 1} > 5$$

$$\frac{-2}{k + 1} = 5 \quad \textit{Solve}$$

$$\frac{-2}{k + 1}(k + 1) = 5(k + 1)$$

$$-2 = 5k + 5$$

$$\frac{-7}{5} = k$$

$$k + 1 = 0 \quad \textit{Set denominator equal to 0}$$

$$k = -1$$

Divide into regions:

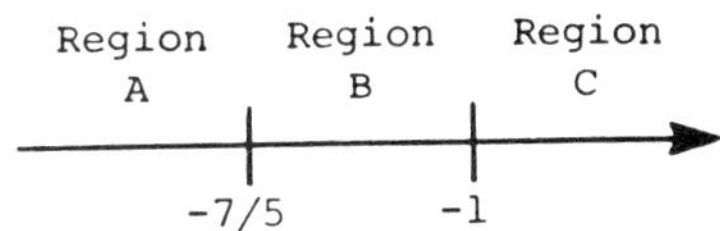


Region A: If k = -2, then

$$\frac{-2}{k + 1} - 5 < 0.$$

Region B: If $k = \frac{-6}{5}$, then

$$\frac{-2}{k + 1} - 5 > 0.$$

Region C: If k = 0, then

$$\frac{-2}{k + 1} - 5 < 0.$$

Region B satisfies the inequality; however -1 cannot be used since it would make the fraction undefined.

Solution set: $(\frac{-7}{5}, -1)$

33. $$\frac{a}{a + 2} \geq 2$$

$$\frac{a}{a + 2}(a + 2) = 2(a + 2) \quad \textit{Solve}$$

$$a = 2a + 4$$

$$-a = 4$$

$$a = -4$$

$$a + 2 = 0 \quad \textit{Set denominator equal to zero}$$

$$a = -2$$

Divide the number line into regions:

Region A | Region B | Region C
-4 | -2

Region A: If a = -5, then

$$\frac{a}{a + 2} - 2 < 0.$$

Region B: If a = -3, then

$$\frac{a}{a - 2} - 2 > 0.$$

Region C: If a = 1, then

$$\frac{a}{a - 2} - 2 < 0.$$

Region B satisfies the inequality; however -2 cannot be used since it would make the fraction undefined.

Solution set: [-4, -2)

37. $$\frac{4k}{2k - 1} < k$$

$$\frac{4k}{2k - 1}(2k - 1) - k(2k - 1) = 0 \quad \textit{Solve}$$

$$4k - 2k^2 + k = 0$$

$$5k - 2k^2 = 0$$

$$k(5 - 2k) = 0$$

$$k = 0 \quad \text{or} \quad 5 - 2k = 0$$

$$k = \frac{5}{2}$$

$$2k - 1 = 0 \quad \textit{Set the denominator equal to 0}$$

$$k = \frac{1}{2}$$

Divide the number line into regions:

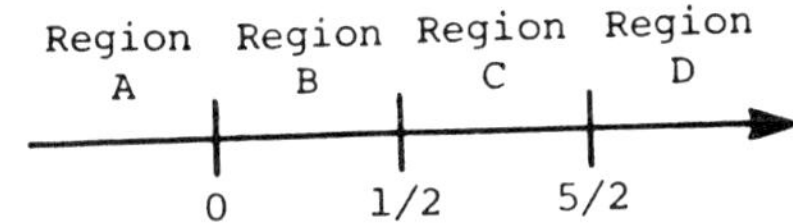


Region A: If k = -1,

$$\frac{4k}{2k - 1} > 0.$$

Region B: If $k = \frac{1}{4}$,

$$\frac{4k}{2k - 1} < 0.$$

Region C: If k = 2,

$$\frac{4k}{2k - 1} > 0.$$

Region D: If k = 3,

$$\frac{4k}{2k - 1} < 0.$$

Regions B and D satisfy the inequality.

Solution set: $(0, \frac{1}{2}) \cup (\frac{5}{2}, +\infty)$

41. $\frac{2x - 3}{x^2 + 1} \geq 0$

$\frac{2x - 3}{x^2 + 1} = 0$ *Solve*

$2x - 3 = 0$ *Multiply both sides by $x^2 + 1$*

$x = \frac{3}{2}$

$x^2 + 1 = 0$ *Set the denominator equal to zero*

$x^2 = -1$

$x = \sqrt{-1}$

Since $\sqrt{-1} = i$, there are no real numbers which would make the denominator equal zero.
Divide the number line into two regions.

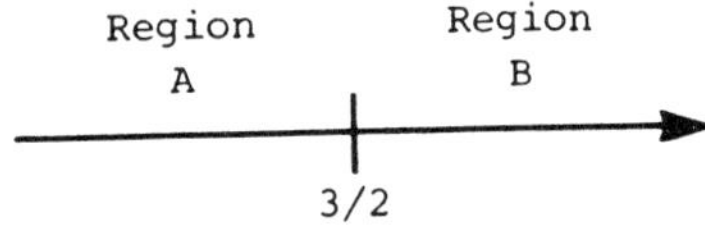


Region A: If x = 1,

$\frac{2x - 3}{x^2 + 1} < 0.$

Region B: If x = 2,

$\frac{2x - 3}{x^2 + 1} > 0.$

Region B and 3/2 satisfy the equation.

Solution set: $[\frac{3}{2}, +\infty)$

45. To show when the company makes a profit we use the inequality

$2x^2 - 25x + 75 > 0.$

Solve the quadratic equation using the quadratic formula.

$2x^2 - 25x + 75 = 0$ *$a = 2$, $b = -25$, $c = 75$*

$x = \frac{-(-25) \pm \sqrt{(-25)^2 - 4(2)(75)}}{2(2)}$

$= \frac{25 \pm 5}{4}$

$x = \frac{30}{4} = \frac{15}{2}$ or $x = \frac{20}{4} = 5$

The numbers 15/2 and 5 divide the number line into three regions.

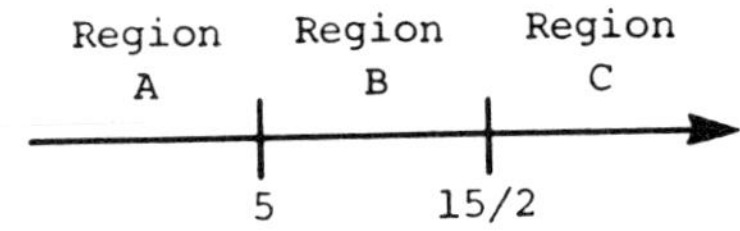


Region A: If x = 4, the profit > 0.

Region B: If x = 6, the profit < 0.

Region C: If x = 8, the profit > 0.

Regions A and C satisfy the inequality. Discard negative values of x since the number of units sold is positive.
Solution set: $[0, 5) \cup (15/2, +\infty)$
The company would make a profit if less than 5 units were sold or if greater than 15/2 or at least 8 units were sold.

49. $3x - 2y = 12$

$3(-2) - 2y = 12$

$-2y = 18$

$y = -9$

**Chapter 6 Review Exercises (page 287)**

1. $t^2 = 16$

$t = \pm\sqrt{16}$

$t = -4$ or $4$

Solution set: $\{-4, 4\}$

5. $(5r + 3)^2 = 72$

$5r + 3 = \sqrt{72}$

$5r + 3 = 6\sqrt{2}$

$5r = -3 + 6\sqrt{2}$

$r = \frac{-3 + 6\sqrt{2}}{5}$

or

$5r + 3 = -\sqrt{72}$

$5r + 3 = -6\sqrt{2}$

$5r = -3 - 6\sqrt{2}$

$r = \frac{-3 - 6\sqrt{2}}{5}$

Solution set: $\{\frac{-3 + 6\sqrt{2}}{5}, \frac{-3 - 6\sqrt{2}}{5}\}$

9. $m^2 + 4m = 21$

$m^2 + 4m + 4 = 21 + 4$

$(m + 2)^2 = 25$

$m + 2 = 5$ or $m + 2 = -5$

$m = 3$ or $m = -7$

Solution set: $\{3, -7\}$

13. $2s^2 + 3s = 10$

$2s^2 + 3s - 10 = 0$ *a = 2, b = 3, c = -10*

$s = \frac{-b \pm \sqrt{b^2 - 4ac}}{2a}$

$= \frac{-3 \pm \sqrt{3^2 - 4(2)(-10)}}{2(2)}$

$= \frac{-3 \pm \sqrt{89}}{4}$

Solution set: $\{\frac{-3 + \sqrt{89}}{4}, \frac{-3 - \sqrt{89}}{4}\}$

17. $ix^2 - 4x + i = 0$ *a = i, b = -4, c = i*

$x = \frac{-(-4) \pm \sqrt{(-4)^2 - 4(i)(i)}}{2(i)}$ *Use quadratic formula*

$= \frac{4 \pm \sqrt{16 - 4(-1)}}{2i}$ $\quad i^2 = -1$

$= \frac{4 \pm 2\sqrt{5}}{2i} = \frac{2 \pm \sqrt{5}}{i}$

Rationalize the denominator.

$\frac{2 \pm \sqrt{5}}{i} \cdot \frac{(i)}{(i)} = \frac{2i \pm i\sqrt{5}}{i^2} = (-1)(2i \pm i\sqrt{5})$

$= -2i + i\sqrt{5}$

or $\quad -2i - i\sqrt{5}$

Solution set: $\{-2i + i\sqrt{5}, -2i - i\sqrt{5}\}$

21. Let t be the time required for the faster pipe. Then t + 1 is the time required for the slower pipe. The faster pipe has a rate of 1/t job per hour and the slower pipe has a rate of $\frac{1}{t + 1}$ job per hour. Working together, they take 3 hours and so this rate is 1/3 job per hour. So, we have

$$\frac{1}{t} + \frac{1}{t + 1} = \frac{1}{3}.$$

$$3t(t + 1)(\frac{1}{t} + \frac{1}{t + 1}) = 3t(t + 1)\cdot\frac{1}{3}$$

$$3(t + 1) + 3t = t(t + 1)$$

$$3t + 3 + 3t = t^2 + t$$

$$6t + 3 = t^2 + t$$

$$t^2 - 5t - 3 = 0$$

*a = 1, b = -5, c = -3*

$t = \frac{-b \pm \sqrt{b^2 - 4ac}}{2a}$

$= \frac{5 \pm \sqrt{(-5)^2 - 4(1)(-3)}}{2(1)}$

$= \frac{5 \pm \sqrt{37}}{2}$

Now $t = \frac{5 - \sqrt{37}}{2}$ is impossible, since time cannot be negative. So $t = \frac{5 + \sqrt{37}}{2} \approx 5.5$ hours. One inlet pipe takes 5.5 hours to fill the tank. The other takes 6.5 hours.

25. $4x^2 = 6x - 9$

$4x^2 - 6x + 9 = 0$

$b^2 - 4ac = (-6)^2 - 4(4)(9)$

$= -108$

Since $\sqrt{-108}$ is not a real number, the solutions of the equation will be (d) two different imaginary numbers.

29. $8p^2 - 35 = 0$

$b^2 - 4ac = 0^2 - 4(8)(-35)$

$= 1120$

which is not a perfect square, so the solutions of the equation will be (c) two different irrational numbers.

33. $x^2 - 4x + 7 = 0$ *a = 1, b = -4, c = 7*

$\frac{-b}{a} = \frac{-(-4)}{1} = 4$ *Sum of the solutions*

$\frac{c}{a} = \frac{7}{1} = 7$ *Product of the solutions*

37. $\{\frac{-2}{5}, \frac{3}{4}\}$

From the solutions given, we write the following equations.

$x = \frac{-2}{5}$ or $x = \frac{3}{4}$

$x + \frac{2}{5} = 0$ or $x - \frac{3}{4} = 0$

$(x + \frac{2}{5})(x - \frac{3}{4}) = 0$ *Zero-factor property*

$x^2 - \frac{3}{4}x + \frac{2}{5}x - \frac{6}{20} = 0$ *Multiply factors (FOIL)*

$20x^2 - 15x + 8x - 6 = 0$ *Multiply by l.c.d.*

$20x^2 - 7x - 6 = 0$

41. $2 - \frac{1}{3 - k} = \frac{15}{(3 - k)^2}$

$2(3 - k)^2 - \frac{1}{3 - k}(3 - k)^2 = \frac{15}{(3 - k)^2}(3 - k)^2.$ *Multiply by l.c.d.*

$2(9 - 6k + k^2) - (3 - k) = 15$

$18 - 12k + 2k^2 - 3 + k = 15$

$2k^2 - 11k = 0$

$k(2k - 11) = 0$

$k = 0$ or $2k - 11 = 0$ *Check both solutions in original equation*

$k = \frac{11}{2}$

Solution set: {0, 11/2}

45. $3p^4 - p^2 - 2 = 0$ *Substitute $x$ for $p^2$*

$3x^2 - x - 2 = 0$

$(3x + 2)(x - 1) = 0$

$3x + 2 = 0$ or $x - 1 = 0$

$x = \frac{-2}{3}$ or $x = 1$

To find p, substitute $p^2$ for x:

$p^2 = \frac{-2}{3}$ $\quad$ $p^2 = 1$

$p = \pm\sqrt{\frac{-2}{3}}$ $\quad$ $p = \sqrt{1}$

$= \pm i\frac{\sqrt{2}}{\sqrt{3}}$ $\quad$ $= \pm 1$

Only real numbers are requested. Solution set: {-1, 1}

49. Let x be John's speed. Then his speed upstream is x - 3 and his speed downstream is x + 3. So his rate upstream is $\frac{10}{x - 3}$ and his rate downstream is $\frac{10}{x + 3}$. Since the total trip took 3 1/2 hours = 7/2 hours, we have

$$\frac{10}{x + 3} + \frac{10}{x - 3} = \frac{7}{2}$$

$$2(x + 3)(x - 3)\left[\frac{10}{x + 3} + \frac{10}{x - 3}\right]$$

$$= 2(x + 3)(x - 3) \cdot \frac{7}{2}$$

$$20(x - 3) + 20(x + 3)$$

$$= 7(x + 3)(x - 3)$$

$$20x - 60 + 20x + 60 = 7(x^2 - 9)$$

$$40x = 7x^2 - 63$$

$$7x^2 - 40x - 63 = 0$$

$$(7x + 9)(x - 7) = 0$$

$$7x + 9 = 0 \quad \text{or} \quad x - 7 = 0$$

$$x = -\frac{9}{7} \quad \text{or} \quad x = 7$$

Now $x = -\frac{9}{7}$ is impossible since the boat cannot have negative speed, therefore the speed is 7 miles per hour.

53. $p = \sqrt{\frac{kL}{g}}$; for k

$$p^2 = \frac{kL}{g}$$

$$\frac{p^2 g}{L} = k$$

57. Let x be the width of the rectangle. Then the length is $5x + 3$. Use $A = LW$.

$$2 = (5x + 3)x = 5x^2 + 3x$$

$$5x^2 + 3x - 2 = 0$$

$$(5x - 2)(x + 1) = 0$$

$$5x - 2 = 0 \quad \text{or} \quad x + 1 = 0$$

$$5x = 2$$

$$x = \frac{2}{5} \quad \text{or} \quad x = -1$$

Now $x - 1$ is impossible and so $x = \frac{2}{5}$. The width is 2/5 centimeters and the length is 5 centimeters.

61. $k^2 + k < 12$

$$k^2 + k - 12 < 0$$

$$k^2 + k - 12 = 0$$

$$(k + 4)(k - 3) = 0$$

$$k = -4 \quad \text{or} \quad k = 3$$

The numbers -4 and 3 divide the number line into three regions.

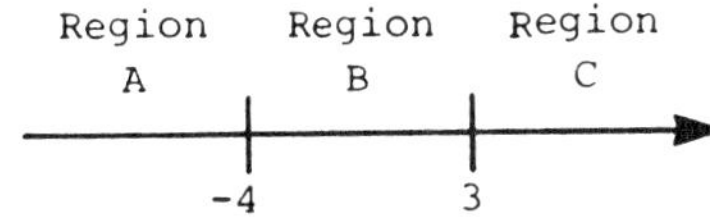


Test a point from each region.

Region A: If $k = -5$, then

$$(k + 4)(k - 3) > 0.$$

Region B: If $k = 0$, then

$$(k + 4)(k - 3) < 0.$$

Region C: If $k = 4$, then

$$(k + 4)(k - 3) > 0.$$

Region B satisfies the inequality $(k + 4)(k - 3) < 0$.

Solution set: $(-4, 3)$

See answer graph in textbook.

65. $3m^2 - 5m < 0$

$3m^2 - 5m = 0$ *Write the quadratic equation*

$$m(3m - 5) = 0$$

$$m = 0 \quad \text{or} \quad m = \frac{5}{3}$$

The numbers 0 and 5/3 divide the number line into three regions.

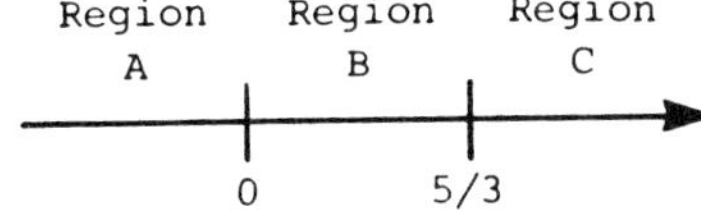


Test a point from each region.

Region A: If $m = -1$, then

$$m(3m - 5) > 0.$$

Region B: If $m = 1$, then

$$m(3m - 5) < 0.$$

Region C: If $m = 2$, then

$$m(3m - 5) > 0.$$

Region B satisfies the inequality $m(3m - 5) < 0$.

Solution set: $(0, 5/3)$

See answer graph in textbook.

69. $\frac{y + 1}{2y - 1} < 3$

$\frac{y + 1}{2y - 1} = 3$ *Write as an equation and solve*

$$\frac{y + 1}{2y - 1}(2y - 1) = 3(2y - 1)$$

$$y + 1 = 6y - 3$$

$$-5y = -4$$

$$y = \frac{4}{5}$$

$2y - 1 = 0$ *Set denominator equal to zero*

$$y = \frac{1}{2}$$

Divide the number line into regions.

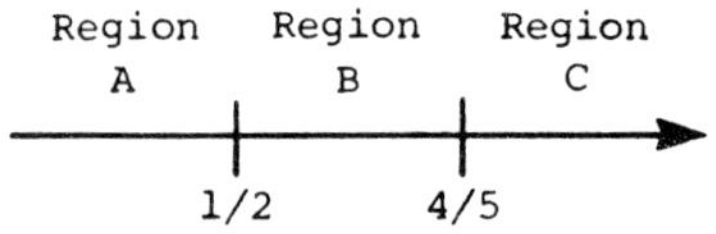


Region A: If $y = 0$,

$$\frac{y + 1}{2y - 1} - 3 < 0.$$

Region B: If $y = \frac{2}{3}$,

$$\frac{y + 1}{2y - 1} - 3 < 0.$$

Region C: If $y = 1$,

$$\frac{y + 1}{2y - 1} - 3 < 0.$$

Regions A and C satisfy the inequality.

Solution set: $(-\infty, \frac{1}{2}) \cup (\frac{4}{5}, +\infty)$

See the answer graph in textbook.

73. $3p^2 - 6p + 4 = 0$   *$a = 3$, $b = -6$, $c = 4$*

$p = \frac{-(-6) \pm \sqrt{(-6)^2 - 4(3)(4)}}{2(3)}$ *Use the quadratic formula*

$$= \frac{6 \pm \sqrt{-12}}{6}$$

$$= \frac{6 \pm i2\sqrt{3}}{6}$$

$$= \frac{3 \pm i\sqrt{3}}{3}$$

Solution set: $\{1 + \frac{\sqrt{3}}{3}i, 1 - \frac{\sqrt{3}}{3}i\}$

77. $(r - 1)(2r + 3)(r + 7) \geq 0$ *Set the polynomial equal to 0*

$$(r - 1)(2r + 3)(r + 7) = 0$$

$r - 1 = 0$ or $2r + 3 = 0$ or $r + 7 = 0$

$r = 1$    $r = \frac{-3}{2}$    $r = -7$

Divide the number line into regions.

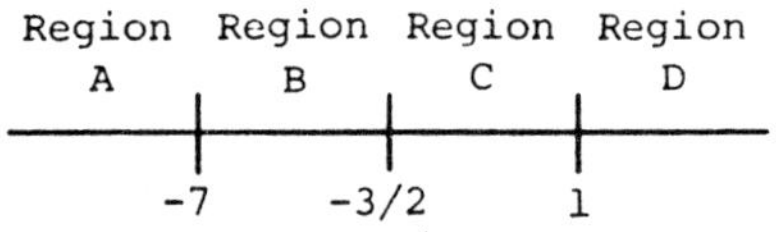


Region A: If $r = -8$,

$$(r - 1)(2r + 3)(r + 7) < 0.$$

Region B: If $r = -5$,

$$(r - 1)(2r + 3)(r + 7) > 0.$$

Region C: If $r = 0$,

$$(r - 1)(2r + 3)(r + 7) < 0.$$

Region D: If $r = 2$,

$$(r - 1)(2r + 3)(r + 7) > 0.$$

Regions B and D satisfy the inequality.
Solution set: $[-7, -3/2] \cup [1, +\infty)$

81. $D = \sqrt{\frac{xy}{3}}$; for y

$D^2 = \frac{xy}{3}$ *Square both sides*

$3D^2 = xy$ *Multiply by 3*

$\frac{3D^2}{x} = y$ *Divide by x*

**Chapter 6 Test (page 289)**

1. $r^2 = 8$

$t = \pm\sqrt{8}$

$= \pm\sqrt{4 \cdot 2}$

$= 2\sqrt{2}$ or $-2\sqrt{2}$

Solution set: $\{2\sqrt{2}, -2\sqrt{2}\}$

2. $(3m + 2)^2 = 4$

$3m + 2 = \sqrt{4}$ *Take the square root of each side*

$3m + 2 = -2$ or $3m + 2 = 2$

$3m = -4$ $\quad$ $3m = 0$

$m = \frac{-4}{3}$ $\quad$ $m = 0$

*Check both solutions in original equation*

Solution set: $\{\frac{-4}{3}, 0\}$

3. $z^2 - 2z = 1$

$z^2 - 2z + 1 = 1 + 1$

$(z - 1)^2 = 2$

$z - 1 = \pm\sqrt{2}$

$z = 1 \pm \sqrt{2}$

Solution set: $\{1 + \sqrt{2}, 1 - \sqrt{2}\}$

4. $2p^2 - 2p + 1 = 0$

$2p^2 - 2p = -1$

$p^2 - 1p = -\frac{1}{2}$

$p^2 - 1p + \frac{1}{4} = -\frac{1}{2} + \frac{1}{4}$

$(p - \frac{1}{2})^2 = -\frac{1}{4}$

$p - \frac{1}{2} = \pm\sqrt{-\frac{1}{4}}$

$p - \frac{1}{2} = \pm i \cdot \frac{1}{2} = \pm\frac{i}{2}$

$p = \frac{1}{2} \pm \frac{i}{2}$

$= \frac{1 \pm i}{2}$

Solution set: $\{\frac{1}{2} + \frac{1}{2}i, \frac{1}{2} - \frac{1}{2}i\}$

5. $2m^2 - 3m + 1 = 0$ $\quad$ $a = 2$, $b = -3$, $c = 1$

$x = \frac{-b \pm \sqrt{b^2 - 4ac}}{2a}$

$= \frac{-(-3) \pm \sqrt{(-3)^2 - 4(2)(1)}}{2(2)}$

$= \frac{3 \pm \sqrt{9 - 8}}{4}$

$= \frac{3 \pm \sqrt{1}}{4}$

$= \frac{3 \pm 1}{4}$

$x = \frac{3 + 1}{4}$ or $x = \frac{3 - 1}{4}$

$x = 1$ $\quad$ $x = \frac{1}{2}$

Solution set: $\{1, 1/2\}$

6. $3r^2 + 4r + 5 = 0$ $\quad$ $a = 3$, $b = 4$, $c = 5$

$x = \frac{-4 \pm \sqrt{(4)^2 - 4(3)(5)}}{2(3)}$

$= \frac{-4 \pm \sqrt{16 - 60}}{6}$

$= \frac{-4 \pm \sqrt{-44}}{6}$

$= \frac{-4 \pm 2i\sqrt{11}}{6}$

$= \frac{2(-2 \pm i\sqrt{11})}{6}$

$= \frac{-2 \pm i\sqrt{11}}{3}$

Solution set: $\{-\frac{2}{3} + \frac{\sqrt{11}}{3}i, -\frac{2}{3} - \frac{\sqrt{11}}{3}i\}$

7. $5m^2 + 2m = 4$

$5m^2 + 2m - 4 = 0$ *a = 5, b = 2, c = -4*

$x = \frac{-2 \pm \sqrt{(2)^2 - 4(5)(-4)}}{2(5)}$

$= \frac{-2 \pm \sqrt{84}}{10}$

$= \frac{-2 \pm 2\sqrt{21}}{10}$

$= \frac{-1 \pm \sqrt{21}}{5}$

Solution set: $\{\frac{-1 + \sqrt{21}}{5}, \frac{-1 - \sqrt{21}}{5}\}$

8. $-2m^2 + 6m + 1 = 0$ *a = -2, b = 6, c = 1*

$m = \frac{-6 \pm \sqrt{(6)^2 - 4(-2)(1)}}{2(-2)}$

$= \frac{-6 \pm \sqrt{44}}{-4}$

$= \frac{-6 \pm 2\sqrt{11}}{-4}$

$= \frac{-3 \pm \sqrt{11}}{-2}$

$m = \frac{-3 + \sqrt{11}}{-2}$ or $m = \frac{-3 - \sqrt{11}}{-2}$

$= \frac{3 - \sqrt{11}}{2}$ or $= \frac{3 + \sqrt{11}}{2}$

Solution set: $\{\frac{3 - \sqrt{11}}{2}, \frac{3 + \sqrt{11}}{2}\}$

9. $2p^2 + 5 = 4p$

$2p^2 - 4p + 5 = 0$ *a = 2, b = -4 c = 5*

$b^2 - 4ac = (-4)^2 - 4(2)(5) = -24$ *Using the discriminant*

Since the discriminant is less than 0, the equation will have two complex solutions.

10. $2p^2 + 5 = 4p$

$2p^2 - 4p + 5 = 0$ *a = 2, b = -4, c = 5*

$\frac{-b}{a} = \frac{-(-4)}{2} = 2$ *Sum of the solutions*

$\frac{c}{a} = \frac{5}{2}$ *Product of the solutions*

11. $\frac{12}{(p - 1)^2} = 1 + \frac{1}{p - 1}$

$\frac{12}{(p - 1)^2}(p - 1)^2 = 1(p - 1)^2 + \frac{1}{p - 1}(p - 1)^2$

$12 = p^2 - 2p + 1 + p - 1$

$0 = p^2 - p - 12$

$(p - 4)(p + 3) = 0$

$p - 4 = 0$ or $p + 3 = 0$

$p = 4$ $\quad$ $p = -3$

Solution set: $\{4, -3\}$

12. $2(\sqrt{m} + 1)^2 + 20 = 13(\sqrt{m} + 1)$

$2(\sqrt{m} + 1)^2 - 13(\sqrt{m} + 1) + 20 = 0$ *Write in quadratic form*

$2x^2 - 13x + 20 = 0$ *Use substitution* $x = \sqrt{m} + 1$

$(2x - 5)(x - 4) = 0$

$2x - 5 = 0$ or $x - 4 = 0$

$x = \frac{5}{2}$ $\quad$ $x = 4$

Substitute $\sqrt{m}+1$ for x.

$$\sqrt{m}+1=\frac{5}{2} \qquad \sqrt{m}+1=4$$
$$\sqrt{m}=\frac{3}{2} \qquad \sqrt{m}=3$$
$$m=\frac{9}{4} \qquad m=9$$

Solution set: {9/4, 9}

13.
$$m=-\sqrt{m+6}$$
$$(m)^2=(-\sqrt{m+6})^2$$
$$m^2=m+6$$
$$m^2-m-6=0$$
$$(m-3)(m+2)=0$$
$$m-3=0 \text{ or } m+2=0$$
$$m=3 \text{ or } m=-2$$

Check:

$$3=-\sqrt{3+6}$$
$$3=-\sqrt{9}$$
$$3=-3 \quad \textit{False}$$

Solution set: {-2}

14.
$$2x=\sqrt{\frac{5x+2}{3}}$$
$$(2x)^2=\left(\sqrt{\frac{5x+2}{3}}\right)^2$$
$$4x^2=\frac{5x+2}{3}$$
$$4x^2(3)=\frac{5x+2}{3}(3)$$
$$12x^2-5x-2=0$$
$$(4x+1)(3x-2)=0$$
$$4x+1=0 \text{ or } 3x-2=0$$
$$x=\frac{-1}{4} \text{ or } x=\frac{2}{3}$$

Check: $2(-\frac{1}{4})=\sqrt{\frac{5(-1/4)+2}{3}}$
$$-\frac{1}{2}=\sqrt{\frac{3/4}{3}}$$
$$-\frac{1}{2}=\sqrt{\frac{1}{4}}$$
$$-\frac{1}{2}=\frac{1}{2} \quad \textit{False}$$

Solution set: {2/3}

15.
$$y^4=6y^2+27$$
$$y^4-6y^2-27=0$$
$$x^2-6x-27=0 \quad \textit{Substitute x for } y^2$$
$$(x-9)(x+3)=0$$
$$x-9=0 \text{ or } x+3=0$$
$$x=9 \qquad x=-3$$

To find y, substitute $y^2$ for x.

$$y^2=9 \qquad y^2=-3$$
$$y=\sqrt{9}=\pm 3 \qquad y=\pm i\sqrt{3}$$

Solution set: $\{3, -3, i\sqrt{3}, -i\sqrt{3}\}$

16.
$$6y^4=13y^2-6$$
$$6y^4-13y^2+6=0$$
$$6x^2-13x+6=0 \quad \textit{Substitute x for } y^2$$
$$x=\frac{-(-13)\pm\sqrt{(-13)^2-4(6)(6)}}{2(6)} \quad \textit{Use the quadratic formula}$$
$$=\frac{13\pm\sqrt{25}}{12}=\frac{3}{2} \text{ or } \frac{2}{3}$$

To find y, substitute $y^2$ for x.

$$y^2=\frac{3}{2} \quad \text{or} \quad y^2=\frac{2}{3}$$
$$y=\pm\sqrt{\frac{3}{2}} \qquad y=\pm\sqrt{\frac{2}{3}}$$
$$=\pm\frac{\sqrt{3}\cdot\sqrt{2}}{\sqrt{2}\cdot\sqrt{2}} \qquad =\pm\frac{\sqrt{2}\cdot\sqrt{3}}{\sqrt{3}\cdot\sqrt{3}}$$
$$=\pm\frac{\sqrt{6}}{2} \qquad =\pm\frac{\sqrt{6}}{3}$$

Solution set: $\{\frac{\sqrt{6}}{2}, -\frac{\sqrt{6}}{2}, \frac{\sqrt{6}}{3}, -\frac{\sqrt{6}}{3}\}$

17. Let x represent the number of hours it takes Mario. So it takes Luis $x - 2$ hours. Their rates per hour, then, are $\frac{1}{x}$ and $\frac{1}{x - 2}$, respectively. They can do the job working together in 5 hours, so their rate per hour working together is 1/5. Adding their individual rates and setting this equal to their rate working together, we get

$$\frac{1}{x} + \frac{1}{x - 2} = \frac{1}{5}.$$

Solve this equation.

$$5x(x - 2) \cdot \frac{1}{x} + 5x(x - 2) \cdot \frac{1}{x - 2} = 5x(x - 2) \cdot \frac{1}{5}$$

$$5(x - 2) + 5x = x(x - 2)$$

$$5x - 10 + 5x = x^2 - 2x$$

$$0 = x^2 - 12x + 10$$

*a = 1, b = -12, c = 10*

$$x = \frac{-b \pm \sqrt{b^2 - 4ac}}{2a}$$

$$= \frac{-(-12) \pm \sqrt{(-12)^2 - 4(1)(10)}}{2(1)}$$

$$= \frac{12 \pm \sqrt{144 - 40}}{2}$$

$$= \frac{12 \pm \sqrt{104}}{2}$$

Since $\sqrt{104} \approx 10.2$ to the nearest tenth, then

$$x = \frac{12 + 10.2}{2} = \frac{22.2}{2} = 11.1$$

or $$x = \frac{12 - 10.2}{2} = \frac{1.8}{2} = .9$$

Now, x represents Mario's time, and $x - 2$ represents Luis' time. But, for $x = .9$ hours, then Luis' time would be $x - 2 = -1.1$, a negative number. So Mario's time would be about 11.1 hours and Luis' time would be about $11.1 - 2 = 9.1$ hours.

18. $S = 4\pi r^2$; for r

$$4\pi r^2 = S$$

$$r^2 = \frac{S}{4\pi}$$

$$r = \pm\sqrt{\frac{S}{4\pi}}$$

$$= \frac{\pm\sqrt{S}}{\sqrt{4\pi}}$$

$$= \frac{\pm\sqrt{S}}{2\sqrt{\pi}} \cdot \frac{\sqrt{\pi}}{\sqrt{\pi}}$$

$$= \frac{\pm\sqrt{\pi S}}{2\pi}$$

19. $A = \pi r\sqrt{r^2 + h^2}$; for h

$$\frac{A}{\pi r} = \sqrt{r^2 + h^2}$$

$$\left(\frac{A}{\pi r}\right)^2 = \left(\sqrt{r^2 + h^2}\right)^2$$

$$\frac{A^2}{\pi^2 r^2} = r^2 + h^2$$

$$\frac{A^2}{\pi^2 r^2} - r^2 = h^2$$

$$\frac{A^2 - r^4(\pi^2)}{\pi^2 r^2} = h^2$$

$$\pm\sqrt{\frac{A^2 - \pi^2 r^4}{\pi^2 r^2}} =$$

or $$h = \frac{\pm\sqrt{A^2 - \pi^2 r^4}}{\sqrt{\pi^2 r^2}}$$

$$h = \frac{\pm\sqrt{A^2 - \pi^2 r^4}}{\pi r}$$

Note: $\sqrt{A^2 - \pi^2 r^4} \neq A - \pi r^2$

20. $r = kp(1 - p)$; for p

$r = kp - kp^2$

$kp^2 - kp + r = 0 \quad a = k,\ b = -k,\ c = r$

$$p = \frac{-b \pm \sqrt{b^2 - 4ac}}{2a}$$

$$= \frac{-(-k) \pm \sqrt{(-k)^2 - 4(k)(r)}}{2(k)}$$

$$= \frac{k \pm \sqrt{k^2 - 4kr}}{2k}$$

21. Let x represent the width of the strip. The pool is rectangular in shape with dimensions of 20 by 24.

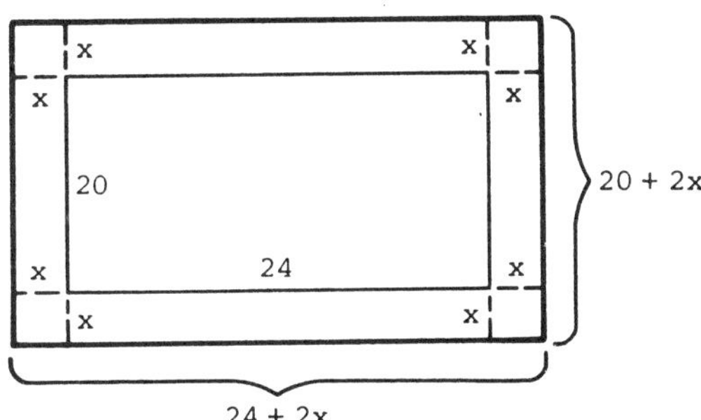


Use A = LW.

(20)(24) = 480 square feet.

By putting a uniform grass strip around the pool, we get a larger rectangular area. Now, the area of the larger rectangle is

| | area of pool | and | area of strip | |
|---|---|---|---|---|
| | ↓ | ↓ | ↓ | |
| (1) | 480 | + | 196 | = 672 square feet. |

By increasing the dimensions of the pool by the uniform width x, the width of the larger rectangle is 20 + 2x and the length 24 + 2x. We know the area of this larger rectangle to be 672 from (1) above.

Again, use A = LW.

$(20 + 2x)(24 + 2x) = 672.$

Solve this equation.

$480 + 40x + 48x + 4x^2 = 672$

$4x^2 + 88x - 192 = 0$

$x^2 + 22x - 48 = 0$

$(x + 24)(x - 2) = 0$

$x + 24 = 0$ or $x - 2 = 0$

$x = -24$ or $x = 2$

Reject -24 since width is positive. So the grass strip has a width of 2 feet.

22. Let x represent the length of the middle side. Then the long side is 110 less than twice the middle side, or 2x - 110. By the Pythagorean formula, $a^2 + b^2 = c^2$ so with the shortest side given as 50, we have

$(50)^2 + (x)^2 = (2x - 110)^2.$

$2500 + x^2 = 4x^2 - 440x + 12100$

$0 = 3x^2 - 440x + 9600$

$0 = (3x - 80)(x - 120)$

$3x - 80 = 0$ or $x - 120 = 0$

$3x = 80$

$x = \frac{80}{3}$ or $x = 120$

$x = \frac{80}{3}$ is not possible solution since the longest side would be $2(\frac{80}{3}) - 110 = -\frac{170}{3}$, a negative number. So, the medium side is 120 meters.

23. $2x^2 + 7x < 15$

$2x^2 + 7x - 15 < 0.$

$2x^2 + 7x - 15 = 0$ *Write the quadratic equation*

$(2x - 3)(x + 5) = 0$

$x = \frac{3}{2}$ or $x = -5$

The numbers -5 and 3/2 divide the number line into three regions.

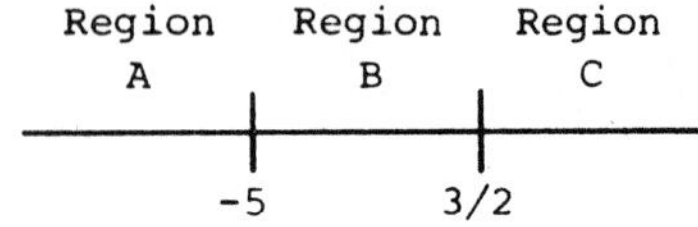


Check a point from each region.

Region A: If x = -6, then

$(2x - 3)(x + 5) > 0.$

Region B: If x = 0, then

$(2x - 3)(x + 5) < 0.$

Region C: If x = 2, then

$(2x - 3)(x + 5) > 0.$

Region B satisfies the inequality $(2x - 3)(x + 5) < 0.$

Solution set: $(-5, 3/2)$

A graph of the solution set is shown in your textbook.

24. Solve $2m^2 - 3m \geq -1$

$2m^2 - 3m + 1 \geq 0$

$2m^2 - 3m + 1 = 0$ *Write the quadratic equation*

$(2m - 1)(m - 1) = 0$

$m = \frac{1}{2}$ or $m = 1$

The numbers 1/2 and 1 divide the number line into three regions.

Region A | Region B | Region C
1/2 | 1

Check a point from each region.

Region A: If m = 0, then

$(2m - 1)(m - 1) > 0.$

Region B: If m = 3/4, then

$(2m - 1)(m - 1) < 0.$

Region C: If m = 2, then

$(2m - 1)(m - 1) > 0.$

If $m = \frac{1}{2}$ or $m = 1$, then

$(2m - 1)(m - 1) = 0.$

The solution set includes points in both Regions A and C (including the endpoints 1/2 and 1).

Solution set: $(-\infty, \frac{1}{2}] \cup [1, +\infty)$

A graph of the solution set is shown in your textbook.

25. Solve $\frac{4}{p - 1} \geq 1$

$\frac{4}{p - 1} - 1 \geq 0$

$\frac{4}{p - 1} - \frac{p - 1}{p - 1} \geq 0$

$\frac{4 - p + 1}{p - 1} \geq 0$

$\frac{5 - p}{p - 1} \geq 0$

The numerator on the left is 0 when p = 5 and the denominator is 0 when p = 1. These two values, 1 and 5, divide the number line into three regions.

Region A | Region B | Region C
1 | 5

Check a test point from each region.

Region A: If p = 0, then

$\frac{5 - p}{p - 1} < 0.$

Region B: If p = 2, then

$\frac{5 - p}{p - 1} > 0.$

Region C: If $p = 6$, then

$$\frac{5 - p}{p - 1} < 0.$$

If $p = 1$, then $\frac{5 - p}{p - 1}$ is undefined.

If $p = 5$, then $\frac{5 - p}{1 - p} = 0$.

Region B (including the endpoint $p = 5$) satisfies the inequality $\frac{5 - p}{p - 1} \geq 0$.

Solution set: $(1, 5]$

(The endpoint $p = 1$ cannot be used as a solution since the denominator equals 0.)

A graph of the solution set is shown in your textbook.

## CHAPTER 7 THE STRAIGHT LINE

### Section 7.1 (page 297)

1. The x- and y-coordinates of a point in quadrant I are both positive. Thus, (1, 5) is in quadrant I.

5. The x-coordinate is positive and the y-coordinate is negative for a point in quadrant IV, so (2, -3) is in quadrant IV.

In Exercises 9-41, see the answer graphs in the textbook.

9. (2, 3) is in the first quadrant. From the origin, go 2 units to the right and 3 units up.

13. (0,5) is on the y-axis. From the origin, go up 5 units.

17. (-2, 0) is on the x-axis. From the origin, go to the left 2 units.

21. $x - y = 4$; (0, ), ( , 0), (2, ), ( , -1)

$0 - y = 4$ *Let x = 0*

$y = -4$

The ordered pair is (0, -4).

$x - 0 = 4$ *Let y = 0*

$x = 4$

(4, 0)

$2 - y = 4$ *Let x = 2*

$y = -2$

(2, -2)

$x - (-1) = 4$ *Let y = -1*

$x = 3$

(3, -1)

25. $3x + 2y = 8$;

(0, ), ( , 0), (2, ) ( , -2)

$3(0) + 2y = 8$ *Let x = 0*

$y = 4$

(0, 4)

$3x + 2(0) = 8$ *Let y = 0*

$x = \frac{8}{3}$

$(\frac{8}{3}, 0)$

$3(2) + 2y = 8$ *Let x = 2*

$y = 1$

(2, 1)

$3x + 2(-2) = 8$ *Let y = -2*

$x = 4$

(4, -2)

29. For the x-intercept, let $y = 0$ in $5x + 6y = 10$.

$5x + 6(0) = 10$

$x = 2$

For the y-intercept, let $x = 0$.

$5(0) + 6y = 10$

$y = \frac{5}{3}$

33. $x = 3y = 2$

$x - 3(0) = 2$ *Let y = 0*

$x = 2$

$0 - 3y = 2$ *Let x = 0*

$y = \frac{-2}{3}$

37. $3x = y$

$3x = 0$ *Let y = 0*

$x = 0$

$3(0) = y$ *Let x = 0*

$0 = y$

Since both intercepts give the same ordered pair (0, 0), another point is needed for graphing. For example, let $x = 1$.

$3(1) = y$

$4 = y$

(1, 4) is on the line.

41. y = 4
y = 4 for every x. If x = 0, y = 4 so the y-intercept is 4.
Since y is always 4, there is no x-intercept. The graph is a horizontal line passing through (0, 4).

45. To find the distance between (-2, 4) and (3, -2), let $x_1 = -2$, $x_2 = 3$, $y_1 = 4$, and $y_2 = -2$.
Use the distance formula.

$$d = \sqrt{(x_2 - x_1)^2 + (y_2 - y_1)^2}$$

$$d = \sqrt{(3 - (-2)^2 + (-2 - 4)^2}$$
$$= \sqrt{25 + 36}$$
$$= \sqrt{61}$$

49. (a, a - b) and (b, a + b)
Using the distance formula

$$d = \sqrt{(x_2 - x_1)^2 + (y_1 + y_2)^2}$$

we have

$$d = \sqrt{(b - a)^2 + [(a + b) - (a - b)]^2}$$
$$= \sqrt{b^2 - 2ab + a^2 + 4b^2}$$
$$= \sqrt{5b^2 - 2ab + a^2}.$$

53. (1, 3), (-2, -4), (1, -2)
To find the perimeter, we need to find the distance between each pair of points determining the triangle. Let A be (1, 3), B be (2, 4) and C be (1, -2).

$$AB = \sqrt{(-2 - 1)^2 + (-4 - 3)^2}$$
$$= \sqrt{9 + 49}$$
$$= \sqrt{58}$$

$$BC = \sqrt{(1 - (-2))^2 + (-2 - (-4))^2}$$
$$= \sqrt{9 + 4}$$
$$= \sqrt{13}$$

$$CA = \sqrt{(1 - 1)^2 + (3 - (-2))^2}$$
$$= \sqrt{25}$$
$$= 5$$

Adding the three distances to get the perimeter, we have

$$\sqrt{58} + \sqrt{13} + 5.$$

57. (0, -4), (2, 2), (5, 11)
A is (0, 4), B is (2, 2), C is (5, 11).

$$AB = \sqrt{(2 - 0)^2 + (2 - (-4))^2}$$
$$= \sqrt{4 + 36}$$
$$= 2\sqrt{10}$$

$$BC = \sqrt{(5 - 2)^2 + (11 - 2)^2}$$
$$= \sqrt{9 + 81}$$
$$= 3\sqrt{10}$$

$$CA = \sqrt{(0 - 5)^2 + (-4 - 11)^2}$$
$$= \sqrt{25 + 225}$$
$$= 5\sqrt{10}$$

Since $2\sqrt{10} + 3\sqrt{10} = 5\sqrt{10}$, the three points lie on the same line.

61. A is (0, 5), B is (5, -5),
C is (10, -14.

$$AB = \sqrt{(5 - 0)^2 + (-5 - 5)^2}$$
$$= \sqrt{25 + 100}$$
$$= 5\sqrt{5}$$

$$BC = \sqrt{(10 - 5)^2 + (-14 - (-5))^2}$$
$$= \sqrt{25 + 81}$$
$$= \sqrt{106}$$

$$CD = \sqrt{(0 - 10)^2 + (5 - (-14))}$$
$$= \sqrt{100 + 361}$$
$$= \sqrt{461}$$

Since $5\sqrt{5} + \sqrt{106} \neq \sqrt{461}$, the three points do not lie on a straight line.

65. A is (-2, -2), B is (8, 4), C is (2, 14).

$$AB = \sqrt{(8 - (-2))^2 + (4 - (-2))^2}$$
$$= \sqrt{100 + 36}$$
$$= \sqrt{136}$$

$$BC = \sqrt{(2 - 8)^2 + (14 - 4)^2}$$
$$= \sqrt{36 + 100}$$
$$= \sqrt{136}$$

$$CA = \sqrt{(-2 - 2)^2 + (-2 - 14)^2}$$
$$= \sqrt{16 + 256}$$
$$= \sqrt{272}$$

Since $(\sqrt{136})^2 + (\sqrt{136})^2 = (\sqrt{272})^2$, this is a right triangle.

69. P(0, 4), Q(-3, -1)
Use the midpoint formulas

$$\overline{x} = \frac{x_1 + x_2}{2} \text{ and } \overline{y} = \frac{y_1 + y_2}{2}.$$

$$\overline{x} = \frac{0 - 3}{2} \qquad \overline{y} = \frac{4 - 1}{2}$$

$$\overline{x} = \frac{-3}{2} \qquad \overline{y} = \frac{3}{2}$$

The midpoint is $(\frac{-3}{2}, \frac{3}{2})$.

73. The x-value is 7 greater than the the y-value.
$x = 7 + y$
To graph, let $x = 0$.

$$0 = 7 + y$$
$$-7 = y$$

(0, -7) is on the graph.
Let $y = 0$.

$$x = 7 + 0$$
$$x = 7$$

(7, 0) is on the graph.
See answer graph in textbook.

77. Let height to top of ladder be h.
Use the Pythagorean formula, with $h = b$.

$$a^2 + b^2 = c^2 = 12^2$$

$$12^2 + h^2 = 20^2$$
$$144 + h^2 = 400$$
$$h^2 = 256$$
$$h = \pm 16$$

Discard negative solution, since height is positive.
The ladder will be 16 feet above the ground.

81. $$\frac{-3 - (-5)}{4 - (-1)} = \frac{-3 + 5}{4 + 1}$$
$$= \frac{2}{5}$$

**Section 7.2 (page 305)**

In Exercises 1-21, see answer graphs in textbook.

1. (2, -3) and (1, 5)

$$m = \frac{\text{change in } y}{\text{change in } x} = \frac{-3 - 5}{2 - 1} = \frac{-8}{1} = -8$$

5. (3, 3) and (-5, -6)

$$m = \frac{3 - (-6)}{3 - (-5)} = \frac{3 + 6}{3 + 5} = \frac{9}{8}$$

9. To graph 2x + 4y = 5, find two points and draw the line through them.

Let x = 0. Then $y = \frac{5}{4}$.

One point is $(0, \frac{5}{4})$.

Let y = 0. Then $x = \frac{5}{2}$.

Another point is $(\frac{5}{2}, 0)$.

The slope is

$$m = \frac{0 - \frac{5}{4}}{\frac{5}{2} - 0} = \frac{-\frac{5}{4}}{\frac{5}{2}} = \frac{-1}{2}.$$

13. 6x - 5y = 30

Replace x with 0 in 6x - 5 = 30 to get y = -6 and replace y with 0 to get x = 5. Using the ordered pairs, (5, 0) and (0, 6), in the slope formula, we get

$$m = \frac{0 - (-6)}{5 - 0} = \frac{6}{5}.$$

17. $m = \frac{-5}{4}$, through (-2, -1)

$$\text{slope} = \frac{\text{change in } y}{\text{change in } x} = \frac{-5}{4}$$

Locate the point (-2, -1); move down 5 units and then 4 units to the right.

21. m = 0, through (2, -5)

A horizontal line has a slope of 0; so we have a horizontal line through y = -5.

25. 3x = y and 2y - 6x = 5

Find the slope of each line by first finding two points on each line.

3x = y If x = 0, y = 0. (0, 0)

If x = 1, y = 3. (1, 3)

$$\text{slope} = m_1 = \frac{3 - 0}{1 - 0} = 3$$

2y - 6x = 5 If x = 0, y = 5/2. (0, 5/2)

If y = 0, x = -5/6. (-5/6, 0)

$$\text{slope} = m_2 = \frac{0 - 5/2}{-5/6 - 0} = \frac{5}{2}\cdot\frac{6}{5} = 3$$

The slopes of $m_1$ and $m_2$ are equal so the lines are parallel.

29. The line through (4, 6) and (-8, 7) and the line through (7, 4) and (-5, 5). The slope of the first line is

$$m_1 = \frac{7 - 6}{-8 - 4} = \frac{1}{-12}.$$

The slope of the second line is

$$m_2 = \frac{5 - 4}{-5 - 7} = \frac{1}{-12}.$$

The two slopes are equal so the lines are parallel.

33. 4x - 3y = 5 and 3x - 4y = 2

First, find the slope of each line:

4x - 3y = 5 If x = 0, $y = \frac{-5}{3}$. $(0, \frac{-5}{3})$

If y = 0, $x = \frac{5}{4}$. $(\frac{5}{4}, 0)$

$$m_1 = \frac{0 - (-\frac{5}{3})}{\frac{5}{4} - 0} = \frac{5}{3}\cdot\frac{4}{5} = \frac{4}{3}$$

$3x - 4y = 2$ If $x = 0$, $y = -\frac{1}{2}$ $(0, -\frac{1}{2})$

If $y = 0$, $x = \frac{2}{3}$ $(\frac{2}{3}, 0)$

$$m_2 = \frac{0 - (-\frac{1}{2})}{\frac{2}{3} - 0} = \frac{1}{2}\cdot\frac{3}{2} = \frac{3}{4}$$

Second, check if the product of the two slopes is -1.

$$\frac{4}{3}\cdot\frac{3}{4} = \frac{12}{12} = 1$$

The lines are not perpendicular.

37. Name the points A(1, 3), B(-2, 9), C(4, -2). Three points lie on the same line if the slopes of 2 segments connecting the points are equal.

Slope of AB $= \frac{9 - 3}{-2 - 1} = \frac{6}{-3} = -2$

Slope of BC $= \frac{-2 - 9}{4 - (-2)} = -\frac{11}{6}$

Since $-2 \neq -\frac{11}{6}$, the points do not lie on the same line.

41. Use the points given to show the opposite sides are parallel.

(-13, -9), (-11, -1)

$$m_1 = \frac{-1 - (-9)}{-11 - (-13)} = \frac{8}{2} = 4$$

(2, -2), (4, 6)

$$m_2 = \frac{6 - (-2)}{4 - 2} = \frac{8}{2} = 4$$

These two lines are parallel since their slopes are equal.

(-11, -1), (4, 6)

$$m_3 = \frac{6 - (-1)}{4 - (-11)} = \frac{7}{15}$$

(2, -2), (-13, -9)

$$m_4 = \frac{-9 - (-2)}{-13 - 2} = \frac{-7}{-15} = \frac{7}{15}$$

These lines are also parallel since their slopes are equal.

45. Let x be miles per hour over the speed limit and y equals the dollars fined. The ordered pairs are (10, 35) and (15, 42).

Use the slope formula.

$$m = \frac{42 - 35}{15 - 10} = \frac{7}{5}$$
$$= 1.40$$

The average rate of change is $1.40 per mile per hour over the speed limit.

49.
$$-3x = 4y - 7$$
$$-3x + 7 = 4y$$
$$\frac{-3x + 7}{4} = y$$
$$y = -\frac{3}{4}x + \frac{7}{4}$$

53.
$$y - (-1) = \frac{5}{3}[x - (-4)]$$
$$y + 1 = \frac{5}{3}(x + 4)$$
$$3(y + 1) = 5(x + 4)$$ *Multiply both sides by 3*
$$3y + 3 = 5x + 20$$
$$3y = 5x + 17$$

**Section 7.3 (page 311)**

1. Put $x + y = 8$ in slope intercept form, $y = mx + b$ by solving for y.

$$y = -x + 8$$

The slope is the coefficient of x, or -1. The y-intercept is 8.

5.
$$2x - 3y = 5$$
$$-3y = -2x + 5$$
$$y = \frac{2}{3}x - \frac{5}{3}$$ *Solve for y*

The slope is $\frac{2}{3}$ and the y-intercept is $-\frac{5}{3}$.

9. $8x + 11y = 9$

$$11y = -8x + 9$$

$$y = \frac{-8}{11}x + \frac{9}{11} \quad \textit{Solve for } y$$

The slope is $-\frac{8}{11}$ and the y-intercept is $\frac{9}{11}$.

13. $m = -\frac{2}{3}$, $b = \frac{1}{2}$

$$y = mx + b$$

$$y = -\frac{2}{3}x + \frac{1}{2}$$

17. $m = 0$, $b = 4$

$$y = 0x + 4$$

$$y = 4$$

21. Use point-slope form

$$y - y_1 = m(x - x_1)$$

with $m = -\frac{3}{4}$, $y_1 = 5$, $x_1 = -2$.

$$y - 5 = -\frac{3}{4}[x - (-2)]$$

$$4(y - 5 = -3(x + 2) \quad \textit{Multiply by 4}$$

$$4y - 20 = -3x - 6$$

$$3x + 4y = 20 - 6$$

$$3x + 4y = 14$$

25. In $y - y_1 = m(x - x_1)$ use

$m = \frac{1}{2}$; $x_1 = 7$; $y_1 = 4$.

$$y - 4 = \frac{1}{2}(x - 7)$$

$$2(y - 4) = x - 7$$

$$2y - 8 = x - 7$$

$$-x + 2y = 1$$

$$x - 2y = -1$$

29. x - intercept 3 means the point (3, 0). Thus, $m = 4$; $x_1 = 3$; $y_1 = 0$.

$$y - 0 = 4(x - 3)$$

$$y = 4x - 12$$

$$-4x + y = -12$$

$$4x - y = 12$$

33. If a line has undefined slope, it is a vertical line with its equation in the form $x = k$ where $k$ is a constant. Thus, the line with undefined slope through (2, 8) is $x = 2$.

37. Find the slope through (3, 4) and (2, 6).

$$m = \frac{6 - 4}{2 - 3} = \frac{2}{-1} = -2$$

Use $m = -2$, $x_1 = 3$, $y_1 = 4$.

Then, $y - 4 = -2(x - 3)$

$$y - 4 = -2x + 6$$

$$2x + y = 10.$$

We would get the same answer if we used $x_1 = 2$, $y_1 = 6$.

41. Find the slope through (1, 1) and (0, -4).

$$m = \frac{1 - (-4)}{1 - 0} = 5$$

Use $m = 5$, $x_1 = 0$, $y_1 = -4$ in the equation $y - y_1 = m(x - x_1)$.

$$y - (-4) = 5(x - 0)$$

$$y + 4 = 5x$$

$$y = 5x - 4$$

So, $5x - y = 4$.

45. Find the slope through (2, 5) and (1, 5).

$$m = \frac{5 - 5}{2 - 1} = \frac{0}{1} = 0$$

Thus, the line is a horizontal line with equation in the form $y = k$. So, $y = 5$.

49. Solve $3x - y = 8$ for y.

$$-y = -3x + 8$$
$$y = 3x - 8$$

The slope is 3, so parallel lines have slope 3.

Let $m = 3$, $x_1 = -7$, $y_1 = 3$ in point-slope form of equation.

$$y - 3 = 3[x - (-7)]$$
$$y - 3 = 3x + 21$$
$$y = 3x + 24$$
$$3x - y = -24$$

53. $2x - y = 4$

$$-y = -2x + 4$$
$$y = 2x - 4$$

The slope is 2. Perpendicular lines have slope $-\frac{1}{2}$ since $(2)(-\frac{1}{2}) = -1$.

Thus, let $m = -\frac{1}{2}$, $x_1 = 8$, $y_1 = 5$

$$y - 5 = -\frac{1}{2}(x_1 - 8)$$
$$y - 5 = -\frac{1}{2}x + 4$$
$$\frac{1}{2}x + y = 9 \text{ or } x + 2y = 18$$

57. (a) We have two points, (10, 7500) and (20, 13,900).

First find the slope.

$$(13{,}900 - 7500) = m(20 - 10)$$
$$6400 = 10m$$
$$640 = m$$

To write the equation, let $m = 640$, $x_1 = 10$, $y_1 = 7500$.

$$y - 7500 = 640(x - 10)$$
$$y - 7500 = 640x - 6400$$
$$y - 1100 = 640x$$
$$y = 640x + 1100$$

(b) To find the cost for 12 generators, substitute 12 for x in the equation.

$$y = 640(12) + 1100$$
$$y = 8780$$

The cost is \$8780.

61. $2x + 5 < 6$

$$2x < 1$$
$$x < \frac{1}{2}$$

Solution set: $(-\infty, \frac{1}{2})$

65. $-5x - 5 > 3$

$$-5x > 8$$
$$x < -\frac{8}{5} \quad \textit{Divide by -5 and reverse inequality}$$

Solution set: $(-\infty, -\frac{8}{5})$

**Section 7.4 (page 316)**

In Exercises 1-45, see answer graphs in textbook.

1. $x + y \leq 2$

First graph the line $x + y = 2$. The x-intercept is 2 and the y-intercept is 2. Draw a solid line since the symbol is $\leq$. Now test a point not on the line, say (0, 0). Substitute $x = 0$ and $y = 0$ into $x + y \leq 2$ to get $0 \leq 2$, a true statement. Shade that side of the line containing the origin.

5. $x + 3y \geq -2$
Graph the line $x + 3y = -2$. The x-intercept is -2 and the y-intercept is $-\frac{2}{3}$. Draw a solid line through these intercepts. Test (0, 0). Substitute $x = 0$ and $y = 0$ into $x + 3y \geq -2$ to get $0 \geq -2$, a true statement. So shade that side of the line containing (0, 0).

9. $4x - 3y < 12$
Graph $4x - 3y = 12$ which has intercepts 3 and -4. Draw a dashed line through these intercepts since the symbol is <. Test (0, 0) and note that the resulting inequality $0 < 12$ is true. So shade the side of the line containing the origin.

13. $x < 1$
Graph $x = 1$. Draw a dashed vertical line. Since $x < 1$ is the desired graph, shade the region to the left of the line, where the x-values are less than 1.

17. $2x - y \geq 1$, $x \geq -4$
Graph $2x - y = 1$ with x-intercept $\frac{1}{2}$ and y-intercept -1. Draw a solid line through these intercepts. Test (0, 0), which gives $0 \geq 1$, a false statement. So shade the side of the line not containing the origin. Graph the solid vertical line $x = -4$ and shade to the right of it. Where the shaded regions overlap is the intersection.

21. $2x - y \leq 4$, $x + 2 \geq 3$
Graph the line $2x - y = 4$ with x-intercept 2 and y-intercept -4. Draw a solid line through these intercepts. Test (0, 0). Substitute $x = 0$, $y = 0$, into $2x - y \leq 4$ to get $0 \leq 4$, a true statement. Shade that side of the line containing (0, 0). Graph the line $x + 2 = 3$ or $x = 1$ as a solid vertical line through (1, 0). Since $x + 2 \geq 3$ or $x \geq 1$, then shade the right side of the line $x = 1$.
The union consists of all points in one graph, the other graph, or where they overlap.

25. $2x + 3y > 6$, $y < 8 + x$
Graph $2x + 3y = 6$ with a dashed line containing x-intercept 3 and y-intercept 2. Notice (0, 0) does not satisfy $2x + 3y > 6$ so shade the side of the line not containing the origin. Graph $y = 8 + x$ with a dashed line containing x-intercept -8 and y-intercept 8. Notice (0, 0) does satisfy $y < 8 + x$ so shade the side of this line containing the origin. All the shaded regions make up the union.

29. Recall that $|y| < 5$ is the same as

$$-5 < y < 5.$$

Draw the line $y = 5$ as the dashed horizontal line through (0, 5), and $y = -5$ as the dashed horizontal line through (0, -5). Since $-5 < y < 5$, shade the region between the horizontal lines.

33. $|x| \geq 2$, with $y \leq 1$

$|x| \geq 2$ means $x \geq 2$ or $x \leq -2$. The graph of $x \geq 2$ consists of a solid vertical line $x = 2$ with shading to the right $x = 2$. The graph of $x \leq -2$ consists of a solid vertical line $x = -2$ with shading to the left of $x = -2$. Both lines and shading are beneath the horizontal solid line $y = 1$.

37. $|y - 3| < 2$ means

$$-2 < y - 3 < 2$$
$$1 < y < 5.$$

The graph consists of the 2 dashed horizontal lines $y = 1$ and $y = 5$ with shading in between.

41. $|x + 2| > y$, $x + y \geq 4$

For $|x + 2| > y$, graph the dashed lines $x + 2 = y$ and $x + 2 = -y$. $y = x + 2$ contains the points (0, 2) and (-2, 0). $-y = x + 2$ contains (0, -2) and (-2, 0). Test a point in each of the four regions. For example, clockwise from the top try (0, 3), (2, 0), (0, -4), (-3, 0) in $|x + 2| > y$. We find that we need to shade the second and fourth regions. That is, the right and left sides. Next, graph $x + y = 4$ as a solid line and test a point on one side in $x + y \geq 4$ to determine which side is shaded. (It turns out to be the right side.) The overlap of the shaded regions is the intersection.

45. $|x - 3| \leq 4$, $|y - 2| \leq 2$

$|x - 3| \leq 4$ can be written as $-4 \leq x - 3 \leq 4$ or $-1 \leq x \leq 7$. The graph consists of 2 solid lines $x = -1$ and $x = 7$ with shading in between.

$|y - 2| \leq 2$ is the same as $-2 \leq y - 2 \leq 2$ or $0 \leq y = 4$. The graph has 2 solid lines at $y = 0$ and $y = 4$ with shading between. The union of the two inequalities includes all the shaded points.

49. ( , -2)

$$6x + 3y = 18$$
$$6x + 3(-2) = 18 \quad \textit{Let } y = -2$$
$$6x - 6 = 18$$
$$6x = 24$$
$$x = 4$$

The ordered pair is (4, -2).

53.

$$4x + 5y = 8$$
$$4x + 5 \cdot 0 = 8 \quad \textit{Let } y = 0$$
$$4x = 8$$
$$x = 2$$
$$4 \cdot 0 + 5y = 8 \quad \textit{Let } x = 0$$
$$5y = 8$$
$$y = \frac{8}{5}$$

**Section 7.5 (page 321)**

1.

$$x = kr$$
$$9 = k(4) \quad \textit{Let } x = 9 \textit{ and } r = 4$$
$$k = \frac{9}{4}$$

So $x = \frac{9}{4}r$. When $r = 10$,

$$x = \frac{9}{4}(10)$$
$$= \frac{45}{2}.$$

5. $z = \frac{k}{w}$

$5 = \frac{k}{1/2} = 2k$ *Let x = 5 and $w = \frac{1}{2}$*

$\frac{5}{2} = k$

So $z = \frac{5}{2w}$. When w = 8,

$z = \frac{5}{2(8)}$

$= \frac{5}{16}$.

9. $r = \frac{kst^2}{q}$

Let r = 50.1291, when s = 10.2483, t = 2.4067, q = 5.3252.

$50.1291 = \frac{k(10.2483)(4.5014)^2}{32.2517}$

k = 7.78566 (rounded)

So $r = \frac{7.78566st^2}{q}$ .

Let x = 8.6321, t = 2.4067, and q = 5.3252.

$r = \frac{(7.78566)(8.6321)(2.4067)^2}{5.3252}$

= 73.1004 (rounded)

13. $p = kd$

$80 = k(30)$ *Let d = 30, p = 80*

$\frac{8}{3} = k$

So $p = \frac{8}{3}d$. When d = 50,

$p = \frac{8}{3}(50)$

$= \frac{400}{3}$.

p = 133.33

The pressure is 133.33 newtons per square centimeter.

17. $d = k\sqrt{h}$

$18 = k\sqrt{144}$ *Let h = 144, d = 18*

$= 12k$

$k = \frac{3}{2}$

So $d = \frac{3}{2}\sqrt{h}$. When h = 1600,

$d = \frac{3}{2}\sqrt{1600}$

$= 60$

The distance is 60 kilometers.

21. $F = kAv^2$

$50 = k(\frac{1}{2})(40)^2$ *Let F = 50, A = $\frac{1}{2}$, v = 40*

$50 = k(\frac{1}{2})(1600) = 800k$

$k = \frac{50}{800} = \frac{5}{80}$

So $F = \frac{5}{80}Av^2$.

If A = 2 and v = 80, then

$F = \frac{5}{80}(2)(80)^2 = \frac{5}{80}(2)(6400) = 800$.

The force is 800 pounds.

25. $L = \frac{kwh^2}{l}$

$360 = \frac{k(.1)(.06)^2}{6}$ *Let L = 360, l = 6, w = .1, h = .06*

$360 = .00006k$

$6{,}000{,}000 = k$

$L = \frac{6{,}000{,}000(.2)(.08)^2}{16}$

*Let l = 16, w = .2, and h = .08*

$= \frac{7680}{16}$

$= 480$

The maximum load is 480 kilograms.

29. $-5(2m + 3) - m = 10$

$-10m - 15 - m = 10$

$-11m = 25$

$m = -\frac{25}{11}$

Solution set: $\{-\frac{25}{11}\}$

33. $-4(\frac{t - 3}{3}) + 2t = 8$

$\frac{-4t + 12}{3} + 2t = 8$

$\frac{-4t + 12}{3}(3) + 2t(3) = 8(3)$

$-4t + 12 + 6t = 24$

$2t = 12$

$t = 6$

Solution set: $\{6\}$

**Chapter 7 Review Exercises (page 325)**

1. $3x + 2y = 10$;

(0, ), (0, ), (2, ), ( , -2)

$3(0) + 2y = 10$ *Let $x = 0$*

$2y = 10$

$y = 5$

(0, 5) is the point.

$3x + 2(0) = 10$ *Let $y = 0$*

$3x = 10$

$x = \frac{10}{3}$

$(\frac{10}{3}, 0)$ is the point.

(2, )

$3 \cdot 2 + 2y = 10$ *Let $x = 2$*

$6 + 2y = 10$

$2y = 4$

$y = 2$

(2, 2) is the point.

( , -2)

$3x + 2 \cdot -2 = 10$ *Let $y = -2$*

$3x - 4 = 10$

$3x = 14$

$x = \frac{14}{3}$

$(\frac{14}{3}), -2)$ is the point.

See answer graph in textbook.

5. $2x + 5y = 20$

$2 \cdot 0 + 5y = 20$ *Let $x = 0$*

$y = 4$ *y-intercept*

$2x + 5 \cdot 0 = 20$ *Let $y = 0$*

$x = 10$ *x-intercept*

Draw a line through (0, 4) and (10, 0).

See answer graph in textbook.

9. Let $(x_1, y_1) = (m, n)$ and $(x_2, y_2) = (m + 1, n - 3)$ in the distance formula.

$d = \sqrt{(x_2 - x_1)^2 + (y_2 - y_1)^2}$

$= \sqrt{(m + 1 - m)^2 + (n - 3 - n)^2}$

$= \sqrt{1^2 + (-3)^2}$

$= \sqrt{10}$

13. If the three points are on the same line, a + b = c, where a, b, c are the distances between the points. From (1, 5) to (-3, -3) is

$a = \sqrt{(1 - (-3))^2 + (5 - (-3))^2}$

$= \sqrt{4^2 + 8^2}$

$= \sqrt{80}$.

From (-3, -3) to (4, 6) is

$c = \sqrt{(-3 - 4)^2 + (-3 - 6)^2}$

$= \sqrt{(-7)^2 + (-9)^2}$

$= \sqrt{130}$.

From (4, 6) to (1, 5) is

$b = \sqrt{(4 - 1)^2 + (6 - 5)^2}$

$= \sqrt{3^2 + 1^2}$

$= \sqrt{10}$.

Since $\sqrt{10} + \sqrt{80} \neq \sqrt{130}$, the points do not lie on the same line.

17. $3x - 4y = 5$

$$-4y = -3x + 5$$

$$y = \frac{3}{4}x - \frac{5}{4} \quad \textit{Solve for } y$$

The slope is $\frac{3}{4}$.

21. Through $(-1, 5)$ and $(-1, -4)$

$$\text{slope} = \frac{5 - (-4)}{-1 - (-1)} = \frac{9}{0} \quad \textit{Undefined}$$

This is a vertical line.

25. Slope $\frac{3}{5}$, y-intercept 2

$m = \frac{3}{5}$, $b = 2$. So, $y = \frac{3}{5}x + 2$

$$\frac{-3}{5}x + y = 2$$

$$3x - 5y = -10.$$

29. Slope 3, through $(-1, 4)$

Let $m = 3$, $x_1 = -1$, and $y_1 = 4$ in

$$y - y_1 = m(x - x_1).$$

$$y - 4 = 3(x - (-1))$$

$$y - 4 = 3(x + 1)$$

$$y - 4 = 3x + 3$$

$$-3x + y = 7$$

or $3x - y = -7$

33. Parallel to $4x - y = 3$ and through $(7, -1)$

$4x - 3 = y$ has slope 4 as do lines parallel to it. The line with slope 4 through $(7, -1)$ is

$$y - (-1) = 4(x - 7)$$

$$y + 1 = 4x - 28$$

$$-4x + y = -29$$

or $4x - y = 29$.

37. $y \leq 2$

Graph $y = 2$ as a solid horizontal line and shade below it to graph $y \leq 2$.

See answer graph in textbook.

41. $2x + y \leq 1$ and $x > 2y$

Graph $2x + y = 1$ as a solid line containing $(\frac{1}{2}, 0)$ and $(0, 1)$ and shade the side containing $(0, 0)$ since $2 \cdot 0 + 0 \leq 1$ means $0 \leq 1$ is true.

Next, graph $x = 2y$ as a dotted line containing $(0, 0)$ and $(2, 1)$ and shade the side containing $(2, 0)$ since $2 > 2(0)$ or $2 > 0$ is true.

The region where the graphs overlap is the intersection.

See answer graph in textbook.

45. $$C = kh^2 l$$

$$1.5 = k(100)^2(600)$$

$$1.5 = 6{,}000{,}000k$$

$$2.5 \times 10^{-7} = k$$

$$C = 2.5 \times 10^{-7}(80)^2(500)$$

$$= .8$$

It would cost .8 million dollars.

**Chapter 7 Test (page 326)**

1. $\text{Slope} = m = \frac{4 - 2}{6 - (-1)} = \frac{2}{7}$

2. $3x - 2y = 20$

To find the x-intercept, let

$$y = 0, \text{ so}$$

$$3x = 20$$

$$x = \frac{20}{3}.$$

To find the y-intercept, let

$$x = 0, \text{ so}$$

$$-2y = 20$$

$$y = -10.$$

Substitute the points $(\frac{20}{3}, 0)$ and $(0, -10)$ into the slope formula:

$$m = \frac{y_2 - y_1}{x_2 - x_1} = \frac{0 - (-10)}{20/3 - 0} = \frac{10}{20/3}$$

$$= 10 \cdot \frac{3}{20} = \frac{3}{2}.$$

3. $x - 5y = 8$

$x - 5(0) = 8$ *Let y = 0*

$x = 8$ *x-intercept*

$0 - 5y = 8$ *Let x = 0*

$y = -\frac{8}{5}$ *y-intercept*

Use the points (8, 0) and (0, -8/5) in the slope formula.

$$m = \frac{-8/5 - 0}{0 - 8} = \frac{-8/5}{-8} = \frac{1}{5}$$

4. The graph of $y = 3$ is a horizontal line through (0, 3). Its y-intercept is 3. It has no x-intercept. The slope of a horizontal line is 0.

5. The graph of $x = -2$ is a vertical line through (-2, 0). Its x-intercept is -2. It has no y-intercept. The slope of a vertical line is undefined.

6. Use $d = \sqrt{(x_2 - x_1)^2 + (y_2 - y_1)^2}$.

Distance from (-2, 3) to (4, 1) is

$$\sqrt{(4 - (-2))^2 + (1 - 3)^2}$$
$$= \sqrt{6^2 + (-2)^2}$$
$$= \sqrt{40}$$

Distance from (4, 1) to (2, -2) is

$$\sqrt{(2 - 4)^2 + (-2 - 1)^2}$$
$$= \sqrt{(-2)^2 + (-3)^2}$$
$$= \sqrt{13}$$

Distance from (2, -2) to (-2, 3) is

$$\sqrt{(-2 - 2)^2 + (3 - (-2))^2}$$
$$= \sqrt{(-4)^2 + (5)^2}$$
$$= \sqrt{41}$$

The perimeter of the triangle is

$\sqrt{13} + \sqrt{40} + \sqrt{41} = \sqrt{13} + 2\sqrt{10} + \sqrt{41}$.

7. Slope through (-6, -6) and (3, -1) is

$$m = \frac{-6 - (-1)}{-6 - 3} = \frac{-5}{-9} = \frac{5}{9}.$$

Slope through (3, -1) and (9, 2) is

$$m = \frac{-1 - 2}{3 - 9} = \frac{-3}{-6} = \frac{1}{2}.$$

Since the slopes are not equal, the points do not lie on the same line.

8. $5x - y = 7$ and $5y = -x + 3$

Find the x- and y-intercepts of $5x - y = 7$.

If $x = 0$, $y = -7$.

If $y = 0$, $x = 3$.

The points (0, -7) and $(\frac{7}{5}, 0)$ are on the line.

$$\text{slope} = \frac{-7}{-7/5} = 7 \cdot \frac{5}{7} = 5$$

Find the x- and y-intercepts of $5y = -x + 3$.

If $x = 0$, $y = \frac{3}{5}$

If $y = 0$, $x = 3$

The points $(0, \frac{3}{5})$ and (3, 0) are on the line.

$$\text{slope} = \frac{3/5}{-3} = \frac{3}{5} \cdot -\frac{1}{3} = -\frac{1}{5}$$

When we multiply the two slopes, we have $5(-\frac{1}{5}) = -1$, so the lines are perpendicular.

9. $2y = 3x + 7$ and $3y = 2x + 9$

Find the x- and y-intercepts of $2y = 3x + 7$.

If $x = 0$, $y = \frac{7}{2}$

If $y = 0$, $x = -\frac{7}{3}$

The points $(0, \frac{7}{2})$ and $(-\frac{7}{3}, 0)$ are on the line.

$$\text{slope} = \frac{7/2}{7/3} = \frac{7}{2} \cdot \frac{3}{7} = \frac{3}{2}$$

Find the x- and y-intercepts of
$3y = 2x + 9$

If $x = 0$, $y = 3$

If $y = 0$, $x = -\frac{9}{2}$

The points (0, 3) and $(-\frac{9}{2}, 0)$ are on the line.

$$\text{slope} = \frac{3}{9/2} = 3\ \frac{2}{9} = \frac{2}{3}$$

The lines are not parallel $(\frac{3}{2} \neq \frac{2}{3})$ and they are not perpendicular:

$$(\frac{3}{2})(\frac{2}{3}) \neq -1.$$

10. $m = -3$, through (4, -1)

Let $m = -3$, $x_1 = 4$ and $y_1 = -1$.

$$y - y_1 = m(x - x_1)$$
$$y - (-1) = -3(x - 4)$$
$$y + 1 = -3x + 12$$
$$y + 3x = 11$$

11. $m = -\frac{5}{8}$, through (-3, 7)

Let $m = -\frac{5}{8}$, $x_1 = -3$, $y_1 = 7$.

$$y - y_1 = m(x - x_1)$$
$$y - 7 = -\frac{5}{8}(x - (-3))$$
$$y - 7 = -\frac{5}{8}x - \frac{15}{8}$$
$$\frac{5}{8}x + y = 7 - \frac{15}{8} = \frac{56}{8} - \frac{15}{8}$$
$$\frac{5}{8}x + y = \frac{41}{8} \quad \text{of} \quad 5x + 8y = 41$$

12. Find slope through (2, 4) and (-2, 6).

$$m = \frac{4 - 6}{2 - (-2)} = -\frac{2}{4} = -\frac{1}{2}$$

Use $m = -\frac{1}{2}$, $x_1 = -2$ and $y_1 = 6$ in

$$y - y_1 = m(x - x_1)$$
$$y - 6 = -\frac{1}{2}(x - (-2))$$
$$y - 6 = -\frac{1}{2}x - 1$$
$$\frac{1}{2}x + y = 5 \quad \text{or} \quad x + 2y = 10$$

13. Horizontal, through (-3, 4)

A horizontal line has the same y-value at every point on the line. The point (-3, 4) has a y-coordinate of 4. Since (-3, 4) is on the line, the equation of the line must be $y = 4$.

14. Perpendicular to $3x + 5y = 12$, through (-7, 2)

$$3x + 5y = 12$$
$$5y = -3x + 12$$
$$y = \frac{-3}{5}x + \frac{12}{5}$$

This line has slope $-\frac{3}{5}$, so a line perpendicular to is has slope $\frac{5}{3}$ since $\frac{5}{3}\cdot\frac{-3}{5} = -1$. The line with slope $\frac{5}{3}$ through (-7, 2) is found by substituting $m = \frac{5}{3}$, $x_1 = -7$, $y = 2$ into $y - y_1 = m(x - x_1)$. Thus,

$$y - 2 = \frac{5}{3}(x - (-7))$$
$$y - 2 = \frac{5}{3}x + \frac{35}{3}$$
$$\frac{-5}{3}x + y = 2 + \frac{35}{3}$$
$$\frac{-5}{3}x + y = \frac{41}{3}$$
$$-5x + 3y = 41.$$

Exercises 15-22, see answer graph in textbook.

15. Find two points on $4x + 3y = 16$, say $(4, 0)$ and $(1, 4)$, and draw the line through them.

16. $y + 3 = 0$ can be written as $y = -3$. The line $y = -3$ is a horizontal line through the point $(0, -3)$.

17. The line through $(-1, 3)$, with slope $\frac{-2}{5}$
Start at $(-1, 3)$, go 2 units down and 5 to the right to $(4, 1)$ and draw a line through these points.

18. Lines with undefined slopes are written $x = k$. Thus, the line with undefined slope through $(2, -4)$ is $x = 2$.

19. $3x + 4y \leq 12$
First graph the solid line $3x + 4y = 12$. Draw the line through the points $(4, 0)$ and $(0, 3)$. Test a point on one side of the line, say $(0, 0)$. Use $x = 0$, $y = 0$ in $3x + 4y \leq 12$ to get $0 \leq 12$, a true statement. So shade that side of the line containing $(0, 0)$.

20. Graph $2x = y - 4$ with a dotted line through $(-2, 0)$ and $(0, 4)$. Test a point not on the line, say $(0, 0)$. Use $x = 0$, $y = 0$ in $2x > y - 4$ to get $0 > -4$, a true statement. So shade that side of the line containing $(0, 0)$.

21. $y < 2x - 1$ and $x - y < 3$
First graph $y = 2x - 1$ with a dotted line through $(2, 3)$ and $(0, -1)$. Test a point not on the line say $(0, 0)$ and use $x = 0$, $y = 0$ in $y < 2x - 1$ to get $0 < -1$, a false statement. So shade the side of the line not containing the origin.
Next, graph $x - y = 3$ with a dashed line through $(3, 0)$ and $(0, -3)$. Test with $x = 0$, $y = 0$ in $x - y < 3$ to get $0 < 3$, a true statement. Shade the side of the line containing the origin. The intersection is the overlap of the shaded regions.

22. $|x| \leq 2$ and $|y| \leq 2$
$|x| \leq 2$ means $-2 \leq x \leq 2$, so graph $x = 2$ and $x = -2$ with solid lines and shade between them.
$|y| \leq 2$ means $-2 \leq y \leq 2$, so graph $y = 2$ and $y = -2$ with solid lines and shade between them. The union is the combination of all the shaded regions.

23. $$y = k\frac{x}{z}$$
$$\frac{3}{2} = k\frac{3}{10} \quad \textit{Let } y = \tfrac{3}{2},\ x = 3, \textit{ and } z = 10$$
$$k = 5$$
$$y = 5\frac{x}{z}$$
When $x = 4$, $z = 12$
$$y = 5\left(\frac{4}{12}\right)$$
$$= \frac{5}{3}.$$

24. $i = kpt$

$360 = k(3000)(2)$ *Let $i = 360$, $p = 3000$, $t = 2$*

$.06 = k$

When $p = 6000$, $t = 5$

$i = (.06)(6000)(5) = 1800$

The interest is \$1800.

25. $I = \frac{k}{R}$ where I is the current and R the resistance.

$80 = \frac{k}{30}$ *Let $I = 80$ and $R = 30$*

$k = 2400$

$I = \frac{2400}{R}$

When $R = 12$,

$I = 200$

The current is 200 amps.

## CHAPTER 8 SYSTEMS OF LINEAR EQUATIONS

### Section 8.1 (page 334)

1. $x + y = 5$; (4, 1)
$x - y = 3$

Substitute 4 for x and 1 for y in each equation.

$4 + 1 = 5$ *True*
$4 - 1 = 3$ *True*

Yes, (4, 1) is a solution of the system.

5. $8x - 3y = 32$; (4, 0)
$4x + 5y = 16$

Substitute 4 for x and 0 for y in each equation.

$8(4) - 3(0) = 32$
$32 - 0 = 32$ *True*

$4(4) + 5(0) = 16$
$16 - 0 = 16$ *True*

Yes, (4, 0) is a solution of the system.

9. $3x + y = 12$
$x + 2y = 4$

Solution set: {(4, 0)}
See answer graph in textbook.

13. $2x - 3y = 3$ *(1)*
$2x + 2y = 8$ *(2)*

$2x - 3y = 3$ *Multiply equation (1) by -1*
$-2x - 2y = -8$ *(2)*
$-5y = -5$ *Add (1) and (2)*
$y = 1$ *Solve for y*

Substitute y = 1 in (1).

$2x - 3(1) = 3$
$2x = 6$
$x = 3$ *Solve for x*

Solution set: {(3, 1)}

17. $7x + 2y = 3$ *(1)*
$-14x - 4y = -6$ *(2)*

$14x + 4y = 6$ *Multiply (1) by 2*
$-14x - 4y = -6$ *(2)*
$0 = 0$ *True*

This true statement means that the equations are equivalent. The solution set is the infinite set of points on the line $7x + 2y = 3$.
Solution set: $\{(x, y) \mid 7x + 2y = 3\}$

21. $5x + 3y = 1$ *(1)*
$-3x - 4y = 6$ *(2)*

$20x + 12y = 6$ *Multiply (1) by 4*
$-9x - 12y = 18$ *Multiply (2) by 3*
$11x = 22$ *Add the equations*
$x = 2$ *Solve for x*

Substitute x = 2 in (1).

$5(2) + 3y = 1$
$3y = -9$
$y = -3$ *Solve for y*

Solution set: {(2, -3)}

25. $.5041x + 1.4932y = 3.19377$ *(1)*
$.2509x - .7532y = -.95427$ *(2)*

Multiply (1) by .2509.
Multiply (2) by -.5041.

$.12647869x + .37464388y = .801315893$
$-.12647869x + .37968812y = .481047507$
$.754332y = 1.2823644$ *Add*
$$\frac{.754332y}{.754332} = \frac{1.2823644}{.754332}$$
$y = 1.700$ *Solve for y*

Substitute y = 1.7 in (1).
$.5041x + 2.53844 = 3.19377$
$.5041x = .65533$
$x = 1.3$ *Solve for x*

Solution set: {(1.3, 1.7)}

29. $\frac{1}{4}x + \frac{5}{4}y = -\frac{1}{2}$ *(1)*

$-x - 5y = 2$ *(2)*

$x + 5y = -2$ *Multiply (1) by 4*

$-x - 5y = 2$ *(2)*

$0 = 0$ *True*

The equations are equivalent. The solution set is the infinite set of points on the line $-x - 5y = 2$.
Solution set: $\{(x, y)\ -x - 5x = 2\}$

33. $2x = y + 6$ *(1)*

$y = 5x$ *(2)*

$2x = 5x + 6$ *Substitute 5x for y in (1)*

$-3x = 6$

$x = -2$ *Solve for x*

$y = 5(-2)$ *Substitute x = -2 in (2), solve for y*

$y = -10$

Solution set: $\{(-2, -10)\}$

37. $5x - 4y = 9$ *(1)*

$3 + x = 2y$ *(2)*

$x = 2y - 3$

$5(2y - 3) - 4y = 9$ *Substitute 2y - 3 for x in (1)*

$10y - 15 - 4y = 9$

$6y = 24$

$y = 4$ *Solve for y*

Substitute $y = 4$ into the bottom equation, solve for x.

$x = 2(4) - 3$

$= 8 - 3$

$= 5$

Solution set: $\{(5, 4)\}$

41. $\frac{x}{2} + \frac{y}{3} = 3$ *(1)*

$y = 3x$ *(2)*

$\frac{x}{2} + \frac{3x}{3} = 3$ *Substitute y = 3x in (1)*

$\frac{x}{2} + x = 3$

$\frac{3x}{2} = 3$

$3x = 6$

$x = 2$ *Solve for x*

Substitute $x = 2$ in (2) and solve.

$y = 3(2) = 6$

Solution set: $\{(2, 6)\}$

45. $11.605x + 16.247y = 2.321$ *(1)*

$9.248x = 4.642y + 37.136$ *(2)*

$x = .5y + 4$ *Divide (2) by 9.284*

$11.605(.5y + 4) + 16.247y = 2.321$ *Substitute x = .5y + 4 in (1)*

$5.8025y + 46.42 + 16.247y = 2.321$

$22.0495y + 46.42 = 2.321$

$22.0495y = -44.099$

$y = -2$

Substitute $y = -2$ in (2).

$x = .5(-2) + 4$

$x = 3$

Solution set: $\{(3, -2)\}$

49. $\frac{2}{x} - \frac{5}{y} = \frac{3}{2}$

$\frac{4}{x} + \frac{1}{y} = \frac{4}{5}$

Let $p = \frac{1}{x}$, $q = \frac{1}{y}$.

$2p - 5q = \frac{3}{2}$ *(1)*

$4p + q = \frac{4}{5}$ *(2)*

$4p - 10q = 3$ *(3) Multiply (1) by 2*

$20p + 5q = 4$ *(4) Multiply (2) by 5*

$$\begin{array}{rcll} 4p - 10q &=& 3 & \\ 40p + 10q &=& 8 & \textit{Multiply (4) by 2} \\ \hline 44p &=& 11 & \textit{Add} \\ p &=& \frac{1}{4} & \end{array}$$

Substitute $p = \frac{1}{4}$ in (2).

$$4(\tfrac{1}{4}) + q = \frac{4}{5}$$
$$1 + q = \frac{4}{5}$$
$$q = -\frac{1}{5}$$

$$x = \frac{1}{p} = \frac{1}{\frac{1}{4}} = 4, \quad y = \frac{1}{q} = \frac{1}{-\frac{1}{5}} = -5$$

Solution set: $\{(4, -5)\}$

53. $x + 2y = 4$ *(1)*

$x - \frac{1}{2}y = 4$ *(2)*

$$\begin{array}{rcll} x + 2y &=& 4 & \textit{(1)} \\ 2x - y &=& 8 & \textit{(3) Multiply (2) by 2} \end{array}$$

$$\begin{array}{rcll} x + 2y &=& 4 & \\ 4x - 2y &=& 16 & \textit{Multiply (3) by 2} \\ \hline 5x &=& 20 & \textit{Add the equations} \\ x &=& 4 & \end{array}$$

Substitute $x = 4$ in (1).

$$4 + 2y = 4$$
$$2y = 0$$
$$y = 0$$

Solution set: $\{(4, 0)\}$

57. $3ax + 2y = 1$ *(1)*

$-ax + y = 2$ *(2)*

$$\begin{array}{rcll} 3ax + 2y &=& 1 & \textit{(1)} \\ -3ax + 3y &=& 6 & \textit{Multiply (2) by 3} \\ \hline 5y &=& 7 & \textit{Add the equations} \\ y &=& \frac{7}{5} & \end{array}$$

Substitute $y = \frac{7}{5}$ in (2).

$$-ax + \frac{7}{5} = 2$$
$$-ax = \frac{3}{5}$$
$$x = \frac{-3}{5a}$$

Solution set: $\{(\frac{-3}{5a}, \frac{7}{5})\}$

61. (a) The point where the graphs cross is (8, 3000). At x = 8 when 800 items are produced, the revenue is $3000.

(b) When x = 4, 400 parts are sold. At x = 4, the graph shows the cost is $2000 and the revenue is $1500. The profit is $1500 - $2000 = -$500, or a loss of $500.

(c) When x = 0, the cost is $1000, so the fixed cost is $1000.

65. Let L be the length. Then L - 4 is the width. Use P = 2L + 2W.

$$2L + 2(L - 4) = 64$$
$$2L + 2L - 8 = 64$$
$$4L = 72$$
$$L = 18$$

The length is 18 units.

**Section 8.2 (page 342)**

1. Let x and y be the two numbers. Then, x + y = 62 and

$$x - y = 16$$

$$\begin{array}{rcll} x + y &=& 62 & \\ x - y &=& 16 & \\ \hline 2x &=& 78 & \textit{Add} \\ &=& 39 & \end{array}$$

Since $x - y = 16$,

$$\begin{aligned} 39 - y &= 16 \\ -y &= -23 \\ y &= 23. \end{aligned}$$

The numbers are 23 and 39.

5. Let the side of the square be x and let the side of the equilateral triangle be y. Since the side of the square is 2 centimeters more than the side of the triangle, then

$$x = y + 2.$$

The perimeter of a square is $4\cdot$(side) or 4x. The perimeter of an equilateral triangle is $3\cdot$(side) or 3y. Since the perimeter of the square is twice the perimeter of the triangle then

$$4x = 2(3y)$$

or

$$4x = 6y.$$

Solve the system:

$$\begin{aligned} x &= y + 2 \\ 4x &= 6y. \end{aligned}$$

Substitute y + 2 for x in 4x = 6y to get:

$$\begin{aligned} 4(y + 2) &= 6y \\ 4y + 8 &= 6y \\ 8 &= 2y \\ 4 &= y. \end{aligned}$$

Now, substitute y = 4 into x = y + 2 to get:

$$x = 4 + 2 = 6.$$

The side of the square is 6 centimeters; the side of the triangle is 4 centimeters.

9. Let x = the number of pounds of peanuts and y = the number of pounds of cashews.

$$\begin{aligned} 5x + 6y &= 70 \quad \textit{(1)} \\ 3x + 7y &= 76 \quad \textit{(2)} \end{aligned}$$

Use elimination to solve.

$$\begin{aligned} -15x - 18y &= -210 \quad \textit{Multiply (1) by (-3)} \\ 15x + 35y &= 380 \quad \textit{Multiply (2) by (5)} \\ \hline 17y &= 170 \quad \textit{Add} \\ y &= 10 \end{aligned}$$

Substitute 10 for y in (1).

$$\begin{aligned} 5x + 6(10) &= 70 \\ 5x &= 10 \\ x &= 2 \end{aligned}$$

The cost of peanuts is \$2; cashews cost \$10 per pound.

13. Let x = the micros to be shipped east and y = the micros to be shipped west. The total is 50 micros, so

$$x + y = 50. \quad \textit{(1)}$$

Adding the costs of each group makes the equation

$$25x + 20y = 1090. \quad \textit{(2)}$$

Solve for x in (1).

$$\begin{aligned} x + y &= 50 \\ x &= 50 - y \end{aligned}$$

Substitute (50 - y) for x in (2).

$$\begin{aligned} 25(50 - y) + 20y &= 1090 \\ 1250 - 25y + 20y &= 1090 \\ -5y &= -160 \\ y &= 32 \end{aligned}$$

$$x = 50 - 32 = 18$$

The number of micros to be shipped west is 32; the number to go east is 18.

17. Let x be the amount of 30% alcohol and y be the amount of 80% alcohol. Complete the table in the textbook by finding the amount of pure alcohol in the 80% alcohol, which is .80y. The equations are written by adding first the second column of the table, then the third.

$$x + y = 5 \quad \textit{(1)}$$
$$.30x + .80y = .50(5) = 2.5 \quad \textit{(2)}$$

$$y = 5 - x \quad \textit{Solve (1) for y}$$
$$.30x + .80(5 - x) = 2.5 \quad \textit{Substitute in (2)}$$
$$.30x + 4 - .80x = 2.5$$
$$-.50x = -1.5$$
$$x = \frac{-1.5}{-.50} = 3$$
$$3 + y = 5 \quad \textit{Substitute for x in (1)}$$
$$y = 2$$

3 gallons of 30% alcohol should be added to 2 gallons of 80% alcohol.

21. Let b = the speed of the boat and c = the speed of the current.

| | d | r | t |
|---|---|---|---|
| Upstream | 18 | b - c | 1 |
| Downstream | 18 | b + c | 3/4 |

Using d = rt, form the two equations from the table.

$$1(b - c) = 18 \quad \textit{(1)}$$
$$\frac{3}{4}(b + c) = 18 \quad \textit{(2)}$$

$$b - c = 18 \quad \textit{(1)}$$
$$b = 18 + c \quad \textit{Solve (1) for b}$$

$$\frac{3}{4}[(18 + c) + c] = 18 \quad \textit{Substitute in (2)}$$
$$3(18 + 2c) = 72 \quad \textit{Multiply by 4}$$
$$54 + 6c = 72$$
$$6c = 18$$
$$c = 3$$

Substitute 3 for c in (1).

$$b - 3 = 18$$
$$b = 21$$

The speed of the current is 3 miles per hour; the speed of the boat is 21 miles per hour.

25. Let x be the amount of pure salt and y be the amount of 10% salt. A table shows the relationships.

| Kind | Amount | Pure salt |
|---|---|---|
| 100% | x | 1.00x = x |
| 10% | y | .10y |
| 20% | 9 | .20(9) |

The equations are written by adding columns.

$$x + y = 9 \quad \textit{(1)}$$
$$x + .10y = .20(9) = 1.8 \quad \textit{(2)}$$

$$y = 9 - x \quad \textit{Solve (1) for y}$$
$$x + .10(9 - x) = 1.8 \quad \textit{Substitute in (2)}$$
$$x + .9 - .10x = 1.8$$
$$.9x = .9$$
$$x = 1$$

$$1 + y = 9 \quad \textit{Substitute in (1)}$$
$$y = 8$$

1 cubic centimeter of salt should be added to 8 cubic centimeters of 10% salt.

29\. Let $x$ be the speed of the slow car and $y$ the speed of the fast car. The fast car travels 15 kilometers per hour faster than the slow car, so

$$y = x + 15.$$

Since $d = rt$, the distance traveled by the fast car after 3 hours is $3y$ and the distance traveled by the slow car after 3 hours is $3x$. So we have

$3x + 3y = 345.$ *(1)*

$y = x + 15$ *(2)*

$3x + 2y = 345$

$3x + 3(x + 15) = 345$ *Substitute for y in (2)*

$6x + 45 = 345$

$6x = 300$

$x = 50$

$y = 50 + 15$ *Substitute for x in (1)*

$= 65$

Answer: 50 kilometers per hour and 65 kilometers per hour.

33\. $-x + 3y - 2z = 18$

$2x - 6y + 4z = -36$ *Multiply by -2*

**Section 8.3 (page 351)**

1\. $2x + y + z = 3$ *(1)*

$3x - y + z = -2$ *(2)*

$4x - y + 2z = 0$ *(3)*

$2x + y + z = 3$ *(1)*

$3x - y + z = -2$ *(2)*

$5x + 2z = 1$ *(5) Add*

$2x + y + z = 3$

$4x - y + 2z = 0$

$6x + 3z = 3$ *Add*

$2x + z = 1$ *(6)*

$5x + 2z = 1$ *(5)*

$2x + z = 1$ *(6)*

$5x + 2z = 1$ *(5)*

$-4x - 2z = -2$ *Multiply equation (6) by -2*

$x = -1$ *Add*

Substitute $x = -1$ in (5).

$5(-1) + 2z = 1$

$-5 + 2z = 1$

$2z = 6$

$z = 3$

Substitute $x = -1$, $z = 3$ in (1).

$-2 + y + 3 = 3$

$y + 1 = 3$

$y = 2$

Solution set: $\{(-1, 2, 3)\}$

5\. $2x + 5y + 2z = 9$ *(1)*

$4x - 7y - 3z = 7$ *(2)*

$3x - 8y - 2z = 9$ *(3)*

$2x + 5y + 2z = 9$

$3x - 8y - 2z = 9$

$5x - 3y = 18$ *(4) Add*

$3x + \frac{15}{2}y + 3z = \frac{27}{2}$ *Multiply equation (1) by 3/2*

$4x - 7y - 3z = 7$

$7x + \frac{1}{2}y = \frac{41}{2}$ *Add*

$14x + y = 41$

So, $5x - 3y = 18$

$42x + 3y = 123$ *Multiply (5) by 3*

$47x = 141$ *Add*

$x = 3$

Substitute $x = 3$ in (5).

$14(3) + y = 41$

$42 + y = 41$

$y = -1$

Substitute x = 3, y = -1 in (1).

$$2(3) + 5(-1) + 2z = 9$$
$$6 - 5 + 2z = 9$$
$$1 + 2z = 9$$
$$2z = 8$$
$$z = 4$$

Solution set: {(3, -1, 4)}

9. $2x + 3y - z = 1$ *(1)*
$x + 2y + 2z = 5$ *(2)*
$x - y + z = 6$ *(3)*

$2x + 3y - z = 1$ *(1)*
$-2x - 4y - 4z = -10$ *Multiply (2) by -2*

$-y - 5z = -9$ *(4) Add*

$2x + 3y - z = 1$ *(1)*
$-2x + 2y - 2z = -12$ *Multiply (3) by -2*

$5y - 3z = -11$ *(5) Add*

$-5y - 25z = -45$ *Multiply (4) by 5*
$5y - 3z = -11$ *(5)*
$-28z = -56$ *Add*
$z = 2$ *Solve for z*

Substitute z = 2 in (4).

$$-y - 5(2) = -9$$
$$-y = 1$$
$y = -1$ *Solve for y*

Substitute z = 2 and y = -1 in (3); solve for x.

$$x - (-1) + 2 = 6$$
$$x = 3$$

The solution is {(3, -1, 2)}

13. $4x + 2y - 3z = 6$ *(1)*
$x - 4y + z = -4$ *(2)*
$-x + 2z = 2$ *(3)*

$x - 4y + z = -4$
$-x + 2z = 2$
$-4y + 3z = -2$ *(4) Add*

$-4x + 8z = 8$ *Multiply (3) by 4*
$4x + 2y - 3z = 6$
$2y + 5z = 14$ *(5) Add*

$4x + 10z = 28$ *Multiply (5) by 2*
$-4y + 3z = -2$
$13z = 26$
$z = 2$

Substitute z = 2 in (5).

$$2y + 5(2) = 14$$
$$2y + 10 = 14$$
$$2y = 4$$
$$y = 2$$

Substitute z = 2 in (3).

$$-x + 2(2) = 2$$
$$-x + 4 = 2$$
$$-x = -2$$
$$x = 2$$

Solution set: {(2, 2, 2)}

17. $-1.46x + .87y = 9.01$ *(1)*
$1.04y - .92z = 11.88$ *(2)*
$1.24x + .87z = -6.83$ *(3)*

Multiply (1) by 1.24.
Multiply (3) by 1.46.

$-1.8104x + 1.0788y = 11.1724$
$1.8104x + 1.2702z = -9.9718$
$1.0788y + 1.2702z = 1.2006$ *(4)*

$1.0788y + 1.2702z = 1.2006$ *(4)*
$1.04y - .92z = 11.88$ *(2)*

Multiply (4) by 1.04.
Multiply (2) by -1.0788.

$1.121952y + 1.321008z = 1.248624$
$-1.121952y + .992496z = -12.816144$
$2.313504z = -11.56752$
$z = -5$

Substitute z = -5 in (2).

$$1.04y - .92(-5) = 11.88$$
$$1.04y + 4.6 = 11.88$$
$$1.04y = 7.28$$
$$y = 7$$

Substitute y = 7 in (1).

$$-1.46x + .87(7) = 9.01$$
$$-1.46x + 6.09 = 9.01$$
$$-1.46x = 2.92$$
$$x = -2$$

Solution set: {(-2, 7, -5)}

21. $x - 2y = 0$ *(1)*
$3y + z = -1$ *(2)*
$4x - z = 11$ *(3)*

$3y + z = -1$ *(2)*
$4x - z = 11$ *(3)*
$4x + 3y = 10$ *(4)*

$-4x + 8y = 0$ *Multiply (1) by -4*
$4x + 3y = 10$ *(4)*
$11y = 10$
$y = \frac{10}{11}$ *Solve for y*

Substitute $y = \frac{10}{11}$ in (2).

$$3(\tfrac{10}{11}) + z = -1$$
$$z = -1 - \frac{30}{11}$$
$$= -\frac{11}{11} - \frac{30}{11}$$
$$= \frac{-41}{11}$$ *Solve for z*

Substitute $y = \frac{10}{11}$ in (1).

$$x - 2(\tfrac{10}{11}) = 0$$
$$x = \frac{20}{11}$$ *Solve for x*

Solution set: $\{(\frac{20}{11}, \frac{10}{11}, \frac{-41}{11})\}$

25. $x - 2y + 4z = -10$ *(1)*
$-3x + 6y - 12z = 20$ *(2)*
$2x + 5y + z = 12$ *(3)*

$3x - 6y + 12z = -30$ *Multiply (1) by 3*
$-3x + 6y - 12z = 20$
$0 = -10$ *False*

Therefore, the system is inconsistent.

Solution set: ∅

29. $2x - 8y + 2z = -10$ *(1)*
$-x + 4y - z = 5$ *(2)*
$3x - 12y + 3z = -15$ *(3)*

$x - 4y + z = -5$ *Multiply (1) by 1/2*
$x - 4y + z = -5$ *Multiply (2) by -1*
$x - 4y + z = -5$ *Multiply (3) by 1/3*

All equations are equivalent.

Solution set:

$\{(x, y, z) | x - 4y + z = -5\}$

33. Let x, y, and z be the three angles. The sum of the angles is 180° and so x + y + z = 180. The second angle is twice as large as the first, so y = 2x. The third angle equals the sum of the other two and so z = x + y.

$x + y + z = 180$ *(1)*
$y = 2x$ *(2)*
$z = x + y$ *(3)*

Substitute (3) in (1).

$$x + y + x + y = 180$$
$$2x + 2y = 180$$
$x + y = 90$ *(4)*

Substitute (2) in (4).

$$x + 2x = 90$$
$$3x = 90$$
$$x = 30$$

Substitute $x = 30$ in (2).

$$y = 2(30) = 60$$

Substitute $x = 30$, $y = 60$ in (3).

$$z = 30 + 60 = 90$$

The three angles are 30°, 60°, and 90°.

37. Let x, y, and z be the number of fives, tens, and twenties, respectively. Since there are 30 bills, $x + y + z = 30$. The value of the bills is $5x + 10y + 20z = 370$, or $x + 2y + 4z = 74$. There are 2 more fives than tens and so $y = x - 2$.

$$\begin{aligned} x + y + z &= 30 && \textit{(1)} \\ x + 2y + 4z &= 74 && \textit{(2)} \\ y &= x - 2 && \textit{(3)} \end{aligned}$$

Substitute (3) in (1).

$$\begin{aligned} x + x - 2 + z &= 30 \\ 2x + z &= 32 && \textit{(4)} \end{aligned}$$

Substitute (3) in (2).

$$\begin{aligned} x + 2(x - 2) + 4z &= 74 \\ x + 2x - 4 + 4z &= 74 \\ 3x + 4z &= 78 && \textit{(5)} \end{aligned}$$

$$\begin{aligned} -8x - 4z &= -128 && \textit{Multiply (4) by -4} \\ 3x + 4z &= 78 \\ \hline -5x &= -50 && \textit{Add} \\ x &= 10 \end{aligned}$$

Substitute $x = 10$ in (4).

$$\begin{aligned} 2(10) + z &= 32 \\ 20 + z &= 32 \\ z &= 12 \end{aligned}$$

Substitute $x = 10$, $z = 12$ in (1).

$$\begin{aligned} 10 + y + 12 &= 30 \\ y + 22 &= 30 \\ y &= 8 \end{aligned}$$

Ken has 10 fives, 8 tens, and 12 twenties.

41. Let C represent the price of "up close" tickets, M represent "in the middle," and F represent "far out." "Up close" cost \$6 more than "in the middle," so

$$C = 6 + M.$$

"In the middle" cost \$3 more than "far out" so

$$M = 3 + F.$$

Twice the cost of an "up close" is \$3 more than 3 times a "far out" or

$$2C = 3 + 3F.$$

Rearrange the terms in the equations above to get the following system.

$$\begin{aligned} C - M \quad\quad &= 6 && \textit{(1)} \\ M - F &= 3 && \textit{(2)} \\ 2C \quad\quad - 3F &= 3 && \textit{(3)} \end{aligned}$$

$$\begin{aligned} C - M \quad &= 6 && \textit{(1)} \\ M - F &= 3 && \textit{(2)} \\ \hline C \quad - F &= 9 && \textit{(4)} \end{aligned}$$

$$\begin{aligned} 2C - 3F &= 3 && \textit{(3)} \\ -2C + 2F &= -18 && \textit{Multiply (4) by -2} \\ \hline -F &= -15 \\ F &= 15 && \textit{Solve for F} \end{aligned}$$

Substitute $F = 15$ in (2).

$$\begin{aligned} M - 15 &= 3 \\ M &= 18 && \textit{Solve for M} \end{aligned}$$

Substitute $M = 18$ in (1).

$$\begin{aligned} C - 18 &= 6 \\ C &= 24 && \textit{Solve for C} \end{aligned}$$

"Up close" cost \$24, "in the middle" cost \$18, and "far out" cost \$15.

45. $x - 2y + 2z + w = 7$ *(1)*

$2x + y + 3z - w = 6$ *(2)*

$-x + y - 2z + 2w = 1$ *(3)*

$x - y - z + w = -4$ *(4)*

$x - 2y + 2z + w = 7$ *(1)*

$-x + y - 2z + 2w = 1$ *(3)*

$-y + 3w = 8$ *(5) Add*

$-x + y - 2z + 2w = 1$ *(3)*

$x - y - z + w = -4$ *(4)*

$-3z + 3w = -3$ *Add*

$-z + w = -1$ *(6)*

$2x + y + 3z - w = 6$ *(2)*

$-2x + 2y - 4z + 4w = 2$ *Multiply (3) by 2*

$3y - z + 3w = 8$ *(7)*

$-y + 3w = 8$ *(5)*

$-z + w = -1$ *(6)*

$3y - z + 3w = 8$ *(7)*

$z - w = 1$ *Multiply (6) by -1*

$3y - z + 3w = 8$ *(7)*

$3y + 2w = 9$ *(8) Add*

$-3y + 9w = 24$ *Multiply (5) by 3*

$3y + 2w = 9$ *(8)*

$11w = 33$

$w = 3$

Substitute w = 3 in (6).

$-z + 3 = -1$

$-z = -4$

$z = 4$

Substitute w = 3 in (5).

$-y + 3(3) = 8$

$-y + 9 = 8$

$-y = -1$

$y = 1$

Substitute y = 1, z = 4, w = 3 in (1).

$x - 2(1) + 2(4) + 3 = 7$

$x - 2 + 8 + 3 = 7$

$x + 9 = 7$

$x = -2$

Solution set: {(-2, 1, 4, 3)}.

49.

| Brand | Nitrogen | Potash | Ammonia |
|---|---|---|---|
| A | .08 | .05 | .10 |
| B | .06 | .10 | .06 |
| C | .10 | .08 | .05 |
| Total | 26.4 | 28 | 26.8 |

$.08A + .06B + .10C = 26.4$ *(1)*

$.05A + .10B + .08C = 28$ *(2)*

$.10A + .06B + .05C = 26.8$ *(3)*

$.08A + .06B + .10C = 26.4$ *(1)*

$-.10A - .06B - .05C = -26.8$ *Multiply (3) by -1*

$-.02A + .05C = -.4$ *(4)*

$.80A + .60B + 1.00C = 264$ *Multiply (1) by 10*

$-.30A - .60B - .48C = -168$ *Multiply (2) by -6*

$.50A + .52C = 96$ *(5)*

$-.50A + 1.25C = -10$ *Multiply (4) by 25*

$.50A + .52C = 96$ *(5)*

$1.77C = 86$

$C = 48.6$

Substitute 48.6 for C in (4).

$-.02A + .05(48.6) = -.4$

$-.02A + 2.43 = -.4$

$-.02A = -2.83$

$A = 141.5$

Substitute the values for A and C in (1).

$$.08(141.5) + .06B + .10(48.6) = 26.4$$
$$11.32 + .06B + 4.86 = 26.4$$
$$.06B = 10.22$$
$$B = 170.3$$

The farmer should use 141.5 pounds of Brand A, 170.3 pounds of B, and 48.6 pounds of C.

53. $3[(-2)(2) - (-3)(1)]$
$= 3[-4 + 3]$
$= 3(-1)$
$= -3$

**Section 8.4 (page 357)**

1. $\begin{vmatrix} 2 & 5 \\ -1 & 4 \end{vmatrix} = (2)(4) - (5)(-1)$
$= 8 + 5$
$= 13$

5. $\begin{vmatrix} -2 & 2 \\ 2 & -2 \end{vmatrix} = (-2)(-2) - (2)(2)$
$= 4 - 4$
$= 0$

9. $\begin{vmatrix} 2.432 & 6.581 \\ -1.235 & 4.209 \end{vmatrix}$
$= (2.432)(4.209) - (6.581)(-1.235)$
$= 18.363823$

13. $\begin{vmatrix} 1 & 0 & 2 \\ 0 & 2 & 3 \\ 1 & 0 & 5 \end{vmatrix}$

$$= 1\begin{vmatrix} 2 & 3 \\ 0 & 5 \end{vmatrix} - 0\begin{vmatrix} 0 & 2 \\ 0 & 5 \end{vmatrix} + 1\begin{vmatrix} 0 & 2 \\ 2 & 3 \end{vmatrix}$$

$= 1[(5)(2) - (0)(3)] - 0 + 1[(0)(3) - (2)(2)]$
$= 10 - 4$
$= 6$

17. $\begin{vmatrix} 2 & 2 & 1 \\ 1 & -1 & -2 \\ 1 & 0 & 2 \end{vmatrix}$

$$= 1\begin{vmatrix} 2 & 1 \\ -1 & -2 \end{vmatrix} - 0\begin{vmatrix} 2 & 1 \\ 1 & -2 \end{vmatrix} + 2\begin{vmatrix} 2 & 2 \\ 1 & -1 \end{vmatrix}$$

Expand about the last row.

$= [2(-2) - (-1)(1)] - 0 + 2[2(-1) - (1)(2)]$
$= -3 - 8$
$= -11$

21. $\begin{vmatrix} 2 & 4 & 0 \\ 3 & -5 & 0 \\ 6 & -7 & 0 \end{vmatrix} = 0$

To understand this result, expand about the third column.
The value of any determinant containing a row or column of all zeros is always 0.

25. $\begin{vmatrix} 3 & 0 & -2 \\ 1 & -4 & 1 \\ 3 & 1 & -2 \end{vmatrix}$

$$= 3\begin{vmatrix} -4 & 1 \\ 1 & -2 \end{vmatrix} - 0\begin{vmatrix} 1 & 1 \\ 3 & -2 \end{vmatrix} - 2\begin{vmatrix} 1 & -4 \\ 3 & 1 \end{vmatrix}$$

*Expand about row 1*

$= 3[(-4)(-2) - (1)(1)] - 0 - 2[(1)(1) - (3)(-4)]$
$= 3[8 - 1] - 2[1 + 12]$
$= 3(7) - 2(13)$
$= -5$

29. $\begin{vmatrix} 3.124 & 0 & -2.531 \\ 1.098 & -4.207 & 1.693 \\ 3.274 & 1.901 & -2.762 \end{vmatrix}$

$$= -0\begin{vmatrix} 1.098 & 1.693 \\ 3.274 & -2.762 \end{vmatrix} -$$

$$4.207\begin{vmatrix} 3.124 & -2.531 \\ 3.274 & -2.762 \end{vmatrix} -$$

$$1.901\begin{vmatrix} 3.124 & -2.531 \\ 1.098 & 1.693 \end{vmatrix}$$ *Expand about column 2*

$= -4.207[(3.124)(-2.762) - (3.274)(-2.531)] - 1.901[(3.124)(1.693) - (1.098)(-2.531)]$

$= -4.207(-8.6285 + 8.2865) - 1.901(5.2889 + 2.7790)$ *Round computations to nearest ten-thousandth*

$= -4.207(-0.3420) - 1.901(8.0679)$

$= 1.4388 - 15.3371$

$= -13.898$ (rounded)

33. $\begin{vmatrix} 1 & 4 & 2 & 0 \\ 0 & 2 & 0 & -1 \\ 3 & -1 & 2 & 0 \\ 1 & 4 & -1 & 2 \end{vmatrix}$

Note that the minors are omitted where the cofactors are 0. Expansion is about row 2 since this row contains 2 zeros.

$$= 0 + 2\begin{vmatrix} 1 & 2 & 0 \\ 3 & 2 & 0 \\ 1 & -1 & 2 \end{vmatrix} - 0 -$$

$$1\begin{vmatrix} 1 & 4 & 2 \\ 3 & -1 & 2 \\ 1 & 4 & -1 \end{vmatrix}$$

Now expansion is about column 3 for the first determinant and column 1 for the second.

$$= 2\left[2\begin{vmatrix} 1 & 2 \\ 3 & 2 \end{vmatrix}\right] -$$

$$1\left[1\begin{vmatrix} -1 & 2 \\ 4 & -2 \end{vmatrix} - 3\begin{vmatrix} 4 & 2 \\ 4 & -1 \end{vmatrix} + 1\begin{vmatrix} 4 & 2 \\ -1 & 2 \end{vmatrix}\right]$$

$= 2[2(2 - 6)] - 1[1(1 - 8) - 3(-4 - 8) + 1(8 + 2)]$

$= -16 - 1[39]$

$= -55$

37. $\begin{vmatrix} x & y & 1 \\ x_1 & y_1 & 1 \\ x_2 & y_2 & 1 \end{vmatrix} = 0$

$$x\begin{vmatrix} y_1 & 1 \\ y_2 & 1 \end{vmatrix} - y\begin{vmatrix} x_1 & 1 \\ x_2 & 1 \end{vmatrix} + 1\begin{vmatrix} x_1 & y_1 \\ x_2 & y_2 \end{vmatrix} = 0$$

*Expand about row 1*

$$x(y_1 - y_2) - y(x_1 - x_2) + 1(x_1x_2 - x_2y_1) = 0$$

$$xy_1 - xy_2 - x_1y + x_2y + x_1y_2 - x_2y_1 = 0$$

The straight line through $(x_1, y_1)$ and $(x_2, y_2)$ has the equation:

$$y - y_1 = \frac{y_2 - y_1}{x_2 - x_1}(x - x_1)$$

$$(y - y_1)(x_2 - x_1) = \frac{y_2 - y_1}{x_2 - x_1}(x - x_1)(x_2 - x_1)$$

*Multiply by $(x_2 - x_1)$*

$$x_2y - x_1y - x_2y_1 + x_1y_1 = xy_2 - x_1y_2 - xy_1 + x_1y_1$$

$$xy_1 - xy_2 - x_1y + x_2y + x_1y_2 - x_2y_1 = 0$$

Thus the equation for the determinant and the straight line are all the same.

**Section 8.5 (page 362)**

1. $8x - 4y = 8$
   $x + 3y = 22$

$$D = \begin{vmatrix} 8 & -4 \\ 1 & 3 \end{vmatrix} = 24 - (-4) = 28$$

$$D_x = \begin{vmatrix} 8 & -4 \\ 22 & 3 \end{vmatrix} = 24 - (-88) = 112$$

$$D_y = \begin{vmatrix} 8 & 8 \\ 1 & 22 \end{vmatrix} = 8(22) - 8 = 176 - 8 = 168$$

$$x = \frac{D_x}{D} = \frac{112}{28} = 4$$

$$y = \frac{D_y}{D} = \frac{168}{28} = 6$$

Solution set: {(4, 6)}

5. $3x + 8y = 3$
   $2x - 4y = 2$

$$D = \begin{vmatrix} 3 & 8 \\ 2 & -4 \end{vmatrix} = -12 - 16 = -28$$

$$D_x = \begin{vmatrix} 3 & 8 \\ 2 & -4 \end{vmatrix} = -12 - 16 = -28$$

$$D_y = \begin{vmatrix} 3 & 3 \\ 2 & 2 \end{vmatrix} = 6 - 6 = 0$$

$$x = \frac{D_x}{D} = \frac{-28}{-28} = 1$$

$$y = \frac{D_y}{D} = \frac{0}{28} = 0$$

Solution set: {(1, 0)}

9. $3x + 5y = -21$
   $-4x - 2y = 14$

$$D = \begin{vmatrix} 3 & 5 \\ -4 & -2 \end{vmatrix} = -6 - (-20) = 14$$

$$D_x = \begin{vmatrix} -21 & 5 \\ 14 & -2 \end{vmatrix} = 42 - 70 = -28$$

$$D_y = \begin{vmatrix} 3 & -21 \\ -4 & 14 \end{vmatrix} = 42 - 84 = -42$$

$$x = \frac{D_x}{D} = \frac{-28}{14} = -2$$

$$y = \frac{D_y}{D} = \frac{-42}{14} = -3$$

Solution set: {(-2, -3)}

13. $2x + 3y + 2z = 10$
    $x - y + 2z = 3$
    $x + 2y - 6z = -15$

$$D = \begin{vmatrix} 2 & 3 & 2 \\ 1 & -1 & 2 \\ 1 & 2 & -6 \end{vmatrix}$$

$$= 2\begin{vmatrix} -1 & 2 \\ 2 & -6 \end{vmatrix} - 1\begin{vmatrix} 3 & 2 \\ 2 & -6 \end{vmatrix} + 1\begin{vmatrix} 3 & 2 \\ -1 & 2 \end{vmatrix}$$

*Expand about column 1*

$$= 2(2) = 1(-22) + 1(8)$$
$$= 34$$

$$D_x = \begin{vmatrix} 10 & 3 & 2 \\ 3 & -1 & 2 \\ -15 & 2 & -6 \end{vmatrix}$$

$$= 10\begin{vmatrix} -1 & 2 \\ 2 & -6 \end{vmatrix} - 3\begin{vmatrix} 3 & 2 \\ 2 & -6 \end{vmatrix} - 15\begin{vmatrix} 3 & 2 \\ -1 & 2 \end{vmatrix}$$

*Expand about column 1*

$$= 20 + 66 - 120$$
$$= -34$$

$$D_y = \begin{vmatrix} 2 & 10 & 2 \\ 1 & 3 & 2 \\ 1 & -15 & -6 \end{vmatrix}$$

$$= 2\begin{vmatrix} 3 & 2 \\ -15 & -6 \end{vmatrix} - 1\begin{vmatrix} 10 & 2 \\ -15 & -6 \end{vmatrix} + 1\begin{vmatrix} 10 & 2 \\ 3 & 2 \end{vmatrix}$$

*Expand about column 1*

$= 24 + 30 + 14$
$= 68$

$$x = \frac{D_x}{D} = \frac{-34}{34} = -1$$

$$y = \frac{D_y}{D} = \frac{68}{34} = 2$$

To find z, substitute x = -1, y = 2 into the second equation x - y + 2z = 3.

$$-1 - 2 + 2z = 3$$
$$2z = 6$$
$$z = 3$$

Solution set: {(-1, 2, 3)}

17. $3x - y + z = 2$
$3x - y - 6z = 5$
$-6x + 2y - 2z = -4$

$$D = \begin{vmatrix} 3 & -1 & 1 \\ 3 & -1 & -6 \\ -6 & 2 & -2 \end{vmatrix}$$

$$= 3\begin{vmatrix} -1 & -6 \\ 2 & -2 \end{vmatrix} - 3\begin{vmatrix} -1 & 1 \\ 2 & -2 \end{vmatrix} - 6\begin{vmatrix} -1 & 1 \\ -1 & -6 \end{vmatrix}$$

*Expand about column 1*

$= 3(2 + 12) - 3(2 - 2) - 6(6 + 1)$
$= 42 - 0 - 42$
$= 0$

D = 0, so Cramer's Rule does not apply.

21. $x + 2y - z = 0$
$2x - y + z = 0$
$3x + 2y - 2z = 0$

$$D = \begin{vmatrix} 1 & 2 & -1 \\ 2 & -1 & 1 \\ 3 & 2 & -2 \end{vmatrix}$$

$$D = 1\begin{vmatrix} -1 & 1 \\ 2 & -2 \end{vmatrix} - 2\begin{vmatrix} 2 & 1 \\ 3 & -2 \end{vmatrix} - 1\begin{vmatrix} 2 & -1 \\ 3 & 2 \end{vmatrix}$$

*Expand about column 1*

$= 1(0) - 2(-7) - 1(7) = 7$

$$D_x = \begin{vmatrix} 0 & 2 & -1 \\ 0 & -1 & 1 \\ 0 & 2 & -2 \end{vmatrix} = 0,$$

$$D_y = \begin{vmatrix} 1 & 0 & -1 \\ 2 & 0 & 1 \\ 3 & 0 & -2 \end{vmatrix} = 0$$

$$x = \frac{D_x}{D} = \frac{0}{7} = 0$$

$$y = \frac{D_y}{D} = \frac{0}{7} = 0$$

$$x + 2y - z = 0$$
$$0 + 2(0) - z = 0$$
$$z = 0$$

Solution set: {(0, 0, 0)}

25. $x - 2y = -4$
$3x + y = -5$
$2x + z = -1$

$$D = \begin{vmatrix} 1 & -2 & 0 \\ 3 & 1 & 0 \\ 2 & 0 & 1 \end{vmatrix}$$

$$= 0\begin{vmatrix} 3 & 1 \\ 2 & 0 \end{vmatrix} - 0\begin{vmatrix} 1 & -2 \\ 2 & 0 \end{vmatrix} + 1\begin{vmatrix} 1 & -2 \\ 3 & 1 \end{vmatrix}$$

*Expand about column 3*

$= 0 - 0 + 1(1 + 6)$
$= 7$

$$D_x = \begin{vmatrix} -4 & -2 & 0 \\ -5 & 1 & 0 \\ -1 & 0 & 1 \end{vmatrix}$$

$$= 0\begin{vmatrix} -5 & 1 \\ -1 & 0 \end{vmatrix} - 0\begin{vmatrix} -4 & -2 \\ -1 & 0 \end{vmatrix} + 1\begin{vmatrix} -4 & -2 \\ -5 & 1 \end{vmatrix}$$

*Expand about column 3*

$= 0 - 0 + 1(-4 - 10)$
$= -14$

$$D_y = \begin{vmatrix} 1 & -4 & 0 \\ 3 & -5 & 0 \\ 2 & -1 & 1 \end{vmatrix}$$

$$= 0\begin{vmatrix} 3 & -5 \\ 2 & -1 \end{vmatrix} - 0\begin{vmatrix} 1 & -4 \\ 2 & -1 \end{vmatrix} + 1\begin{vmatrix} 1 & -4 \\ 3 & -5 \end{vmatrix}$$

*Expand about column 3*

$$= 0 - 0 + 1(-5 + 12)$$
$$= 7$$

$$x = \frac{D_x}{D} = \frac{-14}{7} = -2$$

$$y = \frac{D_y}{D} = \frac{7}{7} = 1$$

To find z substitute x = -2 into the last equation.

$$2x + z = -1$$
$$2(-2) + z = -1$$
$$z = 3$$

Solution set: {(-2, 1, 3)}

29. $$\begin{aligned} 4.5x - 3.7y \qquad &= 34.9 \\ 5.4x + 8.9y + 1.9z &= -57.2 \\ 6.4x \qquad - 2.8z &= 21.2 \end{aligned}$$

$$D = \begin{vmatrix} 4.5 & -3.7 & 0 \\ 5.4 & 8.9 & 1.9 \\ 6.4 & 0 & -2.8 \end{vmatrix}$$

$$= -(-3.7)\begin{vmatrix} 5.4 & 1.9 \\ 6.4 & -2.8 \end{vmatrix} + (8.9)\begin{vmatrix} 4.5 & 0 \\ 6.4 & -2.8 \end{vmatrix}$$

$$= -100.936 - 11.214 = -213.076$$

$$D_x = \begin{vmatrix} 34.9 & -3.7 & 0 \\ -57.2 & 8.9 & 1.9 \\ 21.2 & 0 & -2.8 \end{vmatrix}$$

$$= 34.9\begin{vmatrix} 8.9 & 1.9 \\ 0 & -2.8 \end{vmatrix} + 3.7\begin{vmatrix} -57.2 & 1.9 \\ 21.2 & -2.8 \end{vmatrix}$$

$$= -869.708 + 443.556$$
$$= -426.152$$

$$D_y = \begin{vmatrix} 4.5 & 34.9 & 0 \\ 5.4 & -57.2 & 1.9 \\ 6.4 & 21.2 & -2.8 \end{vmatrix}$$

$$= -1.9\begin{vmatrix} 4.5 & 34.9 \\ 6.4 & 21.2 \end{vmatrix} - 2.8\begin{vmatrix} 4.5 & 34.9 \\ 5.4 & -57.2 \end{vmatrix}$$

$$= 243.124 + 1248.408$$
$$= 1491.532$$

$$x = \frac{D_x}{D} = \frac{-426.152}{-213.076} = 2$$

$$y = \frac{D_y}{D} = \frac{1491.532}{-213.076} = -7$$

Since $6.4x - 2.8z = 21.2$

$$2.8z = (6.4)(2) - 21.2$$

So, $$2.8z = -8.4$$
$$z = -3$$

Solution set: {(2, -7, -3)}

33. $$\begin{aligned} x + y - z + w &= 2 \\ x - y + 2 + w &= 4 \\ -2x + y + 2z - w &= -5 \\ x + 3z + 2w &= 5 \end{aligned}$$

$$D = \begin{vmatrix} 1 & 1 & -1 & 1 \\ 1 & -1 & 1 & 1 \\ -2 & 1 & 2 & -1 \\ 1 & 0 & 3 & 2 \end{vmatrix}$$

$$= 1\begin{vmatrix} -1 & 1 & 1 \\ 1 & 2 & -1 \\ 0 & 3 & 2 \end{vmatrix} - 1\begin{vmatrix} 1 & 1 & 1 \\ -2 & 2 & -1 \\ 1 & 3 & 2 \end{vmatrix} - 1\begin{vmatrix} 1 & -1 & 1 \\ -2 & 1 & -1 \\ 1 & 0 & 2 \end{vmatrix} - 1\begin{vmatrix} 1 & -1 & 1 \\ -2 & 1 & 2 \\ 1 & 0 & 3 \end{vmatrix}$$

*Expand about row 1*

$$= 1\left[-1\begin{vmatrix}2 & -1\\3 & 2\end{vmatrix} - 1\begin{vmatrix}1 & -1\\0 & 2\end{vmatrix} + 1\begin{vmatrix}1 & 2\\0 & 3\end{vmatrix}\right]$$

$$-1\left[1\begin{vmatrix}2 & -1\\3 & 2\end{vmatrix} - 1\begin{vmatrix}-2 & -1\\1 & 2\end{vmatrix} + 1\begin{vmatrix}-2 & 2\\1 & 3\end{vmatrix}\right]$$

$$-1\left[1\begin{vmatrix}1 & -1\\0 & 2\end{vmatrix} + 1\begin{vmatrix}-2 & -1\\1 & 2\end{vmatrix} + 1\begin{vmatrix}-2 & 1\\1 & 0\end{vmatrix}\right]$$

$$-1\left[1\begin{vmatrix}1 & 2\\0 & 3\end{vmatrix} + 1\begin{vmatrix}-2 & 2\\1 & 3\end{vmatrix} + 1\begin{vmatrix}-2 & 1\\1 & 0\end{vmatrix}\right]$$

$$= 1[-1(4 + 3) - 1(2) + 1(3)] -$$
$$1[1(4 + 3) - 1(-4+1) + 1(-6 - 2)]$$
$$- 1[1(2) + 1(-4 + 1) + 1(-1)]$$
$$-1[1(3) + 1(-6 - 2) + 1(-1)]$$
$$= 1(-7 - 2 + 3) - 1(7 + 3 - 8)$$
$$-1(2 - 3 - 1) - 1(3 - 8 - 1)$$
$$= -6 - 2 + 2 + 6$$
$$= 0$$

Cramer's rule does not apply since $D = 0$.

37. $b^2x + a^2y = b^2$

$ax + by = a$

$$D = \begin{vmatrix}b^2 & a^2\\a & b\end{vmatrix} = b^3 - a^3$$

$$D_x = \begin{vmatrix}b^2 & a^2\\a & b\end{vmatrix} = b^3 - a^3$$

$$D_y = \begin{vmatrix}b^2 & b^2\\a & a\end{vmatrix} = ab^2 - ab^2$$

$$x = \frac{D_x}{D} = \frac{b^3 - a^3}{b^3 - a^3} = 1$$

$$y = \frac{D_y}{D} = \frac{ab^2 - ab^2}{b^3 - a^3} = \frac{0}{b^3 - a^3} = 0$$

Solution set: $\{(1, 0)\}$

41. Additive inverse for 27 is $-27$.

Reciprocal for 27 is $\frac{1}{27}$.

**Section 8.6 (page 369)**

1. $$\left[\begin{array}{cc|c}2 & -5 & -1\\3 & 1 & 7\end{array}\right]$$

$$\left[\begin{array}{cc|c}1 & -\frac{5}{2} & -\frac{1}{2}\\3 & 1 & 7\end{array}\right]$$ *Multiply first row by 1/2*

$$\left[\begin{array}{cc|c}1 & -\frac{5}{2} & -\frac{1}{2}\\0 & \frac{17}{2} & \frac{17}{2}\end{array}\right]$$ *Add -3 times first row to the second row*

$$\left[\begin{array}{cc|c}1 & -\frac{5}{2} & -\frac{1}{2}\\0 & 1 & 1\end{array}\right]$$ *Multiply second row by 2/17*

This matrix gives the system

$$x - \frac{5}{2}y = -\frac{1}{2}$$
$$y = 1.$$
$$x - \frac{5}{2}(1) = -\frac{1}{2} \quad \textit{Let } y = 1$$
$$x = -\frac{1}{2} + \frac{5}{2}$$
$$x = 2$$

Solution set: $\{(2, 1)\}$

5. $2x + 4y = 6$

$3x - y = 2$

Put into augmented form:

$$\left[\begin{array}{cc|c}2 & 4 & 6\\3 & -1 & 2\end{array}\right]$$

$$\left[\begin{array}{cc|c}1 & 2 & 3\\3 & -1 & 2\end{array}\right]$$ *Multiply row 1 by 1/2*

$$\left[\begin{array}{cc|c}1 & 2 & 3\\0 & -7 & -7\end{array}\right]$$ *Multiply row 1 by -3 and add to row 2*

$$\left[\begin{array}{cc|c} 1 & 2 & 3 \\ 0 & 1 & 1 \end{array}\right]$$ *Multiply row 2 by -1/7*

This matrix gives the system

$$x + 2y = 3$$
$$y = 1.$$

Substitute $y = 1$ into $x + 2y = 3$ to get $x = 1$.
Solution set: $\{(1, 1)\}$

9. $-4x + 12y = 36$
$x - 3y = 9$

$$\left[\begin{array}{cc|c} -4 & 12 & 36 \\ 1 & -3 & 9 \end{array}\right]$$ *Put into augmented form*

$$\left[\begin{array}{cc|c} -1 & 3 & 9 \\ 1 & -3 & 9 \end{array}\right]$$ *Multiply the first row by 1/4*

$$\left[\begin{array}{cc|c} -1 & 3 & 9 \\ 0 & 0 & 18 \end{array}\right]$$ *Add row 1 to row 2*

The corresponding system is

$$-x + 3y = 9$$
$$0 = 18$$

which has no solution and is inconsistent.
Solution set: ∅

13. $x + y - 3z = 1$
$2x - y + z = 9$
$3x + y - 4z = 8$

$$\left[\begin{array}{ccc|c} 1 & 1 & -3 & 1 \\ 2 & -1 & 1 & 9 \\ 3 & 1 & -4 & 8 \end{array}\right]$$ *Put into augmented form*

$$\left[\begin{array}{ccc|c} 1 & 1 & -3 & 1 \\ 0 & -3 & 7 & 7 \\ 3 & 1 & -4 & 8 \end{array}\right]$$ *Multiply row 1 by -2 and add to row 2*

$$\left[\begin{array}{ccc|c} 1 & 1 & -3 & 1 \\ 0 & -3 & 7 & 7 \\ 0 & -2 & 5 & 5 \end{array}\right]$$ *Multiply row 1 by -3 and add to row 3*

$$\left[\begin{array}{ccc|c} 1 & 1 & -3 & 1 \\ 0 & 1 & -\frac{7}{3} & -\frac{7}{3} \\ 0 & -2 & 5 & 5 \end{array}\right]$$ *Multiply row 2 by -1/3*

$$\left[\begin{array}{ccc|c} 1 & 1 & -3 & 1 \\ 0 & 1 & -\frac{7}{3} & -\frac{7}{3} \\ 0 & 0 & \frac{1}{3} & \frac{1}{3} \end{array}\right]$$ *Multiply row 2 by and add to row 3*

$$\left[\begin{array}{ccc|c} 1 & 1 & -3 & 1 \\ 0 & 1 & -\frac{7}{3} & -\frac{7}{3} \\ 0 & 0 & 1 & 1 \end{array}\right]$$ *Multiply row 3 by 3*

$$z = 1$$
$$y - \frac{7}{3}z = -\frac{7}{3}$$
$$y - \frac{7}{3}(1) = -\frac{7}{3}$$
$$y = 0$$
$$x + y - 3z = 1$$
$$x + 0 - 3(1) = 1$$
$$x = 4$$

Solution set: $\{(4, 0, 1)\}$

17. $x + y - z = 6$
$2x - y + z = -9$
$x - 2y + 3z = 1$

Put into augmented form.

$$\left[\begin{array}{ccc|c} 1 & 1 & -1 & 6 \\ 2 & -1 & 1 & -9 \\ 1 & -2 & 3 & 1 \end{array}\right]$$

$$\left[\begin{array}{ccc|c} 1 & 1 & -1 & 6 \\ 0 & -3 & 3 & -21 \\ 0 & -3 & 4 & -5 \end{array}\right]$$ *Multiply row 1 by -2; add to row 2*
*Multiply row 1 by -1; add to row 3*

$$\left[\begin{array}{ccc|c} 1 & 1 & -1 & -5 \\ 0 & 1 & -1 & 7 \\ 0 & -3 & 4 & -5 \end{array}\right]$$ *Multiply row 2 by -1/3*

$$\left[\begin{array}{ccc|c} 1 & 1 & -1 & 6 \\ 0 & 1 & -1 & 7 \\ 0 & 0 & 1 & 16 \end{array}\right]$$ *Multiply row 2 by 3 and add to row 3*

The system is

$$x + y - z = 6$$
$$y - z = 7$$
$$z = 16$$

Solution set: $\{(-1, 23, 16)\}$

21. $x^2 + 6x - 3 = 0$

$x^2 + 6x + 9 - 3 = 9$

$(x + 3)^2 = 12$

$x + 3 = \sqrt{12}$ or $x + 3 = -\sqrt{12}$

$= 2\sqrt{3}$ $\quad = -2\sqrt{3}$

$x = -3 + 2\sqrt{3}$ or $x = -3 - 2\sqrt{3}$

Solution set: $\{-3 + 2\sqrt{3}, -3 - 2\sqrt{3}\}$

25. $-x^2 - 3x + 2 = 0$

$x^2 + 3x - 2 = 0$

$x^2 + 3x = 2$

$x^2 + 3x + \frac{9}{4} = 2 + \frac{9}{4} = \frac{17}{4}$

$(x + \frac{3}{2})^2 = \frac{17}{4}$

$x + \frac{3}{2} = \frac{\sqrt{17}}{2}$

$x = \frac{-3 + \sqrt{17}}{2}$

or

$x + \frac{3}{2} = \frac{-\sqrt{17}}{2}$

$x = \frac{-3 - \sqrt{17}}{2}$

Solution set: $\{\frac{-3 + \sqrt{17}}{2}, \frac{-3 - \sqrt{17}}{2}\}$

## Chapter 8 Review Exercises (page 371)

1. $5x - 3y = 19$ *(1)*

$4x + y = 5$ *(2)*

$5x - 3y = 19$

$12x + 3y = 15$ *Multiply by 3*

$17x = 34$ *Add*

$x = 2$

Substitute $x = 2$ in (2).

$4(2) + y = 5$

$x + y = 5$

$y = -3$

Solution set: $(2, -3)$.

5. $-3x - y = 4$ *(1)*

$x = \frac{2}{3}y$ *(2)*

Substitute $\frac{2}{3}y$ for x in (1).

$-3(\frac{2}{3}y) - y = 4$

$-2y - y = 4$

$-3y = 4$

$y = \frac{-4}{3}$

Substitute for y in (1).

$-3x - (\frac{-4}{3}) = 4$

$-3x = 4 - \frac{4}{3}$

$-3x = \frac{8}{3}$

$x = \frac{-8}{9}$

Solution set: $\{(\frac{-8}{9}, \frac{-4}{3})\}$

9. $\frac{2}{x} + \frac{4}{y} = 0$ *(1)*

$\frac{2}{x} = \frac{-4}{y}$

$y = -2x$ *Simplify (1)*

$\frac{5}{x} - \frac{6}{y} = -8$ *(2)*

Substitute $y = -2x$ in (2).

$$\frac{5}{x} - \frac{6}{-2x} = -8$$

$$\frac{5}{x} + \frac{3}{x} = -8$$

$$\frac{8}{x} = -8$$

$$\frac{1}{x} = -1$$

Substitute $x = -1$ in (1).

$$y = -2(-1) = 2$$

Solution set: $\{(-1, 2)\}$

13. Let x be the amount of \$2 candy and y the amount of \$1 candy. Then $x + y = 50$. The value of the various candies are $2x$, $1y = y$, and $50(1.30) = 65$, giving

$$2x + y = 65.$$

Solve the system.

$$x + y = 50$$
$$2x + y = 65$$

to get $x = 15$ and $y = 35$. 15 pounds of \$2 candy and 35 pounds of \$1 candy should be used.

17.
$$-x + 2y + 3z = 5 \quad (1)$$
$$2x + 3y + z = 12 \quad (2)$$
$$3x - 6y - 9z = -10 \quad (3)$$

Multiply first equation by 3 and add to third equation.

$$-3x + 6y + 9z = 15 \quad \textit{Multiply (1) by 3}$$
$$3x - 6y - 9z = -10 \quad (3)$$
$$0 = 5 \quad \textit{False}$$

This false result means the system is inconsistant and has no solutions.

Solution set: $\emptyset$

21. Let x be the liters of 8% hydrogen peroxide and y the liters of 20% solution.

| Kind of solution | Liters of solution | Liters of hydrogen peroxide |
|---|---|---|
| 8% | x | .08x |
| 10% | y | .10y |
| 20% | x - 2 | .20(x - 2) |
| 12.5% | 8 | .125(8) |

All of the solution added together is 8, so we have the equation

$$x + y + x - 2 = 8$$
$$y = 10 - 2x. \quad \textit{Solve for y}$$

The liters of hydrogen peroxide in the three solutions should equal the total number of liters of peroxide in the mixture (.125)8.

$$.08x + .10y + .20(x - 2) = (.125)8$$

Substitute $10 - 2x$ for y.

$$.08x + .10(10 - 2x) + .20(x - 2) = (.125)8$$
$$.08x + 1 - .20x + .20x - .40 = 1$$
$$.08x = .40$$
$$x = 5$$

$$y = 10 - 2x = 0$$

The amounts are 5 liters of 8%, none of 10%, and 3 liters of 20%.

25. $$\begin{vmatrix} 10 & 0 \\ -1 & 2 \end{vmatrix} = (10)(2) - (0)(-1) = 20$$

29. $$\begin{vmatrix} 0 & 5 & 3 \\ 7 & 0 & -2 \\ 0 & -1 & 6 \end{vmatrix}$$

$$= 0\begin{vmatrix} 0 & -2 \\ -1 & 6 \end{vmatrix} - 7\begin{vmatrix} 5 & 3 \\ -1 & 6 \end{vmatrix} + 0\begin{vmatrix} 5 & 3 \\ 0 & -2 \end{vmatrix}$$

*Expand by first column*

$= -7(30 + 3)$
$= -231$

33. $3x - 4y = 5$
$2x + y = 8$

$$D = \begin{vmatrix} 3 & -4 \\ 2 & 1 \end{vmatrix} = 3 + 8 = 11$$

$$D_x = \begin{vmatrix} 5 & -4 \\ 8 & 1 \end{vmatrix} = 5 + 32 = 37$$

$$D_y = \begin{vmatrix} 3 & 5 \\ 2 & 8 \end{vmatrix} = 24 - 10 = 14$$

$$x = \frac{D_x}{D} = \frac{37}{11}$$

$$y = \frac{D_y}{D} = \frac{14}{11}$$

Solution set: $\{(\frac{37}{11}, \frac{14}{11})\}$

37. $-x + 3y - 4z = 2$
$2x + 4y + z = 3$
$3x - z = 9$

$$D = \begin{vmatrix} -1 & 3 & -4 \\ 2 & 4 & 1 \\ 3 & 0 & -1 \end{vmatrix}$$

$$= 3\begin{vmatrix} 3 & -4 \\ 4 & 1 \end{vmatrix} - 1\begin{vmatrix} -1 & 3 \\ 2 & 4 \end{vmatrix}$$

$$= 3(19) - (-10) = 67$$

$$D_x = \begin{vmatrix} 2 & 3 & -4 \\ 3 & 4 & 1 \\ 9 & 0 & -1 \end{vmatrix}$$

$$= 9\begin{vmatrix} 3 & -4 \\ 4 & 1 \end{vmatrix} - 1\begin{vmatrix} 2 & 3 \\ 3 & 4 \end{vmatrix}$$

$$= 9(19) - 1(-1) = 172$$

$$D_y = \begin{vmatrix} -1 & 2 & -4 \\ 2 & 3 & 1 \\ 3 & 9 & -1 \end{vmatrix}$$

$$= -1\begin{vmatrix} 3 & 1 \\ 9 & -1 \end{vmatrix} - 2\begin{vmatrix} 2 & 1 \\ 3 & -1 \end{vmatrix} - 4\begin{vmatrix} 2 & 3 \\ 3 & 9 \end{vmatrix}$$

$$= -1(-12) - 2(-5) - 4(9) = -14$$

$$x = \frac{D_x}{D} = \frac{172}{67}$$

$$y = \frac{D_y}{D} = \frac{-14}{67}$$

Since $3x - z = 9$,

$$3(\frac{172}{67}) - z = 9$$

$$\frac{516}{67} - z = 9$$

$$z = \frac{516}{67} - 9.$$

Solution set: $\{(\frac{172}{67}, \frac{-14}{67}, \frac{-87}{67})\}$

41. $x + 2y - z = 1$
$3x + 4y + 2z = -2$
$-2x - y + z = -1$

$$\left[\begin{array}{cc|cc} 1 & 2 & -1 & 1 \\ 3 & 4 & 2 & -2 \\ -2 & -1 & 1 & -1 \end{array}\right]$$

$$\left[\begin{array}{cc|cc} 1 & 2 & -1 & 1 \\ 0 & -2 & 5 & -5 \\ -2 & -1 & 1 & -1 \end{array}\right]$$ *Multiply first row by -3 and add to second row*

$$\left[\begin{array}{ccc|c} 1 & 2 & -1 & 1 \\ 0 & -2 & 5 & -5 \\ 0 & 3 & -1 & 1 \end{array}\right]$$ *Multiply first row by 2 and add to third row*

$$\left[\begin{array}{ccc|c} 1 & 2 & -1 & 1 \\ 0 & 1 & -\frac{5}{2} & \frac{5}{2} \\ 0 & 3 & -1 & 1 \end{array}\right]$$ *Multiply second row by -1/2*

$$\left[\begin{array}{ccc|c} 1 & 2 & -1 & 1 \\ 0 & 1 & -\frac{5}{2} & \frac{5}{2} \\ 0 & 0 & \frac{13}{2} & -\frac{13}{2} \end{array}\right]$$ *Multiply second row by -3 and add to third row*

$$x + 2y - z = 1$$
$$y - \frac{5}{2}z = \frac{5}{2}$$
$$\frac{13}{2}z = -\frac{13}{2}$$
$$z = -1$$

Since $y - \frac{5}{2}z = \frac{5}{2}$,

$$y - \frac{5}{2}(-1) = \frac{5}{2}$$
$$y + \frac{5}{2} = \frac{5}{2}$$
$$y = 0.$$

Since $x + 2y - z = 1$,

$$x + 2(0) - (-1) = 1$$
$$x + 1 = 1$$
$$x = 0.$$

Solution set: $\{(0, 0, -1)\}$

45. $x + 2y + 5z = 9$

$2x - y = 3$

$3x + 2y + z = 9$

Using an augmented matrix:

$$\left[\begin{array}{ccc|c} 1 & 2 & 5 & 9 \\ 2 & -1 & 0 & 3 \\ 3 & 2 & 1 & 9 \end{array}\right]$$

$$\left[\begin{array}{ccc|c} 1 & 2 & 5 & 9 \\ 0 & -5 & -10 & -15 \\ 3 & 2 & 1 & 9 \end{array}\right]$$ *-2 times row 1; add to row 2*

$$\left[\begin{array}{ccc|c} 1 & 2 & 5 & 9 \\ 0 & -5 & -10 & -15 \\ 0 & -4 & -14 & -18 \end{array}\right]$$ *-3 times row 1; add to row 3*

$$\left[\begin{array}{ccc|c} 1 & 2 & 5 & 9 \\ 0 & 1 & 2 & 3 \\ 0 & -4 & -14 & -18 \end{array}\right]$$ *-1/5 times row 2*

$$\left[\begin{array}{ccc|c} 1 & 2 & 5 & 9 \\ 0 & 1 & 2 & 3 \\ 0 & 0 & -6 & -6 \end{array}\right]$$ *4 times row 2; add to row 3*

$$\left[\begin{array}{ccc|c} 1 & 2 & 5 & 9 \\ 0 & 1 & 2 & 3 \\ 0 & 0 & 1 & 1 \end{array}\right]$$ *-1/6 times row 3*

The equations from the last matrix produce the system

$$x + 2y + 5z = 9 \quad (1)$$
$$y + 2z = 3 \quad (2)$$
$$z = 1. \quad (3)$$

Substitute for z.

$$y + 2(1) = 3$$
$$y = 1$$

$$x + 2(1) + 5(1) = 9$$
$$x = 2$$

Solution set: $\{(2, 1, 1)\}$

49. $\frac{2}{3}x + \frac{y}{6} = \frac{19}{2}$ *(1)*

$\frac{1}{3}x - \frac{2}{9}y = 2$ *(2)*

$4x + y = 57$ *Multiply by 6*

$3x - 2y = 18$ *Multiply by 9*

$8x + 2y = 114$ *Multiply by 2*

$3x - 2y = 18$

$11x = 132$ *Add*

$x = 12$

Substitute x = 12 in (1).

$4(12) + y = 57$

$48 + y = 57$

$y = 9$

Solution set: $\{(12, 9)\}$

**Chapter 8 Test (page 373)**

1. $2x + 3y = 10$ *(1)*

$-3x + 2y = 11$ *(2)*

$6x + 9y = 30$ *Multiply (1) by 3*

$-6x + 4y = 22$ *Multiply (2) by 2*

$13y = 52$ *Add*

$y = 4$

$2x + 3y = 10$

$2x + 3(4) = 10$

$2x = -2$

$x = -1$

Solution set: $\{(-1, 4)\}$

2. $3x + 4y = -7$ *(1)*

$2x - y = 10$ *(2)*

$3x + 4y = -7$ *(1)*

$8x - 4y = 40$ *(2) Multiply by 4*

$11x = 33$ *Add*

$x = 3$ *Solve for x*

Substitute $x = 3$ in (1).

$3(3) + 4y = -7$

$4y = -16$

$y = -4$ *Solve for y*

The solution is $\{(3, -4)\}$

3. $3x + 4y = 8$ *(1)*

$6x = 7 - 8y$ *(2)*

$-6x - 8y = -16$ *Multiply (1) by -2*

$6x + 8y = 7$ *(2)*

$0 = -9$ *False*

Therefore, the system has no solutions.

Solution set: $\emptyset$

4. $2x - 3y = -8$ *(1)*

$x = y - 3$ *(2)*

$2x - 3y = -8$ *(1)*

$-2x + 2y = 6$ *Multiply (2) by -2*

$-y = -2$

$y = 2$ *Solve for y*

Substitute $y = 2$ in (2).

$x = 2 - 3$

$x = -1$ *Solve for x*

Solution set: $\{(-1, 2)\}$

5. $\frac{2}{3}x + \frac{1}{3}y = \frac{5}{3}$ *(1)*

$\frac{1}{4}x - \frac{5}{2}y = \frac{17}{2}$ *(2)*

$2x + y = 5$ *Multiply (1) by 3*

$x - 10y = 34$ *Multiply (2) by 4*

$2x + y = 5$ *(4)*

$-2x + 20y = -68$ *Multiply (3) by -2*

$21y = -63$ *Add*

$y = -3$

$2x + y = 5$

$2x - 3 = 5$ *Let y = -3*

$2x = 8$

$x = 4$

Solution set: $\{(4, -3)\}$

6. $\frac{x}{4} + \frac{2y}{3} = -\frac{5}{2}$ *(1)*

$\frac{2x}{5} + \frac{y}{2} = -\frac{23}{10}$ *(2)*

$3x + 8y = -30$ *(3) Multiply (1) by 12*

$4x + 5y = -23$ *(4) Multiply (2) by 10*

$12x + 32y = -120$ *Multiply (3) by 4*

$-12x - 15y = 69$ *Multiply (4) by -3*

$17y = -51$ *Add*

$y = -3$

So, $3x + 8(-3) = -30$

$3x - 24 = -30$

$3x = -6$

$x = -2$

Solution set: $\{(-2, -3)\}$

7. $\frac{1}{x} + \frac{3}{y} = \frac{9}{8}$

$-\frac{2}{x} + \frac{1}{y} = \frac{1}{12}$

Let $p = \frac{1}{x}$, $q = \frac{1}{y}$.

$p + 3q = \frac{9}{8}$ *(1)*

$-2p + q = \frac{1}{12}$ *(2)*

$8p + 24q = 9$ *(3) Multiply (1) by 8*

$-24p + 12q = 1$ *(4) Multiply (2) by 12*

$24p + 72q = 27$ *Multiply (3) by 3*

$-24p + 12q = 1$ *(4)*

$84q = 28$ *Add*

$q = \frac{1}{3}$

So $p + 3(\frac{1}{3}) = \frac{9}{8}$

$p = \frac{1}{8}$

So $x = \frac{1}{p} = \frac{1}{\frac{1}{8}} = 8,$

$y = \frac{1}{q} = \frac{1}{\frac{1}{3}} = 3$

Solution set: $\{(8, 3)\}$

8. $\frac{3}{x} - \frac{2}{y} = \frac{9}{2}$

$\frac{2}{x} + \frac{4}{y} = 7$

Let $p = \frac{1}{x}$, $q = \frac{1}{y}$.

$3p - 2q = \frac{9}{2}$ *(1)*

$2p + 4q = 7$ *(2)*

$6p - 4q = 9$ *Multiply (1) by 2*

$2p + 4q = 7$ *(2)*

$8p = 16$ *Add*

$p = 2$

So $2(2) + 4q = 7$

$4q = 3$

$q = \frac{3}{4}$

So $x = \frac{1}{p} = \frac{1}{2}$, $y = \frac{1}{q} = \frac{1}{\frac{3}{4}} = \frac{4}{3}$

Solution set: $\{(\frac{1}{2}, \frac{4}{3})\}$

9. $2x - y + z = 9$ *(1)*

$3x + y - 2z = 4$ *(2)*

$x + y - 4z = -6$ *(3)*

$2x - y + z = 9$ *(1)*

$-2x - 2y + 8z = 12$ *Multiply (3) by -2*

$-3y + 9z = 21$

$3x + y - 2z = 4$ *(2)*

$-3x - 3y + 12z = 18$ *Multiply (3) by -3*

$-2y + 10z = 22$

$-6y + 18z = 42$ *Multiply (4) by 2*

$6y - 30z = -66$ *Multiply (5) by -3*

$-12z = -24$

$z = 2$ *Solve for z*

Substitute $z = 2$ in (5).

$-2y + 10(2) = 22$

$-2y = 2$ *Solve for y*

$y = -1$

Substitute $y = -1$ and $z = 2$ in (3)

$x + (-1) - 4(2) = -6$ *Solve for x*

$x = 3$

Solution set: $\{(3, -1, 2)\}$

10. $2x - 3y + 2z = -4$ *(1)*

$3x + 4y - 3z = -2$ *(2)*

$-4x - 3y + z = -3$ *(3)*

$4x - 6y + 4z = -8$ *Multiply (1) by 2*

$-4x - 3y + z = -3$ *(3)*

$-9y + 5z = -11$ *(4)*

$-3x + \frac{9}{2}y - 3z = 6$ *Multiply (1) by -3/2*

$3x + 4y - 3z = -2$ *(2)*

$\frac{17}{2}y - 6z = 4$

or $17y - 12z = 8$ *(5)*

$-108y + 60z = -132$ *Multiply (4) by 12*

$85y - 60z = 40$ *Multiply (5) by 5*

$-23y = -92$

$y = 4$

$$-9y + 5z = -11$$
$$-9(4) + 5z = -11 \quad \textit{Let } y = 4$$
$$-36 + 5z = -11$$
$$5z = 25$$
$$z = 5$$

$$2x - 3(4) + 2(5) = -4 \quad \textit{Let } y = 4, z = 5$$
$$2x - 12 + 10 = -4$$
$$2x - 2 = -4$$
$$2x = -2$$
$$x = -1$$

Solution set: $\{(-1, 4, 5)\}$

11. $4x - 2y = -8$ *(1)*
$3y - 5z = 14$ *(2)*
$2x + z = -10$ *(3)*

$4x - 2y = -8$ *(1)*
$-4x - 2z = 20$ *Multiply (3) by -2*
$-2y - 2z = 12$
or $y + z = -6$ *(4)*

$3y - 5z = 14$ *(2)*
$5y + 5z = -30$ *Multiply (4) by 5*
$8y = -16$
$y = -2$

$$y + z = -6$$
$$-2 + z = -6$$
$$z = -4$$

$$4x - 2y = -8$$
$$4x - 2(-2) = -8$$
$$4x + 4 = -8$$
$$4x = -12$$
$$x = -3$$

Solution set: $\{(-3, -2, -4)\}$

12. $\begin{vmatrix} -3 & 2 \\ -1 & 4 \end{vmatrix} = (-3)(4) - (-1)(2)$
$= -12 - (-2)$
$= -10$

13. $\begin{vmatrix} 4 & 6 \\ 6 & 9 \end{vmatrix} = (4)(9) - (6)(6) = 0$

14. $\begin{vmatrix} 2 & -1 & 0 \\ 0 & 3 & 1 \\ -2 & 0 & 1 \end{vmatrix}$

$= 2\begin{vmatrix} 3 & 1 \\ 0 & 1 \end{vmatrix} - (-1)\begin{vmatrix} 0 & 1 \\ -2 & 1 \end{vmatrix}$

$= 2(3) + 1(2) = 8$

15. $\begin{vmatrix} 3 & -17 & 0 \\ -8 & 11 & 0 \\ 9 & -2 & 0 \end{vmatrix} = 0$ *Expand about the third column*

16. Let x be the amount of 30% alcohol needed and y the amount of 80% alcohol.

| Kind | Amount | Pure alcohol |
|---|---|---|
| 30% | x | .30x |
| 80% | y | .80y |
| 50% | 100 | .50(100) = 50 |

The equations are written as sums of the columns.

$x + y = 100$ *(1)*
$.30x + .80y = 50$ *(2)*
$y = 100 - x$ *Solve (1) for y*
$.30x + .80(100 - x) = 50$ *Substitute in (2)*
$.30x + 80 - .80x = 50$
$-.50x = -30$
$x = 60$
$60 + y = 100$ *Substitute in (1)*
$y = 40$

60 gallons of 30% alcohol and 40 gallons of 80% alcohol are needed.

17. Let x and y be the speeds of the two cars. Use $d = rt$ to find the distance traveled by each car and add these expressions in equation (1).

| Distance of slow car | and | distance of fast car |
|---|---|---|
| ↓ | ↓ | ↓ |
| $4x$ | $+$ | $4y$ |

| is | total distance. |
|---|---|
| ↓ | ↓ |
| $=$ | $400$ (1) |

| Speed of slow car | and | 20 | is | speed of fast car. |
|---|---|---|---|---|
| ↓ | ↓ | | | |
| $x$ | $+$ | $20$ | $=$ | $y$ |

$x - y = -20$ (2)

$$\begin{aligned} 4x + 4y &= 400 \quad \textit{(1)} \\ 4x - 4y &= -80 \quad \textit{Multiply (2) by 4} \\ \hline 8x &= 320 \\ x &= 40 \end{aligned}$$

Substitute 40 for x in (2).

$$\begin{aligned} 40 - y &= -20 \\ y &= 60 \end{aligned}$$

The speeds of the cars are 40 miles per hour and 60 miles per hour.

18. Let a be the smallest side, b the medium side, and c the largest side.
Use $P = a + b + c$.

$23 = a + b + c$ *(1)*

$c - a = 4$ *(2) The difference between the largest and smallest side is 4*

$a + b = c + 3$ *The sum of the two smaller sides is 3 more than the largest*

$a + b - c = 3$ *(3)*

$$\begin{aligned} a + b + c &= 23 \quad \textit{(1)} \\ -a - b + c &= -3 \quad \textit{Multiply (3) by -1} \\ \hline 2c &= 20 \\ c &= 10 \end{aligned}$$

Substitute in (2).

$$\begin{aligned} 10 - a &= 4 \\ a &= 6 \end{aligned}$$

Substitute in (1).

$$\begin{aligned} 6 + b + 10 &= 23 \\ b &= 7 \end{aligned}$$

The lengths of the three sides are 6 centimeters, 7 centimeters, and 10 centimeters.

19. $2x + 6y = 3$
$-3x + y = 8$

$$D = \begin{vmatrix} 2 & 6 \\ -3 & 1 \end{vmatrix} = 2 - (-18) = 20$$

$$D_x = \begin{vmatrix} 3 & 6 \\ 8 & 1 \end{vmatrix} = 3 - 48 = -45$$

$$D_y = \begin{vmatrix} 2 & 3 \\ -3 & 8 \end{vmatrix} = 16 - (-9) = 25$$

$$x = \frac{D_x}{D} = \frac{-45}{20} = -\frac{9}{4}$$

$$y = \frac{D_y}{D} = \frac{25}{20} = \frac{5}{4}$$

Solution set: $\{(-\frac{9}{4}, \frac{5}{4})\}$

20. $3x + 5y = 17$
$4x = y - 8$ or $4x - y = -8$

$$D = \begin{vmatrix} 3 & 5 \\ 4 & -1 \end{vmatrix} = -3 - 20 = -23$$

$$D_x = \begin{vmatrix} 17 & 5 \\ -8 & -1 \end{vmatrix} = -17 + 40 = 23$$

$$D_y = \begin{vmatrix} 3 & 17 \\ 4 & -8 \end{vmatrix} = -24 - 68 = -92$$

$$x = \frac{D_x}{D} = \frac{23}{-23} = -1$$

$$y = \frac{D_y}{D} = \frac{-92}{-23} = 4$$

Solution set: $\{(-1, 4)\}$

21. $x + y + z = 4$
$2x - z = -5$
$3y + z = 9$

$$D = \begin{vmatrix} 1 & 1 & 1 \\ 2 & 0 & -1 \\ 0 & 3 & 1 \end{vmatrix}$$

$$= 1\begin{vmatrix} 0 & -1 \\ 3 & 1 \end{vmatrix} - 2\begin{vmatrix} 1 & 1 \\ 3 & 1 \end{vmatrix}$$

$$= 1(3) - 2(-2) = 7$$

$$D_x = \begin{vmatrix} 4 & 1 & 1 \\ -5 & 0 & -1 \\ 9 & 3 & 1 \end{vmatrix}$$

$$= -(-5)\begin{vmatrix} 1 & 1 \\ 3 & 1 \end{vmatrix} - (-1)\begin{vmatrix} 4 & 1 \\ 9 & 3 \end{vmatrix}$$

$$= 5(-2) + 1(3) = -7$$

$$D_y = \begin{vmatrix} 1 & 4 & 1 \\ 2 & -5 & -1 \\ 0 & 9 & 1 \end{vmatrix}$$

$$= 1\begin{vmatrix} -5 & -1 \\ 9 & 1 \end{vmatrix} - 2\begin{vmatrix} 4 & 1 \\ 9 & 1 \end{vmatrix}$$

$$= 1(4) - 2(-5) = 14$$

$$x = \frac{D_x}{D} = \frac{-7}{7} = -1$$

$$y = \frac{14}{7} = 2$$

$2x - z = -5$
$2(-1) - z = -5$ *Let $x = -1$*
$-2 - z = -5$
$-z = -3$
$z = 3$

Solution set: $\{(-1, 2, 3)\}$

22. $3x + 2y = 4$
$5x + 5y = 9$

$$D = \begin{vmatrix} 3 & 2 \\ 5 & 5 \end{vmatrix} = 15 - 10 = 5$$

$$D_x = \begin{vmatrix} 4 & 2 \\ 9 & 5 \end{vmatrix} = 20 - 18 = 2$$

$$D_y = \begin{vmatrix} 3 & 4 \\ 5 & 9 \end{vmatrix} = 27 - 20 = 7$$

$$x = \frac{D_x}{D} = \frac{2}{5}$$

$$y = \frac{D_y}{D} = \frac{7}{5}$$

Solution set: $\{(\frac{2}{5}, \frac{7}{5})\}$

23. $5x = 2y + 6$ or $5x - 2y = 6$
$2x + 3y = 10$

$$D = \begin{vmatrix} 5 & -2 \\ 2 & 3 \end{vmatrix} = 15 - (-4) = 19$$

$$D_x = \begin{vmatrix} 6 & -2 \\ 10 & 3 \end{vmatrix} = 18 - (-20) = 38$$

$$D_y = \begin{vmatrix} 5 & 6 \\ 2 & 10 \end{vmatrix} = 50 - 12 = 38$$

$$x = \frac{D_x}{D} = \frac{38}{19} = 2$$

$$y = \frac{D_y}{D} = \frac{38}{19} = 2$$

Solution set: $\{(2, 2)\}$

24. $x + 3y + 2z = 11$
$3x + 7y + 4z = 23$
$5x + 3y - 5z = -14$

$$D = \begin{vmatrix} 1 & 3 & 2 \\ 3 & 7 & 4 \\ 5 & 3 & -5 \end{vmatrix}$$

$$= 1\begin{vmatrix} 7 & 4 \\ 3 & -5 \end{vmatrix} - 3\begin{vmatrix} 3 & 2 \\ 3 & -5 \end{vmatrix} + 5\begin{vmatrix} 3 & 2 \\ 7 & 4 \end{vmatrix}$$

$$= 1(-47) - 3(-21) + 5(-2)$$
$$= 6$$

$$D_x = \begin{vmatrix} 11 & 3 & 2 \\ 23 & 7 & 4 \\ -14 & 3 & -5 \end{vmatrix}$$

$$= 11\begin{vmatrix} 7 & 4 \\ 3 & -5 \end{vmatrix} - 3\begin{vmatrix} 23 & 4 \\ -14 & -5 \end{vmatrix} + 2\begin{vmatrix} 23 & 7 \\ -14 & 3 \end{vmatrix}$$

$$= 11(-47) - 3(-59) + 2(167)$$
$$= -6$$

$$D_y = \begin{vmatrix} 1 & 11 & 2 \\ 3 & 23 & 4 \\ 5 & -14 & -5 \end{vmatrix}$$

$$= 1\begin{vmatrix} 23 & 4 \\ -14 & -5 \end{vmatrix} - 3\begin{vmatrix} 11 & 2 \\ -14 & -5 \end{vmatrix} + 5\begin{vmatrix} 11 & 2 \\ 23 & 4 \end{vmatrix}$$

$$= 1(-59) - 3(-27) + 5(-2)$$
$$= 12$$

$$x = \frac{D_x}{D} = \frac{-6}{6} = -1$$

$$y = \frac{D_y}{D} = \frac{12}{6} = 2$$

$x + 3y + 2z = 11$
$-1 + 3(2) + 2z = 11$ *Let $x = -1$, $y = 2$*
$5 + 2z = 11$
$2z = 6$
$z = 3$

Solution set: {(-1, 2, 3)}

## CHAPTER 9 CONIC SECTIONS

### Section 9.1 (page 384)

1. Write $y = x^2 - 5$ in $y = a(x - h)^2 + k$ form to get $y = 1(x - 0)^2 - 5$. Thus, the vertex $(h, k) = (0, -5)$.

5. Write $y = x^2 - 18x + 93$ in the form $y = a(x - h)^2 + k$.

   $y = (x^2 - 18x + 81) - 81 + 93$ *Complete the square*

   $= 1(x - 9)^2 + 12$

   The vertex is $(h, k) = (9, 12)$.

9. Write $x = y^2 - 3$ in $x = a(y - k)^2 + h$ form: $x = 1(y - 0)^2 - 3$. Thus, vertex $(h, k) = (-3, 0)$.

13. $x = 4(y - 3)^2 + 1$ is in the proper form to find the vertex $(h, k) = (1, 3)$.

17. Write $y = -5x^2$ in $y = a(x - h)^2 + k$ form to get $y = -5(x - 0)^2 + 0$. Since $a = -5$, the graph opens downward $(a < 0)$.
    To find the discriminant for the quadratic equation resulting when $y = 0$ in $y = -5x^2$, we have

    $a = -5$, $b = 0$, $c = 0$.

    $b^2 - 4ac = 0 - 4(-5)(0) = 0$

    Thus, the equation has exactly one real solution, which is the only x-intercept of the graph of $y = -5x^2$.

21. Write $y = x^2 + 2x + 1$ in $y = a(x - h)^2 + k$ form to get

    $y = 1(x - (-1))^2 + 0$.

    Since $a = 1$, the graph opens upward.
    Using the discriminant of $x^2 + 2x + 1 = 0$, we have $a = 1$, $b = 2$, $c = 1$.

    $b^2 - 4ac = (2)^2 - 4(1)(1) = 0$

    There is one x-intercept.

For Exercises 25-55, see answer graph in textbook.

25. $y = 3x^2$

    Let $x = 0$; y-intercept is 0.
    Let $y = 0$; x-intercept is 0.
    $y = 3x^2$ in the form

    $y = a(x - h)^2 + k$ is

    $y = 3(x - 0)^2 + 0$

    so the vertex $(h, k) = (0, 0)$.
    The x-value of the vertex is 0, so the axis is at $x = 0$.
    Since $a > 0$, the graph opens upward.

    If $x = 1$, $y = 3(1)^2 = 3$, giving $(1, 3)$.

    If $x = -1$, $y = 3(-1)^2 = 3$, giving $(-1, 3)$.

    If $x = 2$, $y = 3(2)^2 = 12$, giving $(2, 12)$.

    If $x = -2$, $y = 3(-2)^2 = 12$, giving $(-2, 12)$.

29. $y = 2x^2 - 2$

When $x = 0$, $y = -2$ *y-intercept*

When $y = 0$, $2x^2 - 2 = 0$

$$2x^2 = 2$$

$$x = \pm 1$$ *x-intercepts*

Using the formula to find the vertex: $y = 2(x - 0)^2 - 2$ and $(h, k) = (0, -2)$.

Since the x-value of the vertex is 0, the axis is $x = 0$.

Since $a > 0$, the graph opens upward.

Let $x = 2$, $y = 2(2)^2 - 2 = 6$, giving $(2, 6)$.

Let $x = -2$, $y = 2(-2)^2 - 2 = 6$, giving $(-2, 6)$.

33. $y = x^2 + 2x - 1$

When $x = 0$, $y = -1$. *y-intercept*

When $y = 0$, $x^2 + 2x - 1 = 0$.

$a = 1$, $b = 2$, $c = -1$ *Use quadratic equation*

$$x = \frac{-(-2) \pm \sqrt{(2)^2 - 4(1)(-1)}}{2(1)}$$

$$= \frac{-2 \pm 2\sqrt{2}}{2}$$

$$= -1 \pm \sqrt{2} = -2.414 \quad \text{or} \quad .414$$

The x-intercepts are -2,414 and .414.

For the vertex, complete the square.

$$y = (x^2 + 2x + 1) - 1 - 1$$

$$= 1(x - (-1))^2 - 2$$

Vertex $(h, k) = (-1, -2)$

The axis is at $x = -1$.

The graph opens upward since $a > 0$.

If $x = 1$, $y = (1)^2 + 2(1) - 1 = 2$, $(1, 2)$.

If $x = -3$, $y = (-3)^2 + 2(-3) - 1 = 2$, $(-3, 2)$.

37. $x = 2 - y^2$ (Reverse the roles for x and y)

To find the x-intercept when $y = 0$, we have $x = 2$.

For the y-intercepts when $x = 0$, we have $0 = 2 - y^2$

$$\pm\sqrt{2} = y = 1.414 \quad \text{or} \quad -1.414$$

To find the vertex, complete the square:

$$x = 2 - y^2$$

$$x = -1(y - 0)^2 + 2$$

Vertex $(h, k) = (2, 0)$.

The y-value of the vertex is 0 so the axis is at $y = 0$.

The graph opens to the left since $a < 0$.

Draw the graph using this info.

Let $y = 1$, $x = 2 - (1)^2 = 1$, $(1, 1)$.

Let $y = -1$, $x = 2 - (-1)^2 = 1$, $(1, -1)$.

Let $y = 2$, $x = 2 - (2)^2 = -2$, $(-2, 2)$.

Let $y = -2$, $x = 2 - (-2)^2 - 2$, $(-2, -2)$.

41. $x = 3(y - 1)^2 + 2$

Since h and k are reversed in the formula, the vertex $(h, k)$ is $(2, 1)$. The graph opens to the right because $a > 0$; the axis is $y = 1$.

Some points for graphing are found:

Let $y = 0$, $x = 5$, $(5, 0)$. *x-intercept, 5*

Let $y = 2$, $x = 5$, $(5, 2)$.

45. $y = x^2 + 4x - 3$

$= (x^2 + 4x + 4) - 4 - 3$ *Complete the square*

$= 1(x + 2)^2 - 7$

The vertex (h, k) is (-2, -7) and the graph opens upward since $a > 0$. The axis is $x = -2$. Plot more points:

Let $x = 0$, $y = -3$, (0, -3).
Let $x = -4$, $y = -3$, (-4, -3).

49. $y = -2x^2 + 4x + 5$

$= -2(x^2 - 2x \quad) + 5$ *Factor out -2*

$= -2(x^2 - 2x + 1) + 2 + 5$ *-2(1) was added inside the parentheses, so add 2*

$= -2(x - 1)^2 + 7$

The vertex at (h, k) is (1, 7) with the axis at $x = 1$. The parabola opens downward since $a < 0$.
Plotting more points:
When $x = 0$, $y = 5$, (0, 5).
When $x = 2$, $y = 5$, (2, 5).

53. $x = 4y^2 + 8y + 2$

$= 4(y^2 + 2y \quad) + 2$ *Factor out 4*

$= 4(y^2 + 2y + 1) - 4 + 2$ *Add 4, subtract 4*

$= 4(y + 1)^2 - 2$ *Complete the square*

The vertex (h, k) is (-2, -1); k and h are reversed in the formula. The axis is $y = -1$ and the graph opens to the right since $a > 0$. Plotting more points:

When $y = 0$, $x = 2$, (2, 0).
When $y = -2$, $x = 2$, (2, -2).

57. $p = 980 - 5x^2$

(a) If $x = 5$, then substitute 5 for x to get

$$p = 980 - 5(5)^2 = 980 - 125 = 855.$$

(b) If $x = 10$,

$$p = 980 - 5(10)^2 = 980 - 500 = 480.$$

(c) If $x = 14$,

$$p = 980 - 5(14)^2 = 980 - 5\cdot 196 = 980 - 980 = 0$$

(d) Write $p = 980 - 5x^2$ in $p = a(x - h)^2 + k$ form.

$$p = -5x^2 + 980$$
$$p = -5(x - 0)^2 + 980$$

The vertex is (0, 980) and the graph opens downward since $a < 0$. Draw the graph using this incormation and the points (5, 855), (10, 480), and (14, 0) from Parts (a), (b), and (c).

61. Let x represent the width of the rectangle and (52 - x) the length.

$A = LW$ so $A = x(52 - x) = 52x - x^2$.

Find the vertex of the parabola $A = 52x - x^2$ to find the width x at the maximum area A.

$$A = 52x - x^2$$
$$= -(x^2 - 52x + 676) + 676$$
$$= -(x - 26)^2 + 676$$

The vertex is (26, 676) so the maximum area of the rectangle is 676 when the width is 26 units.

65. (2, -1) and (4, 3)

$$d = \sqrt{(x_2 - x_1)^2 + (y_2 - y_1)^2}$$
$$= \sqrt{(4 - 2)^2 + (3 - (-1))^2}$$
$$= \sqrt{4 + 16}$$
$$= 2\sqrt{5}$$

69. (x, y) and (-4, -3)

$$d = \sqrt{(x_2 - x_1)^2 + (y_1 - y_2)^2}$$
$$= \sqrt{(-4 - x)^2 + (-4 - y)^2}$$
$$= \sqrt{(x + 4)^2 + (y + 4)^2}$$

**Section 9.2 (page 390)**

1. Center at (2, 4), radius 5
Substitute $h = 2$, $k = 4$, and $r = 5$ into
$$(x - h)^2 + (y - k)^2 = r^2,$$
the equation of the circle with center (h, k) and radius r. Thus,
$$(x - 2)^2 + (y - 4)^2 = 25.$$

5. Center at (0, -1), radius 4
Substitute $h = 0$, $k = -1$, and $r = 4$ into
$$(x - h)^2 + (y - k)^2 = r^2$$
to get
$$(x - 0)^2 + (y - (-1))^2 = (4)^2$$
or $$x^2 + (y + 1)^2 = 16.$$

9. $x^2 + y^2 + 8x + 4y - 29 = 0$

$x^2 + 8x + y^2 + 4y = 29$ *Rearrange terms*

$x^2 + 8x + 16 - 16 + y^2 + 4y + 4 - 4 = 29$ *Complete the square*

$(x + 4)^2 + (y + 2)^2 = 29 + 20$

$(x + 4)^2 + (y + 2)^2 = 49$

$= 7^2$ *Factor and simplify*

The center is (-4, -2) and the radius is 7.

In Exercises 13-33, see answer graph in textbook.

13. Write $x^2 + y^2 = 16$ in $(x - h)^2 + (y - k)^2 = r^2$ form.
$$(x - 0)^2 + (y - 0)^2 = 4^2$$
The center is at (0, 0) and the radius is 4.

17. Write $y^2 = 144 - x^2$ in $(x - h)^2 + (y - k)^2 = r^2$ form.

$x^2 + y^2 = 144$ *Rearrange terms*

$(x - 0)^2 + (y - 0)^2 = (12)^2$

The center is at (0, 0); the radius is 12.

21. $(x + 4)^2 + (y + 5)^2 = 36$ can be written $(x - (-4))^2 + (y - (-5))^2 = 6^2$. The center is at (-4, -5); the radius is 6.

25. $\frac{x^2}{9} + \frac{y^2}{16} = 1$ is of the form
$$\frac{x^2}{a^2} + \frac{y^2}{b^2} = 1.$$
Since $a^2 = 9$, then $a = \pm 3$. And since $b^2 = 16$, then $b = \pm 4$.

The graph intersects the x-axis at 3 and -3, and the y-axis at 4 and -4. Sketch the ellipse through these intercepts.

29. $\frac{x^2}{9/4} + \frac{y^2}{25/16} = 1$ is of the form $\frac{x^2}{a^2} + \frac{y^2}{b^2} = 1$. Since $a^2 = 9/4$, $a = \pm 3/2$. Since $b^2 = \frac{25}{16}$, $b = \pm 5/4$.

The x-intercepts are 3/2 and -3/2. The y-intercepts are 5/4 and -5/4. Sketch the ellipse through these intercepts.

33. $\frac{(x-2)^2}{16} + \frac{(y-1)^2}{9} = 1$ is of the form $\frac{(x-h)^2}{a^2} + \frac{(y-k)^2}{b^2} = 1$,

where $a^2 = 16$ so $a = \pm 4$. The center of the ellipse is (2, 1). Use $a = \pm 4$ and go 4 units right or left from the center, getting (6, 1) and (-2, 1). Since $b^2 = 9$, then $b = \pm 3$. Go 3 units up and down from (2, 1), getting (2, 4) and (2, -2). Draw an ellipse through all these points.

37. (a) (x, y) is on the parabola. The closest point to (x, y) that is on the line $x = -p$ is (-p, y) since the segment joining (x, y) to (-p, y) is perpendicular to $x = -p$. The distance from (x, y) to (-p, y) is

$$d = \sqrt{(x - (-p))^2 + (y - y)^2}$$
$$= \sqrt{(x - p)^2 + 0}$$
$$d = |x + p|.$$

(b) If (p, 0) is the point, then the origin (0, 0) is equidistant from the point and the line, and will be on the parabola. As a matter of fact, (0, 0) will be the vertex of the parabola.

(c) The distance from (x, y) to (p, 0) is

$$d = \sqrt{(x - p)^2 + (y - 0)^2}$$
$$d = \sqrt{x^2 - 2px + p^2 + y^2}.$$

(d) The distance from (x, y) to (p, 0) equals the distance from (x, y) to (-p, y).

$$\sqrt{x^2 - 2px + p^2 + y^2} = |x + p|$$
$$\left(\sqrt{x^2 - 2px + p^2 + y^2}\right)^2 = (|x + p|)^2 \quad \textit{Square both sides}$$
$$x^2 - 2px + p^2 + y^2 = (x + p)^2$$
$$x^2 - 2px + p^2 + y^2 = x^2 + 2xp + p^2 \quad \textit{Simplify}$$
$$y^2 = 4px$$

This is an equation of the parabola.

41.
$$4x - 3y = 5$$
$$4x - 3(0) = 5 \quad \textit{Let } y = 0$$
$$4x = 5$$
$$x = \frac{5}{4} \quad \textit{x-intercept}$$

$$4(0) - 3y = 5 \quad \textit{Let } x = 0$$
$$-3y = 5$$
$$y = -\frac{5}{3} \quad \textit{y-intercept}$$

45. $y = 4$

Since y is always 4, it is never 0 so there is no x-intercept. If $x = 0$, $y = 4$, so 4 is the y-intercept.

**Section 9.3 (page 397)**

For Exercises 1-21, see answer graph in textbook.

1. $\frac{x^2}{25} - \frac{y^2}{9} = 1$ is in the form $\frac{x^2}{a^2} - \frac{y^2}{b^2} = 1$ so it is a hyperbola with $a = 5$, $b = 3$, and x-intercepts $a = 5$ and $-a = -5$, but no y-intercepts. To sketch, draw the rectangle with corners (5, 3), (5, -3), (-5, -3), and (-5, 3). Then draw a hyperbola through the intercepts and with asymptotes that are extension of the diagonals of the rectangle.

5. $\frac{x^2}{64} - \frac{y^2}{100} = 1$ is a hyperbola of the form $\frac{x^2}{a^2} - \frac{y^2}{b^2} = 1$ with $a = 8$, $b = 10$, and x-intercepts 8 and -8. Graph the points (8, 10), (-8, 10), (-8, -10), (8, -10) and draw the diagonals of the rectangle with these points as corners. These lines are the asymptotes of the hyperbola.

9. $\frac{y^2}{4/9} - \frac{x^2}{9/4} = 1$ is a hyperbola in the form $\frac{y^2}{b^2} - \frac{x^2}{a^2} = 1$ with $a = 3/2$, $b = 2/3$, and y-intercepts 2/3 and -2/3. The asymptotes are the extended diagonals of the rectangle with corners at (3/2, 2/3), (3/2, -2/3), (-3/2, -2/3), and (-3/2, 2/3).

13.
$$4x^2 = 16 - 4y^2$$
$$4x^2 + 4y^2 = 16 \quad \textit{Add } 4y^2$$
$$x^2 + y^2 = 4 \quad \textit{Divide by 4}$$
$$(x - 0)^2 + (y - 0)^2 = 2^2$$

This is a circle with center (0, 0) and radius 2.

17.
$$y^2 = 144 - x^2$$
$$x^2 + y^2 = 144$$
$$(x - 0)^2 + (y - 0)^2 = 12^2$$

This is a circle with center (0, 0) and radius 12.

21.
$$25x^2 = 225 + 9y^2$$
$$25x^2 - 9y^2 = 225$$
$$\frac{x^2}{9} - \frac{y^2}{25} = 1$$

This is a hyperbola in the form

$$\frac{x^2}{a^2} - \frac{y^2}{b^2} = 1$$

with $a = 3$, $b = 5$.
The x-intercepts are 3 and -3. The asymptotes are the extended diagonals of the rectangle with corners (3, 5), (3, -5) (-3, -5), and (-3, 5). Sketch the hyperbola with these intercepts and asymptotes.

25. (a) $100x^2 + 324y^2 = 32,400$
$$\frac{x^2}{324} + \frac{y^2}{100} = 1 \quad \textit{Divide by 32,400}$$
$$\frac{x^2}{18^2} + \frac{y^2}{10^2} = 1 \quad \textit{In } \frac{x^2}{a^2} + \frac{y^2}{b^2} = \textit{1 form}$$

The height in the center is the positive y-intercept b, or 10 meters.

(b) The width of the ellipse is the distance between the x-intercepts which are 18 and -18. Thus, the width across the bottom is 36 meters.

29. $$\frac{y^2}{36} - \frac{(x-2)^2}{49} = 1$$

$$\frac{(y-0)^2}{36} - \frac{(x-2)^2}{49} = 1$$

This hyperbola is the hyperbola $\frac{y^2}{36} - \frac{x^2}{49} = 1$ with y-intercepts 6 and -6 and with asymptotes that are the diagonals of the rectangle with corners (7, 6), (7, -6), (-7, -6), and (-7, 6), shifted two units to the right. (The center of the shifted hyperbola is (2, 0)).
See the answer graph in textbook.

33. (a) When a cone is cut by a plane exactly at the tip, the graph is exactly one point.
(b) When a plane cuts the cone on its edge, the graph is a straight line along the side of the cone.
See the drawings in the answers in the textbook.

37. $x^4 - 16 = 0$

$(x^2 - 4)(x^2 + 4) = 0$ *Difference of two squares*

$x^2 - 4 = 0$ or $x^2 + 4 = 0$

$x^2 = 4$ $\qquad x^2 = -4$

$x = \pm 2$ or $x = \pm 2i$

Solution set: {2, -2, 2i, -2i}

**Section 9.4 (page 403)**

1. $y = x^2 + 2x$ *(1)*
$y = x$ *(2)*

$x = x^2 + 2x$ *Substitute x for y in (1)*

$0 = x^2 + x$

$0 = x(x + 1)$

$x = 0$ or $x + 1 = 0$
$\qquad x = -1$

If $x = 0$, $y = x$ *(2)* $y = 0$.

If $x = -1$, $y = x$ *(2)* $y = -1$.

Solution set: {(0, 0), (-1, -1)}

5. $x^2 + y^2 = 1$ *(1)*
$x + 2y = 1$ *(2)*

Solve equation (2) for x.

$x = 1 - 2y$

$(1 - 2y)^2 + y^2 = 1$ *Let x = 1 - 2y in (1)*

$1 - 4y + 4y^2 + y^2 = 1$

$5y^2 - 4y = 0$

$y(5y - 4) = 0$ *Factor*

$y = 0$ or $5y - 4 = 0$
$\qquad y = \frac{4}{5}$

If $y = 0$,
$x = 1 - 2y$
$x = 1 - 2(0)$
$x = 1$.

If $y = \frac{4}{5}$,
$x = 1 - 2y$
$x = 1 - 2(\frac{4}{5})$
$x = \frac{5}{5} - \frac{8}{5}$
$x = -\frac{3}{5}$.

Solution set: $\{(1, 0), (-\frac{3}{5}, \frac{4}{5})\}$

9. $xy = -6$ *(1)*

$x + y = -1$ *(2)*

Solve equation (2) for y.

$$y = -x - 1$$

$$x(-x - 1) = -6$$ *Let $y = -x - 1$ in (1)*

$$-x^2 - x = -6$$

$$-x^2 - x + 6 = 0$$

$$x^2 + x - 6 = 0$$ *Multiply each term by -1*

$$(x + 3)(x - 2) = 0$$ *Factor*

$x + 3 = 0$ or $x - 2 = 0$

$x = -3$ or $x = 0$

If $x = -3$,

$$xy = -6$$
$$-3y = -6$$
$$y = 2.$$

If $x = 2$,

$$xy = -6$$
$$2y = -6$$
$$y = -3.$$

Solution set: $\{(-3, 2), (2, -3)\}$

13. $y = 2x^2 + 4x$ *(1)*

$y = x^2 - 3x - 10$ *(2)*

Substitute $2x^2 + 4x$ for y in (2).

$$2x^2 + 4x = x^2 - 3x - 10$$

$$x^2 + 7x + 10 = 0$$

$$(x + 5)(x + 2) = 0$$ *Factor*

$x + 5 = 0$ or $x + 2 = 0$

$x = -5$ or $x = -2$

If $x = -5$,

$$y = 2x^2 + 4x$$ *(1)*

$$y = 2(-5)^2 + 4(-5)$$
$$= 50 - 20$$
$$= 30.$$

If $x = -2$,

$$y = 2(-2)^2 + 4(-2)$$
$$= 8 - 8$$
$$= 0$$

Solution set: $\{(-5, 30), (-2, 0)\}$

17. $x^2 - xy + y^2 = 0$ *(1)*

$x - 2y = 1$ *(2)*

Solve (2) for x to get

$$x = 1 + 2y$$

and substitute in (1).

$$(1 + 2y)^2 - (1 + 2y)y + y^2 = 0$$

$$1 + 4y + 4y^2 - y - 2y^2 + y^2 = 0$$

$$3y^2 + 3y + 1 = 0$$

Solve using the quadratic formula.

$$y = \frac{-3 \pm \sqrt{(3)^2 - 4 \cdot 3 \cdot 1}}{2 \cdot 3}$$

$$y = \frac{-3 \pm i\sqrt{3}}{6}$$

$$y = -\frac{1}{2} + \frac{\sqrt{3}}{6}i \quad \text{or} \quad y = -\frac{1}{2} - \frac{\sqrt{3}}{6}i$$

If $y = -\frac{1}{2} + \frac{\sqrt{3}}{6}i$,

$$x - 2y = 1$$
$$x - 2\left(-\frac{1}{2} + \frac{\sqrt{3}}{6}i\right) = 1$$
$$x + 1 - \frac{\sqrt{3}}{3}i = 1$$
$$x = \frac{\sqrt{3}}{3}i.$$

If $y = -\frac{1}{2} - \frac{\sqrt{3}}{6}i$,

$$x - 2y = 1$$
$$x - 2\left(-\frac{1}{2} - \frac{\sqrt{3}}{6}i\right) = 1$$
$$x + 1 + \frac{\sqrt{3}}{3}i = 1$$
$$x = \frac{\sqrt{3}}{3}i$$

Solution set:

$$\left\{\left(\frac{\sqrt{3}}{3}i, -\frac{1}{2} + \frac{\sqrt{3}}{6}i\right), \left(-\frac{\sqrt{3}}{3}i, -\frac{1}{2} - \frac{\sqrt{3}}{6}i\right)\right\}$$

21. $2x^2 + 3y^2 = 6$ *(1)*

$x^2 + 3y^2 = 3$ *(2)*

$2x^2 + 3y^2 = 6$ *(1)*

$-x^2 - 3y^2 = -3$ *Multiply (2) by -1*

$x^2 = 3$

$x = \pm\sqrt{3}$

$(\pm\sqrt{3})^2 + 3y^2 = 3$ *Let $x = \pm\sqrt{3}$ in (2)*

$3 + 3y^2 = 3$

$y^2 = 0$

$y = 0$

Solution set: $\{(\sqrt{3}, 0), (-\sqrt{3}, 0)\}$

25. $xy + y^2 = 3$

$2x^2 - xy - y^2 = 5$

$2x^2 = 8$ *Add the 2 equations to eliminate $y^2$ and $xy$*

$x^2 = 4$

$x = \pm 2$

When $x = 2$,

$y^2 + (2)y = 3$

$y^2 + 2y - 3 = 0$

$(y + 3)(y - 1) = 0$

$y = -3$ or $y = 1$.

This gives the ordered pairs $(2, -3)$, $(2, 1)$.

When $x = -2$,

$y^2 + (-2)y = 3$

$y^2 - 2y - 3 = 0$

$(y - 3)(y + 1) = 0$

$y = 3$ or $y = -1$

This gives $(-2, 3)$, $(-2, -1)$.

Solution set:

$\{(2, -3), (2, 1), (-2, 3), (-2, -1)\}$

29. $3x^2 + 2xy - 3y^2 = 5$ *(1)*

$-x^2 - 3xy + y^2 = 3$ *(2)*

$3x^3 + 2xy - 3y^2 = 5$ *Multiply (2) by 3*

$-3x^2 - 9xy + 3y^2 = 9$

$7xy = 14$

$xy = -2$

$y = \frac{-2}{x}$ *(3)*

Substitute the value of y into (1) and solve for x.

$3x^2 + 2x(\frac{-2}{x}) - 3(\frac{-2}{x})^2 = 5$

$3x^2 - 4 - 3(\frac{4}{x^2}) = 5$

$3x^4 - 4x^2 - 12 = 5x^2$ *Multiply each factor by $x^2$*

$3x^4 - 9x^2 - 12 = 0$

$3(x^4 - 3x^2 - 4) = 0$ *Factor out 3*

$3(x^2 - 4)(x^2 + 1) = 0$

$x^2 - 4 = 0$ or $x^2 + 1 = 0$

$x = \pm 2$ $\quad$ $x = \pm i$

Substitute the four values of x in (3) to get the y values.

If $x = 2$, $y = \frac{-2}{2} = -1$, $(2, -1)$.

If $x = -2$, $y = \frac{-2}{-2} = 1$, $(-2, 1)$.

If $x = i$, $y = \frac{-2}{i} = \frac{-2(i)}{i(i)} = \frac{-2i}{-1} = 2i$, $(i, 2i)$.

If $x = -i$, $y = \frac{-2}{-i} = \frac{-2(i)}{-i(i)} = \frac{-2i}{-(-1)}$

$= -2i$, $(-1, -2i)$

Solution set:

$\{(2, -1), (-2, 1), (i, 2i), (-i, -2i)\}$

33. Let x represent the length and y the width.

$A = LW$ $\qquad P = 2L + W$

$84 = xy$ $\qquad 38 = 2x + 2y$

Solve $84 = xy$ for y

$$y = \frac{84}{x}$$

$$2x + 2\left(\frac{84}{x}\right) = 38 \quad \textit{Let } y = \frac{84}{x} \textit{ in } 38 = 2x + 2y$$

$$2x^2 + 168 = 38x \quad \textit{Multiply by } x$$

$$2x^2 - 38x + 168 = 0$$

$$2(x^2 - 19x + 84) = 0$$

$$2(x - 12)(x - 7) = 0$$

$$x = 12 \text{ or } x = 7$$

Substitute the values of x in $y = \frac{84}{x}$, getting $y = 12$ or $y = 7$. Using the larger value for length, the length is 12 feet and the width is 7 feet.

37. Graph $2x - y = 4$ with a solid line through the points (2, 0) and (0, -4). Test (0, 0) in $2x - y \leq 4$ to get $0 \leq 4$, a true statement. So shade the side of the line containing the origin. See answer graph in textbook.

41. Graph $2x = y$ with a solid line through (0, 0) and (1, 2). Since (0, 0) is on the line, we must test a different point, say (5, 0) in $2x \leq y$. Then we have $10 \leq 0$, a false statement, so shade the side of the line not containing the point (5, 0). See answer graph in textbook.

**Section 9.5 (page 407)**

In Exercises 1-45, see the answer graphs in the textbook.

1. $y < x^2$
To graph $y < x^2$, first graph the parabola $y = x^2$ which opens upward, with a dashed curve. The parabola divides the plane into two parts. Test a point not on the parabola, say (3, 0), in $y < x^2$ to get $0 < 3^2$ or $0 < 9$. Since this is a true statement, shade the region of the plane that includes (3, 0), that is the region outside of the parabola.

5. $y^2 \leq 16 + x^2$

Simplify the equation

$$y^2 = 16 + x^2.$$

$$y^2 - x^2 = 16$$

$$\frac{y^2}{16} - \frac{x^2}{16} = 1 \quad \textit{Divide each term by 16}$$

This is a hyperbola of the form

$$\frac{y^2}{a^2} - \frac{x^2}{b^2} = 1$$

intersecting the y-axis at ±4. The graph is a solid curve, because of the $\leq$ sign. Now test a point not on the curve, such as (0, 0), in $y^2 \leq 16 + x^2$.

$$0^2 \leq 16 + 0^2$$

or $$0 \leq 16 \quad \textit{True}$$

Because (0, 0) is between the two branches of the hyperbola, and (0, 0) produced a true statement above, the shading is between the two branches.

9. $9x^2 > 4y^2 - 36$

Simplify the equation

$$9x^2 = 4y^2 - 36.$$

$$36 = 4y^2 - 9x^2$$

$$1 = \frac{y^2}{9} - \frac{x^2}{4} \quad \textit{Divide each term by 36}$$

or $$\frac{y^2}{9} - \frac{x^2}{4} = 1.$$

This is a hyperbola of the form $\frac{y^2}{a^2} - \frac{x^2}{b^2} = 1$, intersecting the y-axis at ±3. The graph is a dashed curve because of the $>$ sign. Now, test a point not on the curve, such as (0, 0),

$$9x^2 > 4y^2 - 36.$$

$$9 \cdot 0^2 > 4 \cdot 0^2 - 36$$

$$0 > -36 \quad \textit{True}$$

Because (0, 0) is between the two branches of the hyperbola, and (0, 0) produced a true statement above, shade the region between the two branches.

13. $2x^2 + 2y^2 \geq 98$

Simplify

$$2x^2 + 2y^2 = 98.$$

$$x^2 + y^2 = 49 \quad \textit{Divide by 2}$$

This is a solid circle of radius 7 with center at the origin. Test a point not on the circle, say (0, 0) in the inequality to get $0 \geq 98$, a false statement. Since (0, 0) is inside the circle and produces a false statement, shade the region outside the circle.

17. $y^2 - 9x^2 \geq 9$

Simplify the equation

$$y^2 - 9x^2 = 9.$$

$$\frac{y^2}{9} - \frac{x^2}{1} = 1 \quad \textit{Divide each term by 9}$$

This is a hyperbola of the form

$$\frac{y^2}{a^2} - \frac{x^2}{b^2} = 1,$$

intersecting the y-asis at ±3. The graph is a solid curve because of the $\geq$ sign. Now, test a point not on the curve, such as (0, 0), in

$$y^2 - 9x^2 \geq 9.$$

$$0^2 - 9 \cdot 0^2 \geq 0$$

$$0 \geq 9 \quad \textit{False}$$

Because (0, 0) is between the two branches of the hyperbola, and (0, 0) produced a false statement, the shading is in the region above and below the two branches.

21. $x < y^2 - 3,\ x < 0$

$$x = y^2 - 3$$

$$x = (y - 0)^2 - 3.$$

This is a parabola of the form $x = a(y - k)^2 + h$ with the vertex at (-3, 0), opening to the right with the same shape as $x = y^2$. The graph is a dashed curve because of the $<$ sign. Test a point not on the curve, such as (0, 0), in

$$x < y^2 - 3.$$

$$0 < 0^2 - 3$$

$$0 < -3 \quad \textit{False}$$

Because (0, 0) is inside the parabola, and (0, 0) produced a false statement, then the shading will be outside the parabola. Note that by including the added conditions that $x < 0$, we only consider the graph and shading to the left of, but not including, the y-axis.

25. $x^2 + 4y^2 \geq 1$, $x \geq 0$, $y \geq 0$

Simplify the equation

$$x^2 + 4y^2 = 1,$$

$$\frac{x^2}{1} + \frac{y^2}{1/4} = 1.$$

The graph is a solid ellipse with x-intercepts 1 and -1 and y-intercepts 1/2 and -1/2. Test a point not on the ellipse, say (0, 0) in the inequality.

$$0^2 + 4(0)^2 \geq 1$$

$$0 \geq 1 \quad \textit{False}$$

Thus, shade the region outside the ellipse. The restrictions $x \geq 0$ and $y \geq 0$ give exactly the first quadrant and the positive x- and y-axes. The graph is the intersection of this region and the region on the ellipse and outside of it.

29. $3x + 2y < 12$
$x - 3y < 6$

For $3x + 2y < 12$ graph a dashed line through the points (4, 0) and (0, 6). Test (0, 0) to get $0 < 12$, a true statement. So shade the side of the line containing the origin. For $x - 3y < 6$, graph the dashed line through (6, 0) and (0, -2). Test (0, 0) to get $0 < 6$, a true statement. Shade the side of this line containing the origin. The overlap of the shaded regions is the intersection.

33. $x \leq 4$
$y \leq 6$

To graph $x \leq 4$ graph $x = 4$ as a solid vertical line and shade to the left. To graph $y \leq 6$, graph $y = 6$ as a solid horizontal line and shade below. The overlap of these 2 regions in the intersection.

37. $x^2 - y^2 \geq 1$
$\frac{x^2}{9} + \frac{y^2}{4} \leq 1$

To graph $x^2 - y \geq 1$, first graph $x^2 - y^2 = 1$ as a solid hyperbola through (1, 0) and (-1, 0). Test (0, 0) in the inequality to get $0 \geq 1$, a false statement. Since (0, 0) lies between the 2 branches of the hyperbola and produces a false statement, do not shade between the branches, but shade the outer regions.

To graph $\frac{x^2}{9} + \frac{y^2}{4} \leq 1$, graph $\frac{x^2}{9} + \frac{y^2}{4} =$ 1 as a solid ellipse with x-intercepts $\pm 3$ and y-intercepts $\pm 2$ and and centered at the origin. Test (0, 0) in the inequality to get $0 \leq 1$, a true statement. Since (0, 0) is inside the ellipse and produces a true statement, shade inside the ellipse. The intersection is the overlapping regions that are shaded.

41. $x > 1$
$y < 2$
$x^2 + y^2 < 4$

$x < 1$ is the right side of the dashed vertical line $x = 1$. To graph $y < 2$, graph $y = 2$ as a dashed horizontal line and shade below. To graph $x^2 + y^2 < 4$, draw the dashed circle $x^2 + y^2 = 4$ with center (0, 0) and radius 2. Test (0, 0) in the inequality to get $0 < 4$, a true statement. Since (0, 0) is inside the circle and yields a true statement, shade the inside of the circle. The region of overlap of all 3 shaded portions is the intersection.

45. $1.2x^2 + 1.2y^2 \geq 7.5$
$5.1x^2 + 2.8y^2 \geq 14.4$

$1.2x^2 + 1.2y^2 = 7.5$
$x^2 + y^2 = 2.5$ *Divide by 1.2*

Graph this equation as a solid circle with radius $\sqrt{2.5} \approx 1.6$, center at the origin. Testing (0, 0) gives a false statement in the inequality. Shade outside the circle.

$$5.1x^2 + 2.8y^2 = 14.4$$
$$\frac{x^2}{2.82} + \frac{y^2}{5.14} = 1$$

Graph as a solid ellipse with x-intercepts $\pm\sqrt{2.82}$ and y-intercepts $\pm\sqrt{5.14}$ centered at the origin. Testing (0, 0) in $5.1x^2 + 2.8y^2 \leq 14.4$ gives $0 \leq 14.4$, a true statement. So shade inside the ellipse. The shaded regions don't overlap, so there are no ordered pairs that are solutions to both inequalities.

49. $9^{3/2} = (\sqrt{9})^3 = 27$

53. $(\frac{16}{81})^{-3/4} = (\frac{81}{16})^{3/4} = \frac{(81)^{3/4}}{(16)^{3/4}}$
$= \frac{(\sqrt{81})^3}{(\sqrt{16})^3} = \frac{(3)^3}{(2)^3} = \frac{27}{8}$

**Chapter 9 Review Exercises (page 409)**

For Exercises 1-9, see answer graphs in textbook.

1. Write $y = -5x^2$ in $y = a(x - h)^2 + k$ form $y = -5(x - 0)^2 + 0$. The vertex $(h, k) = (0, 0)$; since $a < 0$, the graph opens downward. Find additional points for graphing:

If $x = 1$, $y = -5(1)^2 = -5$, (1, -5).
If $x = -1$, $y = -5(-1)^2 = -5$, (-1, -5).

5. $y = (x - 3)^2 - 7$ is in $y = a(x - h)^2 + k$ form where $a = 1$. Thus, the vertex is (3, -7). The graph opens upward since $a > 0$. The axis is $x = 3$.
Find additional points for graphing:

If $x = 0$, $y = 2$, (0, 2).
If $x = 6$, $y = 2$, (6, 2).

9. Write $x = 2y^2 + 3$ in $x = a(y - k)^2 + h$ form. $x = 2(y - 0)^2 + 3$ has vertex $(h, k) = (3, 0)$. Since $a > 0$ the graph opens to the right. The axis is $x = 0$. Find additional points for graphing:

When $y = 1$, $x = 5$, $(5, 1)$.
When $y = -1$, $x = 5$, $(5, -1)$.

13. Write $s = -16t^2 + 160t$ in $s = a(t - h)^2 + k$ form.

$$s = -16(t^2 - 10t)$$
$$s = -16(t^2 - 10t + 25 - 25)$$
$$s = -16[(t - 5)^2 - 25]$$
$$= -16(t - 5)^2 + 400$$

This is a parabola which opens downward. Thus, its maximum occurs at the vertex $(h, k) = (5, 400)$. That is at time 5 seconds.

17. The circle with center at $(-3, 4)$ and radius 5 has equation

$$(x - (-3))^2 + (y - 4)^2 = 5^2$$

or $$(x + 3)^2 + (y - 4)^2 = 25.$$

21. Write $x^2 + y^2 + 8x - 8y + 23 = 0$ in $(x - h)^2 + (y - k)^2 = r^2$ form.

$$x^2 + 8x + y^2 - 8y = -23$$ *Rearrange terms*

$$x^2 + 8x + 16 - 16 + y^2 - 8y + 16 - 16 = -23$$ *Complete the square*

$$(x + 4)^2 + (y - 4)^2 = -23 + 32$$
$$(x + 4)^2 + (y - 4)^2 = 9$$

This is a circle with center $(-4, 4)$ and radius 3.

25. $\frac{x^2}{9} + \frac{y^2}{25} = 1$

The given equation is in $\frac{x^2}{a^2} + \frac{y^2}{b^2} = 1$ form. $\frac{x^2}{9} + \frac{y^2}{25} = 1$ is an ellipse with x-intercepts $\pm 3$ and y-intercepts $\pm 5$. Draw the ellipse throught these points. See answer graph in textbook.

29. $y = x^2 - 2$ can be written $y = (x - 0)^2 - 2$ which is the equation of a parabola.

33. Write $x^2 - y^2 = 16$ in the form

$$\frac{x^2}{a^2} - \frac{y^2}{b^2} = 1.$$

$$\frac{x^2}{16} - \frac{y^2}{16} = 1$$ *Divide by 16*

The graph is a hyperbola.

37. $\frac{x}{2} + \frac{y}{3} = 1$

$3x + 2y = 6$ *Multiply each term by 6*

This is the equation of a straight line.

41. $x + y^2 = 2$

Write in the form $x = a(y - k) + h$.

$$x = -1(y - 0)^2 + 2$$

This is the equation of a parabola.

45. Write $8x^2 + 2y^2 = 5$ in the form $\frac{x^2}{a^2} + \frac{y^2}{b^2} = 1$.

$$\frac{8x^2}{5} + \frac{2y^2}{5} = \frac{5}{5} \quad \textit{Divide each term by 5}$$

$$\frac{x^2}{5/8} + \frac{y^2}{5/2} = 1$$

This is the equation of an ellipse.

49. $\frac{(x - 5)^2}{4} - \frac{(y - 2)^2}{9} = 1$ is a hyperbola of the form $\frac{(x - h)^2}{a^2} - \frac{(y - k)^2}{b^2} = 1$.

The center $(h, k) = (5, 2)$, $a = \pm 2$, $b = \pm 3$. Since the hyperbola opens to the right and left, it passes through the points 2 units right and left of (5, 2). Plot the points 3 units up from (5, 2) and 3 units below (5, 2). Sketch the rectangle and asymptotes from these points.
See answer graph in textbook.

53.
$$y = 2x - x^2 \quad (1)$$
$$x + 2y = -3 \quad (2)$$
$$x + 2(2x - x^2) = -3 \quad \textit{Substitute } y = 2x - x^2 \textit{ in (2)}$$
$$x + 4x - 2x^2 = -3$$
$$2x^2 - 5x - 3 = 0$$
$$(2x + 1)(x - 3) = 0$$
$$2x + 1 = 0 \quad \text{or} \quad x - 3 = 0$$
$$x = -\frac{1}{2} \qquad x = 3$$

When $x = -\frac{1}{2}$,
$$y = 2(-\tfrac{1}{2}) - (-\tfrac{1}{2})^2$$
$$= -\frac{5}{4}.$$

When $x = 3$,
$$y = 2(3) - (3)^2$$
$$= -3.$$

Solution set: $\{(-\frac{1}{2}, -\frac{5}{4}), (3, -3)\}$

57. $4x^2 + 3y^2 = 24 \quad (1)$
$2x^2 - y^2 = 12 \quad (2)$

Substitute the value of $y^2$ in (2) into (1).

$$2x^2 - y^2 = 12$$
$$y^2 = 2x^2 - 12$$
$$4x^2 + 3(2x^2 - 12) = 24$$
$$10x^2 = 60$$
$$x = \pm\sqrt{6}$$

When $x = \pm\sqrt{6}$
$$4(\pm\sqrt{6})^2 + 3y^2 = 24$$
$$24 + 3y^2 = 24$$
$$3y^2 = 0$$
$$y = 0.$$

Solution set: $\{(\sqrt{6}, 0), (-\sqrt{6}, 0)\}$

61.
$$\begin{array}{rl} x^2 + y^2 = & 8 \\ 2x^2 - y^2 = & 4 \\ \hline 3x^2 \quad = & 12 \end{array} \quad \textit{Add the equations to eliminate } y^2$$
$$x^2 = 4$$
$$x = \pm 2$$

When $x = 2$,

$$(2)^2 + y^2 = 8$$
$$y^2 = 4$$
$$y = \pm 2.$$

When $x = -2$,

$$(-2)^2 + y^2 = 8$$
$$y^2 = 4$$
$$y = \pm 2.$$

Solution set:
$\{(2, 2), (2, -2), (-2, 2), (-2, -2)\}$

65. $x^2 + y^2 \le 25$

Graph $x^2 + y^2 = 25$ as a solid circle centered at the origin with radius 5. Test the point $(0, 0)$, $0 \le 25$ is a true statement. Shade the inside of the circle.
See answer graph in the textbook.

69. $3x^2 + y^2 = 36$

Write the equation in the form $\frac{x^2}{a^2} + \frac{y^2}{b^2} = 1$ which is an ellipse.

$$\frac{3x^2}{36} + \frac{y^2}{36} = \frac{36}{36} \quad \textit{Divide by 36}$$
$$\frac{x^2}{12} + \frac{y^2}{36} = 1$$

The x-intercepts are $+2\sqrt{3}$ and $-2\sqrt{3}$.
The y-intercepts are $\pm 6$.
Plot the points $(2\sqrt{3}, 0)$, $(-2\sqrt{3}, 0)$ $(0, 6)$ and $(0, -6)$ and draw the ellipse.
See answer graph in textbook.

73. $9x^2 + 49y^2 \le 441$, $y \ge 0$

Simplify the equation.

$$9x^2 + 49y^2 = 441.$$
$$\frac{x^2}{49} + \frac{y^2}{9} = 1 \quad \textit{Divide each term by 441}$$

Graph as a solid ellipse with x-intercepts at 7 and -7 and y-intercepts 3 and -3. Test the point $(0, 0)$ in the inequality. $0 \le 441$ is a true statement.
The inside of the ellipse is shaded, but only the area above and including $y = 0$ because of the restriction $y \ge 0$.
See answer graph in the textbook.

**Chapter 9 Test (page 411)**

1. Simplify $2x^2 + 2y^2 = 36$ by dividing each term by 2 to get

$$x^2 + y^2 = 18.$$

This is an equation of a circle.

2. Simplify $6x^2 - 4y^2 = 12$.

$$\frac{6x^2}{12} - \frac{4y^2}{12} = 1 \quad \textit{Divide each term by 12}$$
$$\frac{x^2}{2} - \frac{y^2}{3} = 1 \quad \textit{Simplify}$$

This is an equation of a hyperbola of the form

$$\frac{x^2}{a^2} - \frac{y^2}{b^2} = 1$$

with $a = \sqrt{2}$, $b = \sqrt{3}$.

3. Write $3y^2 + 3x = 8$ as

$$3x = -3y^2 + 8.$$

Divide each term by 3 to get

$$x = -y^2 + \frac{8}{3}$$

or $\quad x = -(y - 0)^2 + \frac{8}{3}$.

This is an equation of a parabola in $x = a(y - k)^2 + h$ form, with $h = 8/3$ and $k = 0$.

4. Simplify $9x^2 + 25y^2 = 225$.

$$\frac{9}{225}x^2 + \frac{25}{225}y^2 = 1 \quad \textit{Divide each term by 225}$$

$$\frac{x^2}{25} + \frac{y^2}{9} = 1 \quad \textit{Simplify}$$

This is an equation of an ellipse in the form

$$\frac{x^2}{a^2} + \frac{y^2}{b^2} = 1,$$

with $a = 5$, $b = 3$.

5. Write $2x = 9 - y^2$ as $2x = -y^2 + 9$. Divide each term by 2 and write in $x = a(y - k)^2 + h$ form.

$$x = -\frac{1}{2}(y - 0)^2 + \frac{9}{2}$$

This is an equation of a parabola, with $h = 9/2$ and $k = 0$.

6. $(y - 2)^2 + (x + 3)^2 = 5$ can be written $(y - 2)^2 + (x - (-3))^2 = 5$. This is an equation of a circle.

7. Simplify $x^2 + y^2 - 4x + 2y - 2 = 0$.

$$x^2 - 4x + y^2 - 2y = 2$$

$$x^2 - 4x + 4 - 4 + y^2 - 2y + 1 - 1 = 2 \quad \textit{Complete the square}$$

$$(x^2 - 4x + 4) + (y^2 - 2y + 1) = 2 + 5$$

$$(x - 2)^2 + (y - 1)^2 = 7$$

This is an equation of a circle.

8. Simplify $5y = 2(x + 3)^2 - 4$.

$$y = \frac{2}{5}(x + 3)^2 - \frac{4}{5} \quad \textit{Divide by 5}$$

$$y = \frac{2}{5}(x - (-3))^2 - \frac{4}{5}$$

This is an equation of a parabola in the form $y = a(x - h)^2 + k$, with $h = -3$ and $k = -4/5$.

For Exercises 9-14, see the answer graph in the textbook.

9. $y = x^2 - 4x + 5$

$$y = x^2 - 4x + 4 - 4 + 5 \quad \textit{Complete the square}$$

$$y = 1(x - 2)^2 + 1$$

This is a parabola in form $y = a(x - h)^2 + k$ with vertex (2, 1) and opening upward.
Plot additional points:
When $x = 0$, $y = 5$, (0, 5).
When $x = 4$, $y = 5$, (4, 5).

10. $x = 3 - y^2$

$$x = -y^2 + 3$$

$$x = -1(y - 0)^2 + 3$$

This is a parabola in $x = a(y - k)^2 + h$ form, with vertex (3, 0), opening to the left.
Plot additional points:
When $y = 2$, $x = -1$, (-1, 2).
When $y = -2$, $x = -1$, (-1, -2).

11. $(x + 1)^2 + (y - 4)^2 = 3$ can be written $(x - (-1))^2 + (y - 4)^2 = 3$. This is a circle of radius $\sqrt{3}$ centered at (-1, 4).

12.
$$x^2 + y^2 - 6x - 8y = 0$$

$$x^2 - 6x + y^2 = 0 \quad \textit{Rearrange terms}$$

$$x^2 - 6x + 9 - 9 + y^2 - 8y + 16 - 16 = 0 \quad \textit{Complete the square}$$

$$(x - 3)^2 + (y - 4)^2 = 25$$

This equation is a circle with center (3, 4) and radius $\sqrt{25} = 5$.

13. Simplify $9x^2 + 16y^2 = 144$.

$$\frac{9x^2}{144} + \frac{16y^2}{144} = 1 \quad \textit{Divide each term by 144}$$

$$\frac{x^2}{16} + \frac{y^2}{9} = 1$$

This equation is of the form

$$\frac{x^2}{a^2} + \frac{y^2}{b^2} = 1,$$

an ellipse intersecting the x-axis at $a = \pm 4$, and intersecting the y-axis at $b = \pm 3$.

14. $4x^2 - 25y^2 = 100$

$$\frac{x^2}{25} - \frac{y^2}{4} = 1 \quad \textit{Divide each term by 100}$$

This equation is of the form

$$\frac{x^2}{a^2} - \frac{y^2}{b^2} = 1,$$

a hyperbola with $a = \pm 5$, $b = \pm 2$ and x-intercepts 5 and -5.

15. $(x + 6)^2 + (y - 2)^2 = 144$

$$(x - (-6))^2 + (y - 2)^2 = 12^2.$$

This is the equation of a circle with center (-6, 2) and radius 12.

16. $x^2 - 2x + y^2 - 6y = 6$

$$x^2 - 2x + 1 - 1 + y^2 - 6y + 9 - 9 = 6 \quad \textit{Complete the square}$$

$$(x - 1)^2 + (y - 3)^2 = 6 + 10$$

$$(x - 1)^2 + (y - 3)^2 = 16$$

This is a circle centered at (1, 3) with radius $\sqrt{16} = 4$.

17. Simplify $d = -2t^2 + 12t + 32$.

$$d = -2(t^2 - 6t) + 32$$
$$= -2(t^2 - 6t + 9 - 9) + 32$$
$$= -2(t^2 - 6t + 9) + 18 + 32$$
$$= -2(t - 3)^2 + 50$$

This is a parabola with vertex (3, 50), opening downward. Since the graph of the parabola has a maximum value at the vertex (3, 50), the maximum distance the object reaches is 50 feet. The object reaches that distance in 3 seconds.

18. $x + 4y = 10$ *(1)*

$xy = 4$ *(2)*

Solve (1) for x.

$$x = 10 - 4y$$

$$(10 - 4y)(y) = 4 \quad \textit{Let } x = 10 - 4y \textit{ in (2)}$$

$$10y - 4y^2 = 4$$

$$0 = 4y^2 - 10y + 4 \quad \textit{Rearrange terms}$$

$$0 = 2y^2 - 5y + 2 \quad \textit{Divide by 2}$$

$$0 = (2y - 1)(y - 2) \quad \textit{Factor}$$

$2y - 1 = 0$ or $y - 2 = 0$

$y = \frac{1}{2}$ or $y = 2$

$$x = 10 - 4y$$
$$x = 10 - 4(\tfrac{1}{2}) \quad \textit{Let } y = \tfrac{1}{2}$$
$$= 10 - 2$$
$$= 8$$

Thus, one solution is $(8, \frac{1}{2})$.

Using $y = 2$ in $x = 10 - 4y$, we have

$x = 10 - 4 \cdot 2$ *Let y = 2*
$= 10 - 8$
$= 2$

The other solution is (2, 2)

Solution set: $\{(8, \frac{1}{2}), (2, 2)\}$

19. $y + 2 = 3x$ *(1)*
$x^2 + y^2 = 2$ *(2)*
Solve (1) for y.
$y = 3x - 2$

$x^2 + (3x - 2)^2 = 2$ *Let y = 3x - 2 in (2)*

$x^2 + 9x^2 - 12x + 4 = 2$

$10x^2 - 12x + 2 = 0$

$5x^2 - 6x + 1 = 0$ *Divide by 2*

$(5x - 1)(x - 1) = 0$ *Factor*

$5x - 1 = 0$ or $x - 1 = 0$

$x = \frac{1}{5}$ or $x = 1$

$y = 3x - 2$

$y = 3(\frac{1}{5}) - 2$ *Let $x = \frac{1}{5}$*

$= \frac{3}{5} - \frac{10}{5}$

$= -\frac{7}{5}.$

One solution is $(\frac{1}{5}, -\frac{7}{5})$.

$y = 3 \cdot 1 - 2$ *Let x = 1*
$= 3 - 2$
$= 1.$

The other solution is (1, 1).

Solution set: $\{(\frac{1}{5}, -\frac{7}{5}), (1, 1)\}$

20. $x^2 + y^2 = 16$
$2x^2 - y^2 = 8$
$3x^2 = 24$
$x^2 = 8$ *Divide by 3*
$x = \pm\sqrt{8}$
$= \pm 2\sqrt{2}$

$(\pm 2\sqrt{2})^2 + y^2 = 16$ *Let $x = 2\sqrt{2}$ in (1)*

$8 + y^2 = 16$

$y^2 = 8$

$y = \pm 2\sqrt{2}$

Solution set:
$\{(2\sqrt{2}, 2\sqrt{2}), (2\sqrt{2}, -2\sqrt{2}),$
$(-2\sqrt{2}, 2\sqrt{2}), (-2\sqrt{2}, -2\sqrt{2})\}$

For Exercises 21-24, see answer graphs in the textbook.

21. $y \le x^2 + 2x - 1$

Simplify $y = x^2 + 2x - 1$.

$y = x^2 + 2x + 1 - 1 - 1$
*Complete the square*

$y = (x + 1)^2 - 2$

$y = [x - (-1)]^2 - 2.$

This is a parabola in the form $y = a(x - h)^2 + k$. It has vertex $(h, k) = (-1, -2)$ and opens upward. Draw it with a solid curve since there is a $\le$ sign. Test (0, 0), not on the parabola, in $y \le x^2 + 2x - 1$.

$0 \le 0^2 + 2(0) - 1$
$0 \le -1$ *False*

Thus, shade the side of the parabola not containing (0, 0).

22. $x^2 \geq 9 + y^2$

Simplify $x^2 = 9 + y^2$.

$$x^2 - y^2 = 9$$

$$\frac{x^2}{9} - \frac{y^2}{9} = 1 \quad \textit{Divide each term by 9}$$

This is a hyperbola of the form $\frac{x^2}{a^2} - \frac{y^2}{b^2} = 1$ with $a = \pm 3$, $b = \pm 3$, passing through (3, 0) and (-3, 0). Draw this hyperbola with solid curves. Test a point, say (0, 0) in the inequality $x^2 \geq 9 + y^2$ to get

$$0^2 \geq 9 + 0^2$$

or $0 \geq 9$ which is false.

Since (0, 0) lies between the 2 branches and leads to a false statement, shade the outside of the branches.

23. To graph $2x - 5y \geq 12$, graph $2x - 5y = 12$ as a solid line through (6, 0) and $(0, -\frac{12}{5})$. Test (0, 0) in $2x - 5y \geq 12$ to get $0 \geq 12$. Since this is false, shade the side of the line not containing the origin. To graph $3x + 4y \leq 12$, graph $3x + 4y = 12$ as a solid line through (4, 0) and (0, 3). Test (0, 0) in $3x + 4y \leq 12$ to get $0 \leq 12$. Since this is true, shade the side of the line containing (0, 0). The overlapping shaded region is the intersection.

24. $3x^2 + 2y^2 \leq 24$

Simplify $3x^2 + 2y^2 = 24$.

$$\frac{x^2}{8} + \frac{y^2}{12} = 1 \quad \textit{Divide by 24}$$

Draw this ellipse with x-intercepts $\pm\sqrt{8} = \pm 2\sqrt{2}$ and y-intercepts $\pm\sqrt{12} = \pm 2\sqrt{3}$. Since (0, 0) is inside the ellipse and

$$3(0)^2 + 2(0)^2 \leq 24$$

$$0 \leq 24$$

is a true statement, shade the interior of the ellipse. Draw the ellipse with a solid curve because of the $\leq$ sign.

$$x^2 + y^2 \geq 4$$

Graph $x^2 + y^2 = 4$ as a solid circle of radius 2 with center at the origin. Since (0, 0) is inside the circle and

$$0^2 + 0^2 \geq 4$$

$$0 \geq 4$$

is a false statement, shade the outside of the circle. The overlapping shaded region is the intersection.

## CHAPTER 10 FUNCTIONS

### Section 10.1 (page 419)

1. {(5, 1), (3, 2), (4, 9), (7, 6)}
   Domain: {3, 4, 5, 7}
   Range: {1, 2, 6, 9}

   This is a set of ordered pairs in which no first components are repeated. It is a function.

5. {(-3, 1), (4, 1), (-2, 7)}
   Domain: {-3, -2, 4}
   Range: {1, 7}
   Each domain value is assigned to exactly one range value; ordered pairs can have the same y-value in a function. Thus, it is a function.

9. Both domain and range are the set of all real numbers.
   For each value of x in the equation of the line, there is only one value of y.
   It is a function.

13. Domain is $\{x \mid -4 \leq x \leq 4\}$; range is $\{y \mid -3 \leq y \leq 3\}$.

    Use the vertical line test; the line intersects the graph of the circle in more than one point so it is not a function.

17. Domain is set of all real numbers; range is $\{y \mid y \geq -6\}$.

    Any value of x leads to one value of y; a vertical line intersects the graph at only one point. It is a function.

21. $x = y^2$
    $\{x \mid x \geq 0\}$

    If x = 4, y equals 2 or -2. Since one value of x leads to two values of y, y is not a function of x.

25. $x = y^3$
    Set of all real numbers

    If x = 16, y = 2; $(-2)^3 = -16$. Thus for each value of x, there is one corresponding value of y. It is a function.

29. $y \geq 3x$
    Set of all real numbers
    One value of x leads to many values of y. For example when x = 1, the ordered pairs (1, 3), (1, 4), (1, 5), and soon satisfy the inequality. It is not a function.

33. $y = \frac{x + 1}{x - 7}$

    Since 7 makes the denominator equal 0, the domain is $\{x \mid x \neq 7\}$. For each value of x, there is exactly one value of y. It is a function.

37. $y = \frac{-4}{x^2 + 1}$

    Set of all real numbers
    Any value of x leads to exactly one value of y, so this equation defines a function.

Exercises 41-53, see answer graphs in textbook.

41. $y = 3x - 2$
    Both domain and range are the set of all real numbers.

Since this is the equation of a straight line, each value of x corresponds to one value of y. Also, the vertical line test determines it is a function.

45. $x^2 + y^2 = 16$

Domain: $\{x|-4 \le x \le 4\}$
Range: $\{y|-4 \le y \le 4\}$
This is the graph of a circle; when $x = 0$, $y^2 = 16$ or $y = \pm 4$. Thus one value of x corresponds to two values of y. Also, the vertical line test determines it is not a function.

49. $x = 1 - y^2$

Domain: $\{x|x \le 1\}$ Range: set of all real numbers
Some vertical lines intersect the graph in two points. It is not a function.

53. $\frac{x^2}{4} - \frac{y^2}{9} = 1$

Domain: $\{x|x \le -2 \text{ or } x \ge 2\}$
Range: set of all real numbers
A hyperbola has more than one value of y for each x; some vertical lines intersect the graph at two points. It is not a function.

57. $xy = 1$ Solve for y: $y = \frac{1}{x}$.

All real numbers except 0 can be used for x so the domain is $\{x|x \ne 0\}$. This is a function because each domain value (x) corresponds to exactly one range value (y).

61. (a) Use the vertical line test to see that all are functions.
(b) Domain is set of years from 1975 to 1985; range is $\{y|\$6.35 \le y \le \$12.82\}$.
(c) Domain is set of years from 1975 to 1985; range is $\{y|\$3.75 \le y \le \$7.50\}$.

65. (a) $2x^2 - 4x + 1$

$= 2\cdot 1^2 - 4\cdot 1 + 1$ *Let $x = 1$*
$= 2 - 4 + 1$
$= -1$

(b) $2x^2 - 4x + 1$

$= 2\cdot(-2)^2 - 4(-2) + 1$ *Let $x = -2$*
$= 8 + 8 + 1$
$= 17$

69. (a) $\frac{3x + 1}{2x - 5} = \frac{3\cdot 1 + 1}{2\cdot 1 - 5}$ *Let $x = 1$*

$= \frac{3 + 1}{2 - 5} = \frac{4}{-3}$
$= -\frac{4}{3}$

(b) $\frac{3x + 1}{2x - 5} = \frac{3\cdot(-2) + 1}{3\cdot(-2) - 5}$ *Let $x = -2$*

$= \frac{-6 + 1}{-4 - 5} = \frac{-5}{-9}$
$= \frac{5}{9}$

**Section 10.2 (page 426)**

In Exercises 1-13, $f(x) = 3 + 2x$ and $g(x) = x^2 - 2$.

1. $f(1) = 3 + 2(1) = 5$

5. $g(-1) = (-1)^2 - 2 = 1 - 2 = -1$

9. $f(-2) + f(-3)$
$f(-2) = 3 + 2(-2) = -1$
$f(-3) = 3 + 2(-3) = -3$
$f(-2) + f(-3) = -1 + (-3) = -4$

13. $g(5) + f(-1)$

$g(5) = 5^2 - 2 = 25 - 2 = 23$

$f(-1) = 3 + 2\cdot(-1) = 3 - 2 = 1$

$g(5) + f(-1) = 23 + 1 = 24$

In Exercises 17-25, $f = \{(-1, 2), (0, 1), (1, 5), (2, 5)\}$ and $g = \{(0, 4), (1, 3), (2, 2), (6, 1)\}$

17. $f(-1) = 2$

21. $g(6) + f(2) = 1 + 5 = 6$

25. $(f + g)(1)$

$f(1) = 5,\ g(1) = 3,$

$(f+g)(1) = f(1)+g(1) = 5+3 = 8$

29. $4x + y = 9$

$y = -4x + 9$

$f(x) = -4x + 9 \quad y = f(x)$

$f(-2) = -4(-2) + 9$

$= 17$

$f(1) = -4(1) + 9$

$= 5$

33. $x^2 - y = 4$

$-y = -x^2 + 4$

$y = x^2 - 4$

$f(x) = x^2 - 4 \quad y = f(x)$

$f(-2) = (-2)^2 - 4$

$= 0$

$f(1) = (-1)^2 - 4$

$= -3$

In Exercises 37-45, $f(x) = x^2 + 2x$ and $g(x) = 2 + 7x$.

37. $(f + g)(2) = f(2) + g(2)$

$= (2)^2 + 2(2) + 2 + 7(2)$

$= 8 + 16$

$= 24$

41. $(f - g)(-3)$

$= f(-3) - g(-3)$

$= (-3)^2 + 2(-3) - [2 + 7(-3)]$

$= 3 - (-19)$

$= 22$

45. $(f + g)(r) = f(r) + g(r)$

$= (r^2 + 2r) + (2 + 7r)$

$= r^2 + 9r + 2$

In Exercises 49-61, $f(x) = -x^2 + 2x + 1$ and $g(x) = 4x - 3$.

49. $f(r) = -r^2 + 2r + 1$

53. $g(2q - 5) = 4(2q - 5) - 3$

$= 8q - 23$

57. $(f\circ g)(1)$

$= f(g(1))$

$= f(4(1) - 3)$

$= f(1)$

$= -1^2 + 2(1) + 1 = 2$

61. $(g\circ f)(r)$

$= g(f(r))$

$= g(-r^2 + 2r + 1)$

$= 4(-r^2 + 2r + 1) - 3$

$= -4r^2 + 8r + 1$

65. $f(x) = 9x - 1$

(a) $f(x + h) = 9(x + h) - 1$

$= 9x + 9h - 1$

(b) $f(x + h) - f(x)$

$= (9x + 9h - 1) - (9x - 1)$

$= 9h$

(c) $\dfrac{f(x + h) - f(x)}{h} = \dfrac{9h}{h} = 9$

69. $f(x) = -4x^2 + 6x$

(a) $f(x + h)$

$= -4(x + h)^2 + 6(x + h)$

$= -4(x^2 + 2xh + h^2) + 6x + 6h$

$= -4x^2 - 8xh - 4h^2 + 6x + 6h$

(b) $f(x + h) - f(x)$

$= (-4x^2 - 8xh - 4h^2 + 6x + 6h) - (-4x^2 + 6x)$

$= -4x^2 - 8xh - 4h^2 + 6x + 6h + 4x^2 - 6x$

$= -8xh - 4h^2 + 6h$

(c) $\dfrac{f(x+h) - f(x)}{h} = \dfrac{-8xh - 4h^2 + 6h}{h}$

$= -8x - 4h + 6$

73. $C(x) = 7x$

77. Write the formula as

$f(x) = 35 + .20x.$

(a) $f(8) = 35 + .20(8)$

$= 36.6$

$= \$36.60$

(b) $f(27\frac{1}{2})$

$f(28) = 35 + .20(28)$ *Round fraction of a mile*

$= 40.6$

$= \$40.60$

(c) $f(27\frac{3}{4})$

$f(28) = \$40.60$ *Round fraction to 28*

(d) $f(15) = 35 + .20(15)$

$= 38$

$= \$38.00$

(e) See answer graph in your textbook.

(f) Yes

81. $y = 1 - x^2$ is a parabola with vertex at (0, 1) and opening downward.
See graph in your textbook.

**Section 10.3 (page 432)**

For Exercises 1-45, see the answer graphs in the textbook.

1. $f(x) = 4 - x$
Linear function; graph by letting $x = 0$ to get $y = 4$; letting $x = 4$ gives $y = 0$. Locate (0, 4) and (4, 0) and draw a line through them.

5. $f(x) = x^2$
Quadratic function; the graph is a parabola opening upward and having vertex (0, 0).

9. $f(x) = (x - 1)^2$
Quadratic function; graph is a parabola with vertex at (1, 0) and opening upward.

13. $f(x) = x^2 - 2x + 3$
Quadratic function; complete the square to get

$f(x) = (x^2 - 2x \quad ) + 3$

$= (x^2 - 2x + 1) + 3 - 1$

$= (x - 1)^2 + 2,$

a parabola with vertex (1, 2) and opening upward.

17. $f(x) = x^3$

Polynomial function; find some points satisfying $y = x^3$:

| x | -2 | -1 | -1/2 | 0 | 1/2 | 1 | 2 |
|---|---|---|---|---|---|---|---|
| y | -8 | -1 | -1/8 | 0 | 1/8 | 1 | 8 |

Locate these points and draw a smooth curve through them.

21. $f(x) = -x^3 + 2x^2 + 5x - 6$

Polynomial function; find some points satisfying the function:

| x | -2 | -1 | 0 | 1 | 2 | 3 |
|---|---|---|---|---|---|---|
| y | 0 | -8 | -6 | 0 | 4 | 0 |

Locate these points and draw a smooth curve through them.

25. $f(x) = x^4 + 2x^3 - 3x^2 - 3x - 5$

Polynomial function; find some points satisfying the function:

| x | -3 | -2 | -1 | 0 | 1 | 2 |
|---|---|---|---|---|---|---|
| y | 94 | -11 | -6 | -5 | -8 | 9 |

Locate these points and draw a smooth curve through them.

29. $f(x) = -\sqrt{9 - x}$

Square root function; square both sides of $y = -\sqrt{9 - x}$ to get

$$y^2 = 9 - x,$$

or
$$x = 9 - y^2,$$

a parabola with vertex (0, 3) and opening to the left.
However, y is negative since $y = -\sqrt{9 - x}$, so only the bottom half is graphed.

33. $y = \sqrt{16 - x^2}$

Square both sides to get $y^2 = 16 - x^2$ which can be written in the standard form of an equation of a circle, $x^2 + y^2 = 16$. The circle has a radius of 4 centered at the origin. Since y equals a positive square root, only the top half of the circle is on the graph.
The vertical line test shows that $y = \sqrt{16 - x^2}$ is a function.

37. $$\frac{x}{3} = \sqrt{1 + \frac{y^2}{9}}$$

$$\frac{x^2}{9} = 1 + \frac{y^2}{9}$$ *Square both sides*

$$\frac{x^2}{9} - \frac{y^2}{9} = 1$$ *Write in the form of a hyperbola*

This hyperbola has x-intercepts (3, 0) and (-3, 0) with corners of the rectangle at (3, 3), (-3, 3), (3, -3), and (-3, -3). Since x/3 equals a positive square root, only the half of the hyperbola with the positive x-intercept is graphed.

This is not a function since some vertical lines intersect graph in two points.

41. $x = \sqrt{36 + y^2}$, $y \geq 0$

$$x^2 = 36 + y^2$$ *Square both sides*

$$x^2 - y^2 = 36$$

$$\frac{x^2}{36} - \frac{y^2}{36} = 1$$ *Write in the form of a hyperbola*

Graph only the half of the hyperbola with the positive x-intercept (6, 0) since x equals a positive square root. Adding the restriction $y \geq 0$, we include only the part above $y = 0$. This graph is a function.

45. $f(x) = -|x - 2|$

Plotting points:

$f(-3) = -|-3 - 2| = -5 \quad (-3, -5)$
$f(-2) = -|-2 - 2| = -4 \quad (-2, -4)$
$f(-1) = -|-1 - 2| = -3 \quad (-1, -3)$
$f(0) = -|0 - 2| = -2 \quad (0, -2)$
$f(1) = -|1 - 2| = -1 \quad (1, -1)$
$f(2) = -|2 - 2| = 0 \quad (2, 0)$
$f(3) = -|3 - 2| = -1 \quad (3, -1)$

49. $A = kr^2$

Since the formula for the area of a circle is $A = \pi r^2$, k equals $\pi$.

53. $5x - 2y = 9$

$5x = 9 + 2y$

$x = \frac{1}{5}(9 + 2y)$

or $x = \frac{9 + 2y}{5}$

57. $\frac{2x - 1}{4} = -2y$

$2x - 1 = -8y$

$2x = 1 - 8y$

$x = \frac{1 - 8x}{2}$

**Section 10.4 (page 438)**

1. $\{(3, 5), (2, 9), (4, 7)\}$

Exchange x and y in each ordered pair:

$\{(5, 3), (9, 2), (7, 4)\}$.

5. $y = f(x) = 2x$

$x = 2y$ *Exchange x and y*

$\frac{x}{2} = y$ *Solve for y*

$f^{-1}(x) = \frac{x}{2}$

9. $2y = x + 1$

$2x = y + 1$ *Exchange x and y*

$y = 2x - 1$ *Solve for y*

$f^{-1}(x) = 2x - 1$ *Write y as an inverse function*

13. $y = \sqrt{x}$ *Note that* $y \geq 0$

$x = \sqrt{y}$ *Exchange x and y*

$x^2 = y$ *Solve for y; square both sides*

$f^{-1}(x) = x^2 \quad x \geq 0$

This is an inverse function if we add the restriction $x \geq 0$.

17. $2y = x^2 + 1$

$2x = y^2 + 1$ *Exchange x and y*

$2x - 1 = y^2$

$y = \sqrt{2x - 1}$ or $y = -\sqrt{2x - 1}$ *Solve for y*

Not a one-to-one function.

21. This is not a one-to-one function since the inverse of the parabola (reflected about $y = x$) fails the vertical line test and is not a function.

25. $f(x) = 2x$

The inverse is $f^{-1}(x) = \frac{x}{2}$.

These two lines are reflected about the line $y = x$.
See answer graph in textbook.

29. $2y = x + 1$

$y = \frac{x + 1}{2}$

Draw this function as a solid line.

The inverse is $f^{-1}(x) = 2x - 1$, drawn as a dashed line.
See answer graph in the textbook.

33. $y = \sqrt{x}$

This function is the top half of the parabola $x = y^2$.

$f^{-1}(x) = x^2$
The inverse is the right side of the parabola $y = x^2$.
See answer graph in the textbook.

37. $f(3) = 2^3 = 8$

41. $f^{-1}(8) = 3$ because $2^3 = 8$.

45. $f[f^{-1}(x)]$

$= 3[f^{-1}(x)] - 1$

$= 3[\frac{x + 1}{3}] - 1$

$= x + 1 - 1$

$= x$

$f^{-1}[f(x)]$

$= \frac{f(x) + 1}{3}$

$= \frac{3x - 1 + 1}{3}$

$= \frac{3x}{3}$

$= x$

49. $100^{3/2} = (\sqrt{100})^3 = 10^3 = 1000$

53. $27^{-2/3} = (\sqrt[3]{27})^{-2} = (3)^{-2} = \frac{1}{3^2} = \frac{1}{9}$

**Chapter 10 Review Exercises (page 440)**

1. $3y = 2x - 1$

Set of all real numbers
For each value of x that is put into the equation, there will be one value of y. It is a function.

5. $y \leq 2x + 3$

Set of all real numbers

If $x = 1$, y can have the values 5, 4, 3, 2, etc. so y is not a function of x.

9. $\{x | x \leq -4 \text{ or } x \geq 4\}$

The vertical line test fails, so it is not a function.

13. $f(x) = x^2 + 2x - 1$

$f(3) = 3^2 + 2(3) - 1$
$= 9 + 6 - 1$
$= 14$

17. $f(x) = x^2 + 2x - 1$

$f(x + k) = (x + k)^2 + 2(x + k) - 1$
$= x^2 + 2xk + k^2 + 2x + 2k - 1$

21. Function; use the vertical line test.

For Exercises 25-37, see the answer graphs in the textbook.

25. $f(x) = -4$

Constant function or linear function; graph the straight line through such points as (0, -4) and (2, -4).

29. $f(x) = -(x + 2)^2 - 1$

Quadratic function; the graph is a parabola with vertex at (-2, -1), and opening downward.

33. $f(x) = x^3 - 3x^2 + 2x$

Polynomial function; locate some points to get the graph:

| x | -3 | -2 | -1 | 0 | 1 | 2 |
|---|---|---|---|---|---|---|
| y | -60 | -24 | -6 | 0 | 0 | 0 |

37. $f(x) = -\sqrt{2 - x}$

Square root function; square both sides to get

$$y^2 = 2 - x$$

or $$x = 2 - y^2,$$

a parabola opening to the left with vertex (2, 0). Because of the negative sign in front of the radical, use only the bottom half of the graph.

41.

$$3y = x - 1$$
$$3x = y - 1 \quad \textit{Exchange } x \textit{ and } y$$
$$y = 3x + 1 \quad \textit{Solve for } y$$

$f^{-1}(x) = 3x + 1$

45.

$$y = f(x) = x^2 - 4$$
$$x = y^2 - 4 \quad \textit{Exchange } x \textit{ and } y$$
$$x + 4 = y^2$$
$$y = \sqrt{x + 4} \quad \text{or} \quad y = -\sqrt{x + 4}$$

This is not a one-to-one function since there are 2 y-values for each choice of x.

49.

$$y = f(x) = \sqrt{2x + 5}$$
$$x = \sqrt{2y + 5} \quad \textit{Exchange } x \textit{ and } y$$
$$x^2 = 2y + 5$$
$$\frac{1}{2}(x^2 - 5) = y \quad \textit{Solve for } y$$
$$f^{-1}(x) = \frac{x^2 - 5}{2}$$

53. $3y = x - 1$; $f^{-1}(x) = 3x + 1$

Draw the graph of $3y = x - 1$, using the points (0, -1/3), and (1, 0). Then reflect the graph in the line $y = x$ to get the inverse, using the points (0, 1) and (-1/3, 0). See answer graph in the textbook.

**Chapter 10 Test (page 442)**

1. {(1, 2), (1, 3), (-2, 2), (-2, 3)}
Domain: {1, -2}
Range: {2, 3}

There are two y-values that correspond to the x-values of 1 and -2, so it is not a function.

2. $y = x^2 - 4$

Domain: set of all real numbers
Range: $\{y | y \geq -4\}$

Each x-value has only one y-value. For example, when $x = 2$, $y = 0$; when $x = -2$, $y = 0$. Thus it satisfies the definition even though $x = 2$ and $x = -2$ have the same y-value. It is a function.

3. $y = \dfrac{1}{x - 2}$

Domain: $\{x | x \neq 2\}$ Range: $\{y | y \neq 0\}$

Given any value of x, we get exactly one value of y, so it is a function.

4. Domain: $\{x | x \leq 2\}$ Range: set of all real numbers
The vertical line test through the parabola determines that it is not a function.

5. Domain: set of all real numbers
Range: set of all real numbers
Any value of x corresponds to many values of y, so it is not a function.

6. $g(x) = -x^2 + 2x - 1$
$g(0) = -(0) + 2(0) - 1$
$= -1$

7. $f(x) = 3x - 1$

$f(1) = 3(1) - 1$

$= 2$

8. $(f \circ g)(2) = f[g(2)]$

$= f[-2^2 + 2(2) - 1]$

$= f(-1)$

$= 3(-1) - 1$

$= -4$

9. $(f + g)(-2)$

$= f(-2) + g(-2)$

$= 3(-2) - 1 + [-(-2)^2 + 2(-2) - 1]$

$= -7 + (-9)$

$= -16$

10. $(\frac{f}{g})(3) = \frac{f(3)}{g(3)}$

$= \frac{3(3 - 1)}{-3^2 + 2(3) - 1}$

$= \frac{8}{-4}$

$= -2$

11. $f(x) = 5 - 2x$

Linear function; let $x = 0$ to get $y = 5$; let $x = 3$ to get $y = -1$; draw the line through (0, 5) and (3, -1).

See graph in your textbook.

12. $f(x) = -x^2 - 4x + 1$

Quadratic function; complete the square to get

$y = -x^2 - 4x + 1$

$= -(x^2 + 4x \quad ) + 1$

$= -(x^2 + 4x + 4) + 1 + 4$

$= -(x + 2)^2 + 5,$

the equation of a parabola with vertex (-2, 5) and opening downward.

See graph in your textbook.

13. $f(x) = -x^3 + 2x^2 + x - 2$

Polynomial function; complete some points:

| x | -2 | -1 | 0 | 1 | 2 | 3 |
|---|---|---|---|---|---|---|
| y | 12 | 0 | -2 | 0 | 0 | -8 |

See graph in your textbook.

14. $f(x) = -\sqrt{3 - x}$

Square root function; square both sides to get

$y^2 = 3 - x,$

or $\quad x = 3 - y^2$

a parabola opening sideways with vertex (3, 0). Because of the negative sign in front of the radical, use only the bottom half of the graph.

See graph in your textbook.

15. $2y = x + 3$

$2x = y + 3$ *Exchange x and y*

$2x - 3 = y$ *Solve for y*

$f^{-1}(x) = 2x - 3$

16. $y = \sqrt[3]{4 - x}$

$y^3 = 4 - x$ *Cube both sides*

$x^3 = 4 - y$ *Exchange x and y*

$y = 4 - x^3$ *Solve for y*

$f^{-1}(x) = 4 - x^3$ *Write as an inverse function*

17. $3y = x^2 + 1$

$3x = y^2 + 1$ *Exchange x and y*

$3x - 1 = y^2$

$y = \sqrt{3x - 1}$ or $y = -\sqrt{3x - 1}$

*Solve for y*

This is not a one-to-one function; for one value of x there can be two values of y.

18. This is not a one-to-one function since the parabola drawn as the inverse (opening to the right) does not pass the vertical line test.
See answer graph in the textbook.

19. Reflect the graph of the line about the line $y = x$. Exchange the x and y in each point plotted for the line to find points for the inverse.
See answer graph in the textbook.

20. Reflect the graph of the top half of the parabola about the line $y = x$. Exchange the x and y in each plotting point to find points for the inverse.
See answer graph in the textbook.

# CHAPTER 11 EXPONENTIAL AND LOGARITHMIC FUNCTIONS

## Section 11.1 (page 448)

For Exercises 1-13, see answer graphs in the textbook.

1. $y = 3^x$

Make a table of values.

| x | -2 | -1 | 0 | 1 | 2 |
|---|---|---|---|---|---|
| y | 1/9 | 1/3 | 1 | 3 | 9 |

Note that $3^{-2} = \frac{1}{3^2} = \frac{1}{9}$.

Now, plot the points given by the ordered pairs in the table above, and draw the graph through them.

5. $y = 2^{2x}$

Make a table of values.

| x | -2 | -1 | 0 | 1 | 2 |
|---|---|---|---|---|---|
| y | 1/16 | 1/4 | 1 | 4 | 16 |

Note that for $x = -2$, $y = 2^{2(-2)}$ $= 2^{-4} = \frac{1}{2^4} = \frac{1}{16}$.

Now, plot the points given by the ordered pairs in the table above, and draw the graph through them.

9. $y = 3^x - 3$

Make a table of values.

| x | -2 | -1 | 0 | 1 | 2 |
|---|---|---|---|---|---|
| y | -26/9 | -8/3 | -2 | 0 | 6 |

Plot these points and draw a smooth curve through them.

13. $y = 2.718^x$

Make a table of values.

| x | -2 | -1 | 0 | 1 | 2 |
|---|---|---|---|---|---|
| y | .135 | .368 | 1 | 2.718 | 7.388 |

Plot the points given by the ordered pairs in the table above and draw the graph through them.

17. $2^x = \frac{1}{8}$

$2^x = 2^{-3}$ *Write each side using the same base*

$x = -3$ *Since the bases are equal, their exponents are equal*

Solution set: {-3}

21. $4^x = 8$

$(2^2)^x = 2^3$ *Write each side using the same base*

$2^{2x} = 2^3$

$2x = 3$ *Since the bases are equal, their exponents are equal*

$x = \frac{3}{2}$ *Solve for x*

Solution set: {3/2}

25. $2^{3x+2} = 16$

$2^{3x+2} = 2^4$ *Write each side using the same base*

$3x + 2 = 4$ *Equate the exponents*

$3x = 2$

$x = \frac{2}{3}$ *Solve for x*

Solution set: {2/3}

29. (a) In 1980, $x = 0$.

$y = 5000(2)^{.1x}$

$y = 5000(2)^{(.1)(0)}$

$y = 5000(2)^{0}$

$y = 5000(1)$

$= 5000$

The population in 1980 was 5000.

(b) In 1990, $x = 10$.

$y = 5000(2)^{(.1)(10)}$

$= 5000(2)^{1}$

$= 5000(2)$

$= 10{,}000$

The population in 1990 will be 10,000.

(c) In the year 2000, $x = 20$

$y = 5000(2)^{(.1)(20)}$

$y = 5000(2)^{2}$

$y = 5000(4)$

$= 20{,}000$

The population in 2000 will be 20,000.

(d) Plot the points solved for in (a), (b) and (c) and draw the graph through them.
See answer graph in the textbook.

33. $y = 3^{-|x|}$

| x | -2 | -1 | 0 | 1 | 2 |
|---|---|---|---|---|---|
| y | 1/9 | 1/3 | 1 | 1/3 | 1/9 |

Plot the points given by the ordered pairs in the table above and draw the graph through them.
See answer graph in the textbook.

37. $25^{-2x} = 3125$

$5^{2(-2x)} = 5^5$ *Write each side using the same base*

$5^{-4x} = 5^5$

$-4x = 5$ *Equate the exponents*

$x = \frac{-5}{4}$

Solution set: {-5/4}

41. $2^{-3} = \frac{1}{2^3} = \frac{1}{8}$

45. $3^{5/2} = (\sqrt{3})^5$

$= \sqrt{243}$ or $9\sqrt{3}$

**Section 11.2 (page 454)**

1. $3^4 = 81$ becomes $\log_3 81 = 4$ in logarithmic form.

5. $(\frac{1}{2})^{-2} = 4$ becomes $\log_{1/2} 4 = -2$ in logarithmic form.

9. $\log_2 8 = 3$ becomes $2^3 = 8$ in exponential form.

13. $\log_4 16 = 2$ becomes $4^2 = 16$ in exponential form.

17. $\log_{1/2} 4 = -2$ becomes $(\frac{1}{2})^{-2} = 4$ in exponential form.

21. Let $y = \log_3 \frac{1}{9}$ and write in expoential form as an equation and solve it:

$3^y = \frac{1}{9}$.

$3^y = 3^{-2}$ *Write each side with the same base*

$y = -2$ *Equate exponents, since the bases are equal*

25. Let $y = \log_{36} 6$ and write in exponential form:

$36^y = 6.$

$(6^2)^y = 6$ *Write each side with the same base*

$6^{2y} = 6^1$ *Simplify*

$2y = 1$ *Equate the exponents since the bases are equal*

$y = \frac{1}{2}$ *Solve for y*

29. Let $y = \log_3 \sqrt{3^5}$ and write in exponential form:

$3^y = \sqrt{3^5}$

$3^y = 3^{5/2}.$ $\quad 3^{5/2}$

$\sqrt{3^5} = (3^5)^{1/2}$

$y = \frac{5}{2}$ *Since the bases are equal, the exponents are equal*

33. Let $\log_9 \sqrt{3^3} = y.$

$9^y = \sqrt{3^3}$ *In exponential form*

$9^y = 3^{3/2}$

$(3^2)^y = 3^{3/2}$ *Same base on both sides*

$3^{2y} = 3^{3/2}$

$2y = \frac{3}{2}$

$y = \frac{3}{4}$

37. Write $y = \log_6 216$ in exponential form:

$6^y = 216$

$6^y = 6^3$ *Get each side to the same base*

$y = 3$ *Since the bases are equal, the exponents are equal*

Solution set: {3}

41. Write $\log_x 64 = -6$ in exponential form:

$x^{-6} = 64.$

$(\frac{1}{x})^6 = 2^6$ *Write each using equal exponents*

$\frac{1}{x} = 2$ *Since exponents are equal, their bases must be equal*

$x = \frac{1}{2}$ *Solve for x*

Solution set: {1/2}

45. Write $\log_m 3 = \frac{1}{2}$ in exponential form:

$m^{1/2} = 3.$

$m^{1/2} = 9^{1/2}$ *Since $3 = \sqrt{9} = 9^{1/2}$*

$m = 9$ *Since the exponents are equal, the bases must be equal*

Solution set: {9}

49. Write $\log_{12} P = 1$ in exponential form:

$12^1 = P$

$P = 12$

Solution set: {12}

53. Write $y = \log_{1/4} x$ in exponential form:

$$\left(\frac{1}{4}\right)^y = x.$$

Use this exponential equation to complete the table:

| x | 64 | 16 | 4 | 1 | 1/4 | 1/16 | 1/64 |
|---|---|---|---|---|---|---|---|
| y | -3 | -2 | -1 | 0 | 1 | 2 | 3 |

Note that for x = 64, then y = -3, since $\left(\frac{1}{4}\right)^{-3} = 4^3 = 64$.

Graph these points.
See answer graph in textbook.

57. $2^4 \cdot 2^5 = 2^{4+5} = 2^9$

61. $(2^4)^3 = 2^{4\cdot 3} = 2^{12}$

**Section 11.3 (page 460)**

1. $\log_6 \frac{3}{4} = \log_6 3 - \log_6 4$ *Quotient rule for logarithms*

5. $\log_3 3^2 = 2$ *Since $\log_b b^x = x$*

9. $\log_2 \frac{5\sqrt{7}}{3}$

$= \log_2 5\sqrt{7} - \log_2 3$ *Quotient rule for logarithms*

$= \log_2 5 + \log_2 \sqrt{7} - \log_2 3$ *Product rule for logarithms*

$= \log_2 5 + \log_2 7^{1/2} - \log_2 3$

$= \log_2 5 + \frac{1}{2}\log_2 7 - \log_2 3$ *Power rule for logarithms*

13. $\log_5 (9x + 4y) \neq \log_5 9x + \log_5 4y$ *So it is not possible to rewrite using the properties*

Note, however, that

$\log_5 (9x)(4y) = \log_5 9x + \log_5 4y$

17. $4 \log_b m - \log_b n$

$= \log_b m^4 - \log_b n$ *Power rule*

$= \log_b \frac{m^4}{n}$ *Quotient rule*

21. $3 \log_b 4 - 2 \log_b 3$

$= \log_b 4^3 - \log_b 3^2$ *Power rule*

$= \log_b 64 - \log_b 9$

$= \log_b \frac{64}{9}$ *Quotient rule*

25. $\log_p \frac{5}{2} + \log_p \frac{5}{4} - (\log_p \frac{3}{4} + \log_p 2)$

$= \log_p (\frac{5}{2} \cdot \frac{5}{4}) - (\log_p \frac{3}{4} \cdot 2)$ *Product rule*

$= \log_p \frac{25}{8} - \log_p \frac{3}{2}$

$= \log_p \frac{25/8}{3/2}$ *Quotient rule*

$= \log_p \frac{25}{8} \cdot \frac{2}{3}$

$= \log_p \frac{25}{12}$

29. $\log_{10} 6 = \log_{10} (2\cdot 3)$

$= \log_{10} 2 + \log_{10} 3$ *Product rule*

$= .3010 + .4771$

$= .7781$

33. $\log_{10} 16 = \log_{10} 2^4$

$= 4 \log_{10} 2$ *Power rule*

$= 4(.3010)$

$= 1.2040$

37. $\log_b AB = \log_b A + \log_b B$ *Product rule*

$= 2 + (-4)$

$= -2$

$(\log_b A)\cdot(\log_b B) = 2\cdot(-4) = -8$

The values -2 and -8 are unequal, so $\log_b AB \neq (\log_b A)\cdot(\log_b B)$.

41. $\log_{10} 2 + \log_{10} 3 = \log_{10} 2\cdot 3$

$= \log_{10} 6$

$\log_{10} (2 + 3) = \log_{10} 5$

Since $\log_{10} 6 \neq \log_{10} 5$

$\log_{10} 2 + \log_{10} 3 = \log_{10} (2 + 3)$

is false.

45. $\dfrac{\log_{10} 8}{\log_{10} 16} = \dfrac{\log_{10} 2^3}{\log_{10} 2^4}$

$= \dfrac{3 \log_{10} 2}{4 \log_{10} 2}$

$= \dfrac{3}{4}$

Since $3/4 \neq 1/2$,

$$\frac{\log_{10} x}{\log_{10} 16} = \frac{1}{2}$$

is false.

49. $482 = 4.82 \times 10^2$ since we must move the decimal point 2 spaces left.

53. $4.23 \times 10^3 = 4230$; we move the decimal point 3 spaces right.

**Section 11.4 (page 466)**

1. Since $278 = 2.78 \times 10^2$, the characteristic is 2. From Table 3, the mantissa of 2.78 is .4440. Therefore

$\log 278 = 2 + .4440 = 2.4440.$

5. Since $.327 = 3.27 \times 10^{-1}$, the characteristic is -1. From Table 3, the mantissa of 3.27 is .5145. Therefore,

$\log .327 = .5145 - 1$ or $-.4855$.

9. Since $675{,}000 = 6.75 \times 10^5$, the characteristic is 5. From Table 3, the mantissa of 6.75 is .8293. Therefore,

$\log 675{,}000 = 5 + .8293 = 5.8293.$

13. We want the number whose logarithm is 2.6571.
Look in the table for a mantissa of .6571. The number whose mantissa is .6571 is 4.54.
Since the characteristic is 2, the antilog of 2.6571 is $4.54 \times 10^2 = 454$.

17. We want the number whose logarithm is .6017 + -3.
Look in the table for a mantissa of .6107. The number whose mantissa is .6107 is 4.08.
Since the characteristic is -3, the antilog of 0.6107 - 3 is $4.08 \times 10^{-3} = .00408$.

21. (37.3)(9.72)

$\log 37.3 = \log (3.73 \times 10^1)$

$= 1 + .5717$

$= 1.5717$

$\log 9.72 = \log (9.72 \times 10^0)$
$= 0.9877$

Thus, $\log (37.3)(9.72)$
$= \log 37.3 + \log 9.72$
$= 1.5717 + 0.9877$
$= 2.5594$

In Table 3, the number whose mantissa is closest to .5594 is 3.63. Since the characteristic is 2, we have $(37.3)(9.72) \approx 3.63 \times 10^2$
$= 363.$

25. $\frac{65.4}{129}$

$\log 65.4 = 1.8156$
$\log 129 = 2.1106$

Thus,

$\log \frac{6.54}{129} = \log 65.4 - \log 129$
$= 1.8156 - 2.1106.$
$= 1.8156 - 1.1106 - 1$
*Write -2.1106 as -1.1106 - 1*
$= .7050 - 1.$

The antilog of .7050 - 1 is $5.07 \times 10^{-1} = .507.$

29. $\sqrt{694}$

$\log \sqrt{694} = \log 694^{1/2}$
$= \frac{1}{2} \log 694$

In Table 3, $\log 694 = 2.8414$.

Thus,

$\frac{1}{2} \log 694 = \frac{1}{2}(2.8414)$
$= \frac{2.8414}{2}$
$= 1.4207.$

The antilog of 1.4207
$= 2.63 \times 10^1$
$= 26.3$

33. $\sqrt[3]{\frac{(15.8)(62.1)}{12}}$

$\log \sqrt[3]{\frac{(15.8)(62.1)}{12}}$
$= \log \left(\frac{(15.8)(62.1)}{12}\right)^{1/3}$
$= \frac{1}{3}[\log \frac{(15.8)(62.1)}{12}]$
$= \frac{1}{3}[\log (15.8) + \log (62.1) - \log 12]$

In Table 3, $\log 15.8 = 1.1987$; $\log 62.1 = 1.7931$; $\log 12 = 1.0792$. Thus,

$\frac{1}{3}[\log (15.8) + \log(62.1) - \log (12)]$
$= \frac{1}{3}(1.1987 + 1.7931 - 1.0792)$
$= \frac{1}{3}(1.9126)$
$= .6375.$

The antilog of .6375 = 4.34.

37. $\log 53.89$

Since $53.89 = 5.389 \times 10^1$, the characteristic is 1.
Find log 5.389 as follows.

$$10\left\{9\left\{\begin{array}{l}\log 5.380 = .7308\\ \log 5.389 = \end{array}\right.\\ \log 5.390 = .7316\right\}.008$$

Since 5.389 is 9/10 of the way from 5.380 to 5.390, take 9/10 of .0008, the difference between the two logarithms.

$\frac{9}{10}(.0008) \approx .0007$

$\log 5.389 = .7308 + .0007$
$= .7315$

Finally, $\log 53.89 = 1.7315$

41. 0.8342 - 1

Find the antilog of .8342 as follows.

$$6\left\{ \begin{array}{l} 4\left\{ \begin{array}{l} .8338 = \log 6.820 \\ .8342 \end{array} \right. \\ .8344 = \log 6.830 \end{array} \right\} .010$$

$$\frac{4}{6}(.010) \approx .007$$

antilog .8342 = 6.820 + .007
= 6.827

Finally, antilog .8342 - 1
$= 6.827 \times 10^{-1}$
= .6827.

45. $\dfrac{(.0008214)(9461)}{8162}$

Find log 8.214 as follows.

$$10\left\{ \begin{array}{l} 4\left\{ \begin{array}{l} \log 8.210 = .9143 \\ \log 8.214 \end{array} \right. \\ \log 8.220 = .9149 \end{array} \right\} .0006$$

$$\frac{4}{10}(.0006) \approx .0002$$

log 8.214 = .9143 + .0002
= .9145

Find log 9.461 in the same way.

$$10\left\{ \begin{array}{l} 1\left\{ \begin{array}{l} \log 9.460 = .9759 \\ \log 9.461 \end{array} \right. \\ \log 9.470 = .9763 \end{array} \right\} .0004$$

$$\frac{1}{10}(.0004) \approx 0$$

log 9.461 = .9759 + .0000
= .9759

Find log 8.162 in the same way.

$$10\left\{ \begin{array}{l} 2\left\{ \begin{array}{l} \log 8.160 = .9117 \\ \log 8.162 \end{array} \right. \\ \log 8.170 = .9122 \end{array} \right\} .0005$$

$$\frac{2}{10}(.0005) = .0001$$

log 8.162 = .9117 + .0001
= .9118

Using the preceding information

$$\log \frac{(.0008214)(9461)}{8162}$$

= log .0008214 + log 9461 - log 8162

$= \log 8.124 \times 10^{-4} + \log 9.461 \times 10^{-3} -$

$\log 8.162 \times 10^{3}$

= .9145 - 4 + .9759 - 3 - 3.9118

= .9786 - 4

The antilog of .9786 is 9.520, so the answer is

$$9.520 \times 10^{-4},$$

or .0009520.

A calculator gives a more accurate answer, .0009521.

49. $\sqrt[3]{\dfrac{(57.32)(3.681)}{.684}}$

By interpolation, we find log 57.32 = 1.7583
(that is, $.7582 + \frac{2}{10}(.0017)$).

log 3.681 = .5659
(that is , $.5658 + \frac{1}{10}(.0012)$)

log .684 = .8351 - 1

Thus,

$\log \sqrt[3]{\frac{(57.32)(3.681)}{.684}}$

$= \log \left(\frac{(57.32)(3.681)}{.684}\right)^{1/3}$

$= \frac{1}{3} \log \left(\frac{(57.32)(3.681)}{.684}\right)$

$= \frac{1}{3}[\log 57.32 + \log 3.681 - \log .684]$

$= \frac{1}{3}[1.7583 + .5659 - (.8351 - 1)]$

$= \frac{1}{3}[1.7583 + .5659 - .8351 + 1]$

$= \frac{1}{3}[2.4891]$

$= .8297.$

Find the antilog of .8297.

$$.001\left\{\begin{array}{l}\log 6.750 = .8293 \\ \qquad\qquad\quad .8297 \\ \log 6.760 = .8299\end{array}\right\} \quad (4 \text{ between } .8293 \text{ and } .8297;\ 6 \text{ between } .8293 \text{ and } .8299)$$

$\frac{4}{6}(.001) = .007$

antilog $8.297 = 6.750 + .007$
$= 6.757$

53. $pH = -\log [H_3O^+]$

$= -\log 4.1 \times 10^{-5}$

$= -(\log 4.1 + \log 10^{-5})$

$= -[.6128 - 5]$

$= 5 - .6128$

$= 4.3872$

or 4.4 to the nearest tenth.

57. $pH = -\log [H_3O^+]$

$3.8 = -\log [H_3O^+]$

So, $\log [H_3O^+] = -3.8$
$= -4 + .2.$

The antilog of .2000 is 1.59.

So the antilog of

$.2 - 4 = 1.59 \times 10^{-4}$
$= .000159.$

Thus, $[H_3O^+] = 1.6 \times 10^{-4}$

61. With $a = 9$, $b = 10$, $c = 11$, then

$s = \frac{a + b + c}{2} = \frac{9 + 10 + 11}{2} = 15.$

Area $= \sqrt{15(15 - 9)(15 - 10)(15 - 11)}$
$= \sqrt{15(6)(5)(4)}$
$= \sqrt{1800}$

$\log \sqrt{1800} = \log (1800)^{1/2}$
$= \frac{1}{2} \log 1800$
$= \frac{1}{2}(3.2553)$
$= 1.6277$

The antilog of

$1.6277 = 4.24 \times 10^1$
$= 42.4.$

The area is 42.4 square units.

65. $(3m + 2)(4) = \left(\frac{4m - 1}{5}\right)(2)$

$5(3m + 2)(4) = 5\left(\frac{4m - 1}{5}\right)(2)$

$60m + 40 = 8m - 2$ *Multiply both sides by 5*

$52m = -42$ *Simplify*

$m = \frac{-42}{52} = \frac{-21}{26}$

Solution set: $\{-\frac{21}{26}\}$

69. $\sqrt{k + 2} = 7$

$(\sqrt{k + 2})^2 = 7^2$ *Square both sides*

$k + 2 = 49$

$k = 47$

Solution set: $\{47\}$

**Section 11.5 (page 471)**

1. $25^x = 125$ Since both 25 and 125 are both powers of 5, take base 5 logarithms of both sides.

$$\log_5 25^x = \log_5 125$$
$$\log_5 (5^2)^x = \log_5 5^3$$
$$\log_5 5^{2x} = \log_5 5^3$$
$$2x = 3$$
$$x = \frac{3}{2}$$

Solution set: {3/2}

5. $6^{y+1} = 8$

$\log 6^{y+1} = \log 8$ *Take the log of each side*

$(y + 1) \log 6 = \log 8$

$y + 1 = \frac{\log 8}{\log 6}$ *Divide each side by log 6*

$y = \frac{\log 8}{\log 6} - 1$ *Subtract 1*

$y = \frac{.9031}{.7782} - 1$ *Find log 8 and log 6*

$y = 1.16 - 1$

$= .16$ *Simplify*

Solution set: {.16}

9. $5^{1-n} = 8$

$\log 5^{1-n} = \log 8$ *Take the log of each side*

$(1 - n) \log 5 = \log 8$

$1 - n = \frac{\log 8}{\log 5}$ *Divide by log 5*

$n = 1 - \frac{\log 8}{\log 5}$

$n = 1 - \frac{.9031}{.6990}$ *Calculate log 8 and log 5*

$= 1 - 1.2920$

$= -.29$

Solution set: {-.29}

13. $\log (x + 2) = \log (3x - 6)$

$x + 2 = 3x - 6$ *Property 4*

$-2x = -8$

$x = 4$

Solution set: {4}

17. $\log 4x = \log 2 + \log (x - 3)$

$\log 4x = \log 2(x - 3)$ *Product rule*

$4x = 2(x - 3)$ *Property 4*

$2x = -6$

$x = -3$

The solution cannot be -3 since by the definition of logarithm ($y = \log_a x$), x must be a positive real number. Substituting -3 for x in the original equation gives a negative x.
Solution set: ∅

21. $\log_2 x = 3$

$2^3 = x$ *Write in exponential form*

$8 = x$

Solution set: {8}

25. $\log_a 5 = \frac{-3}{4}$

$a^{-3/4} = 5$ *Change to exponential form*

$\frac{1}{a^{3/4}} = 5$ *Write with a positive exponent*

$5a^{3/4} = 1$ *Multiply each side by $a^{3/4}$*

$a^{3/4} = \frac{1}{5}$

$a = (\frac{1}{5})^{4/3}$ *Take both sides to 4/3 power*

$= \sqrt[3]{(\frac{1}{5})^4}$

$= \sqrt[3]{\frac{1}{625}} = \frac{1}{5\sqrt[3]{5}}$

$= \frac{\sqrt[3]{25}}{25}$

$= .117$ *Table 2*

Solution set: {.117}

29. $A = P(1 + i)^t$

$1000 = P(1 + .05)^{10}$

$1000 = P(1.05)^{10}$

$\log 1000 = \log P(1.05)^{10}$ *Take the log of each side*

$\log 1000 = \log P + \log (1.05)^{10}$ *Product rule*

$\log 1000 = \log P + 10 \log 1.05$ *Power rule*

$3 = \log P + 10(.02119)$ *Calculate each log*

$3 = \log P + .2119$

$3 - .2119 = \log P$

$2.7881 = \log P$

$613.9 = P$ *Find the antilog*

The amount of money that must be deposited today is \$613.90. (By calculator, the amount is \$613.91.)

33. $7^{2y-1} = 1$

$7^{2y-1} = 7^0$

$2y - 1 = 0$ *Since the bases are equal, the exponents are equal*

$2y = 1$

$y = \frac{1}{2}$ *Solve for y*

Solution set: {1/2}

37. $2^{2x^2+1} = 8^x$

$2^{2x^2+1} = 2^{3(x)}$

$2x^2 + 1 = 3x$ *Property 2*

$2x^2 - 3x + 1 = 0$

$(2x - 1)(x - 1) = 0$ *Factor*

$2x - 1 = 0$ or $x - 1 = 0$

$x = \frac{1}{2}$ or $x = 1$

Solution set: {1/2, 1}

41. $\log_3 x + \log_3 (2x + 5) = 1$

$\log_3 x(2x + 5) = 1$ *Product rule*

$x(2x + 5) = 3^1$ *Write in exponential form*

$2x^2 + 5x = 3$

$2x^2 + 5x - 3 = 0$

$(2x - 1)(x + 3) = 0$ *Factor*

$2x - 1 = 0$ or $x + 3 = 0$

$x = \frac{1}{2}$ or $x = -3$

However, log -3 does not exist.

So, $x = \frac{1}{2}$ is the only solution.

Solution set: {1/2}

45. $\dfrac{\log_2 4}{\log_2 8}$

Let $x = \log_2 4$ and $y = \log_2 8$.

$2^x = 4$ *In exponential form*

$x = 2$

$2^y = 8$ *In exponential form*

$y = 3$

Thus, $\dfrac{\log_2 4}{\log_2 8} = \dfrac{x}{y} = \dfrac{2}{3}$.

**Section 11.6 (page 476)**

1. $\log_5 11 = \dfrac{\log_{10} 11}{\log_{10} 5}$

$= \dfrac{1.0414}{.6990}$

$= 1.49$

5. $\log_8 9.63 = \dfrac{\log_{10} 9.63}{\log_{10} 8}$

$= \dfrac{.9836}{.9031}$

$= 1.09$

9. $\log_{50} 31.3 = \dfrac{\log_{10} 31.3}{\log_{10} 50}$

$= \dfrac{1.4955}{1.6990}$

$= .88$

13. $\ln 4.83 = \dfrac{\log 4.83}{.4343}$ *Use formula* $\ln x = \dfrac{\log x}{.4343}$

$= \dfrac{.6839}{.4343}$

$= 1.57$

17. $P = 1{,}000{,}000\ e^{.02t}$

(a) If $t = 0$,

$P = 1{,}000{,}000\ e^{.02(0)}$

$= 1{,}000{,}000\ e^{(0)}$

$= 1{,}000{,}000(1)$

$= 1{,}000{,}000.$

(b) If $t = 2$,

$P = 1{,}000{,}000\ e^{(.02)(2)}$

$P = 1{,}000{,}000\ e^{.04}.$

Now,

$\log e^{.04} = .04 \log e$

$= .04(.4343)$

$= .017372$

$e^{.04} = 1.0408$ *Take antilogs*

Thus, when $t = 2$,

$P = (1{,}000{,}000)(1.0408)$

$P = 1{,}041{,}000$ (rounded to nearest 1,000).

(c) If $t = 4$,

$P = 1{,}000{,}000\ e^{.02(4)}$

$P = 1{,}000{,}000\ e^{.08}.$

Now, $\log e^{.08} = .08 \log e$

$= .08(.4343)$

$= .0347$

$e^{.08} = 1.083.$ *Take antilogs*

So, when $t = 4$,

$P = (1{,}000{,}000)(1.083)$

$= 1{,}083{,}000.$

(d) when $t = 10$,

$P = 1{,}000{,}000\ e^{.02(10)}$

$= 1{,}000{,}000\ e^{.2}$

Now, $\log e^{.2} = .2 \log e$

$= .2(.4343)$

$= .086866.$

Thus, $e^{.2} = 1.221.$

$P = (1{,}000{,}000)(1.221)$

$= 1{,}221{,}000$

21. At time $t = 0$, the number of ants is

$y = 300\ e^{.4(0)}$

$= 300\ e^{0}$

$= 300(1)$

$= 300.$

To find when they triple, we want to find the time such that the number of ants is 900.

$y = 300\ e^{.4t}$

$900 = 300\ e^{.4t}$ *Let $y = 900$*

$3 = e^{.4t}$ *Divide each side by 300*

$\log 3 = \log e^{.4t}$ *Take the log of each side*

$.4771 = (.4t) \log e$

$.4771 = (.4)t(.4343)$

$.4771 = 0.1737t$

$\dfrac{.4771}{.1737} = t$

$2.75 = t$

Thus, it takes 2.75 days for the number of ants to triple.

25. Find N(.5).
Use the formula N = -5000 ln r where r = .5.

$$N = -5000 \ln .5$$
$$\approx -5000\left(\frac{\log .5}{.4343}\right)$$
$$\approx -5000\left(\frac{-.30103}{.4343}\right)$$
$$\approx -5000(-.6931)$$
$$\approx 3465.5$$
$$\approx 3500 \text{ years (rounded to nearest 100)}$$

29. $\log e^2 = 2 \log e$
$= 2(.4343)$
$= .8686$
$= .869$
since
$\log e \approx .4343$

33. $\frac{n-1}{n+1}$

(a) $\frac{1-1}{1+1} = \frac{0}{2} = 0$ *Let n = 1*

(b) $\frac{2-1}{2+1} = \frac{1}{3}$ *Let n = 2*

(c) $\frac{3-1}{3+1} = \frac{2}{4} = \frac{1}{2}$ *Let n = 3*

(d) $\frac{4-1}{4+1} = \frac{3}{5}$ *Let n = 4*

**Chapter 11 Review Exercises (page 478)**

1. $y = 4^x$
Make a table of values.

| x | -2 | -1 | 0 | 1 | 2 |
|---|---|---|---|---|---|
| y | 1/16 | 1/4 | 1 | 4 | 16 |

Plot these ordered pairs and draw the graph through them.
See answer graph in textbook.

5. Write $y = \log_4 x$ in exponential form. $x = 4^y$. Make a table of values.

| x | 1/16 | 1/4 | 1 | 4 | 16 |
|---|---|---|---|---|---|
| y | -2 | -1 | 0 | 1 | 2 |

Plot the ordered pairs given by the table and draw the graph through them.
See answer graph in textbook.

9. $\log_8 2 = \frac{1}{3}$ becomes $8^{1/3} = 2$ in exponential form.

13. $\log_5 \frac{3}{10}x$
$= \log_5 \frac{3}{10} + \log_5 x$
$= \log_5 3 - \log_5 10 + \log_5 x$

17. $\log_a 2x + \log_a 3 = \log_a (2x)(3)$
$= \log_a 6x$

21. $4 \log_2 x - 3 \log_2 x = 1 \log_2 x$
$= \log_2 x$

25. $\log \sqrt[3]{16.2} = \log (16.2)^{1/3}$
$= \frac{1}{3} \log 16.2$
$= \frac{1}{3}(1.2095)$
$= .4032$
$\sqrt[3]{16.2} = 2.53$ *Take antilogs*

29. $4^{m+1} = 5$

$\log 4^{m+1} = \log 5$ *Take the log of each side*

$(m + 1)\log 4 = \log 5$

$m + 1 = \frac{\log 5}{\log 4}$

$m = \frac{\log 5}{\log 4} - 1$

$= \frac{.6990}{.6021} - 1$

$= 1.161 - 1$

$= .161$

Solution set: {.161}

33. $\log_m 125 = 3$ becomes $m^3 = 125$ in exponential form.

$m^3 = 5^3$ since $5^3 = 125$.

Thus, m = 5. Since the exponents are equal, the bases must be equal.
Solution set: {5}

37. $S = C(1 - r)^n$, C = 50,000, n = 10, r = .10

$S = 50{,}000(1 - .10)^{10}$

$S = 50{,}000(.9)^{10}$

Note:

$\log(.9)^{10} = 10 \log .9$

$= 10(.9542 - 1)$

$= 10(.0457575)$ *Use calculator to get more accurate answer*

$= -.457575$

$= .542425 - 1$

Thus, $.9^{10} = 3.4867836 \times 10^{-1}$

$= .34867836.$

$S = 50{,}000(.9)^{10}$

$= 50{,}000(.34867836)$

$= \$17{,}433.92.$

41. $\ln 7 = \log_e 7$

$= \frac{\log_{10} 7}{\log_{10} e}$

$= \frac{.8451}{.4343}$

$= 1.95$

45. $P = 1000\, e^{.05t}$

(a) If t = 2, then

$P = 1000\, e^{(.05)(2)}$

$= 1000\, e^{.1}.$

Now

$\log e^{.1} = .1 \log e$

$= (.1)(.4343)$

$= .0434.$

Taking antilogs gives $e^{.1} = 1.105$

So, $P = 1000\, e^{.1}$

$= 1000(1.105)$

$= 1105.$

(b) If t = 5,

$P = 1000\, e^{(.05)(5)}$

$= 1000\, e^{.25}.$

$\log e^{.25} = .25 \log e$

$= .25(.4343)$

$= .1086$

Thus, $e^{.25} = 1.284.$

So, $P = 1000\, e^{.25}$

$= 1000(1.284)$

$= 1284.$

(c) If t = 10,

$P = 1000\, e^{(.05)(10)}$

$= 1000\, e^{.5}.$

$$\log e^{.5} = .5 \log e$$
$$= \frac{1}{2}(.4343)$$
$$= .2171$$

So, $e^{.5} = 1.649$ and

$$P = 1000\, e^{.5}$$
$$= 1000(1.649)$$
$$= 1649.$$

**Chapter 11 Test (page 480)**

In Exercises 1-4, see the answer graphs in your textbook.

1. $y = 2^x$

Make a table of values.

| x | -2 | -1 | 0 | 1 | 2 |
|---|---|---|---|---|---|
| y | 1/4 | 1/2 | 1 | 2 | 4 |

Plot the ordered pairs given by the table above and draw a graph through them.

2. $y = (\frac{1}{3})^x$

Make a table of values.

| x | -2 | -1 | 0 | 1 | 2 |
|---|---|---|---|---|---|
| y | 9 | 3 | 1 | 1/3 | 1/9 |

Plot the ordered pairs given by the table above and draw a graph through them.

3. Write $y = \log_3 x$ in exponential form.

$$x = 3^y$$

Complete the following table of values.

| x | 1/27 | 1/9 | 1/3 | 1 | 3 | 9 | 1/27 |
|---|---|---|---|---|---|---|---|
| y | -3 | -2 | -1 | 0 | 1 | 2 | 3 |

Plot the ordered pairs given in the table above, then graph the curve through them.

4. Write $y = \log_{1/2} x$ in exponential form.

$$x = (\frac{1}{2})^y$$

Make a table of values.

| x | 4 | 2 | 1 | 1/2 | 1/4 |
|---|---|---|---|---|---|
| y | -2 | -1 | 0 | 1 | 2 |

Plot the ordered pairs given by the table and draw a graph through them.

5. $\log_6 36 = \log_6 6^2 = 2$

6. $\log_2 (\frac{1}{64}) = \log_2 2^{-6} = -6$

7. Let $\log_{81} \frac{1}{27} = y$. Then,

$$81^y = \frac{1}{27}.$$

$(3^4)^y = 3^{-3}$ *Write both sides as powers of 3*

$$3^{4y} = 3^{-3}$$
$$4y = -3$$
$$y = -\frac{3}{4}$$

8. $\log_5 5 = 1$ since $5^1 = 5$.

9. $\log 246 = 2.3909$

10. $\log .000317 = .5011 - 4$ or $-3.4989$

11. $\log_2 10 = \dfrac{\log_{10} 10}{\log_{10} 2}$

$= \dfrac{1}{.30103}$

$= 3.3219$

12. $\ln 89.1 = \log_e 89.1$

$= \dfrac{\log_{10} 89.1}{\log_{10} e}$

$= \dfrac{1.9499}{.4343}$

$= 4.4898$

13. $\ln 17 = \dfrac{\log_{10} 17}{\log_{10} e}$

$= \dfrac{1.23045}{.4343}$

$= 2.8332$

14. $\log \dfrac{(6.49)(28.3)}{(71.5)^2}$

$= \log 6.49 + \log 28.3 - 2 \log 71.5$

$= 0.8122 + 1.4518 - 3.7086$

$= 2.2640 - 3.7086$

$= -1.4446$

$= .5554 - 2$

Antilog $.5554 - 2 = 3.59 \times 10^{-2}$

$= .0359.$

15. $\log (8.1)^{.81} = .81 \log 8.1$

$= .81(.9085)$

$= .7359$

Antilog $.7359 = 5.44.$

16. Write $\log_9 x = \frac{5}{2}$ in exponential form.

$9^{5/2} = x$

$x = 9^{5/2}$

$x = (9^{1/2})^5$

$= (\sqrt{9})^5$

$= 3^5$

$= 243$

Solution set: {243}

17. $\log_{1/8} x = \frac{2}{3}$ in exponential form becomes

$x = (\frac{1}{8})^{2/3}$

$= [(\frac{1}{8})^{1/3}]^2$

$= (\sqrt[3]{\frac{1}{8}})^2$

$= (\frac{\sqrt[3]{1}}{\sqrt[3]{8}})^2$

$= (\frac{1}{2})^2 = \frac{1}{4}.$

Solution set: {1/4}

18. $2^m = 14$

$\log 2^m = \log 14$ *Take the log of each side*

$m \log 2 = \log 14$

$m = \dfrac{\log 14}{\log 2}$

$= \dfrac{1.1461}{.3010}$

$= 3.808$

$= 3.81$ (rounded)

Solution set: {3.81}

19. $\log_2 (x + 3) + \log_2 (x - 1)$

$$= \log_2 5$$

$$\log_2 (x + 3)(x - 1) = \log_2 5$$

$$(x + 3)(x - 1) = 5$$

$$x^2 + 2x - 3 = 5$$

$$x^2 + 2x - 8 = 0$$

$$(x + 4)(x - 2) = 0$$

$x + 4 = 0$ or $x - 2 = 0$

$x = -4$ or $x = 2$

$\log_2 (-4 + 3) = \log_2 (-1)$, which does not exist.

The only solution is 2.

Solution set: $\{2\}$

20. $C(t) = 8000\, e^{.4t}$

(a) If $t = 0$, we have

$$C(0) = 8000\, e^{.4(0)}$$
$$= 8000\, e^0$$
$$= 8000(1)$$
$$= 8000 \text{ bacteria.}$$

(b) If $t = 5$, we have

$$C(5) = 8000\, e^{.4(5)}$$
$$= 8000\, e^2$$
$$= 8000(2.718)^2$$
$$= 59{,}100 \text{ bacteria.}$$

(c) The culture will contain 16,000 bacteria when $C(t) = 16{,}000$. So we have

$$16{,}000 = 8000\, e^{.4t}.$$

$2 = e^{.4t}$ *Divide each side by 8000*

$\log 2 = \log e^{.4t}$ *Take logs*

$\log 2 = .4t \log e$

$.3010 = .4(.4343)t$ *Find logs*

$.3010 = .1737t$

$t \approx 1.7$ hours. *Solve for t*

## CHAPTER 12 SEQUENCES AND SERIES

### Section 12.1 (page 483)

1. $a_n = n - 1$

$a_1 = n - 1 = 0$

$a_2 = 2 - 1 = 1$

$a_3 = 3 - 1 = 2$

$a_4 = 4 - 1 = 3$

$a_5 = 5 - 1 = 4$

5. $a_n = 2^n$

$a_1 = 2^1 = 2$

$a_2 = 2^2 = 4$

$a_3 = 2^3 = 8$

$a_4 = 2^4 = 16$

$a_5 = 2^5 = 32$

9. $a_n = \frac{1}{n^2}$

$a_1 = \frac{1}{1^2} = 1$

$a_2 = \frac{1}{2^2} = \frac{1}{4}$

$a_3 = \frac{1}{3^2} = \frac{1}{9}$

$a_4 = \frac{1}{4^2} = \frac{1}{16}$

$a_5 = \frac{1}{5^2} = \frac{1}{25}$

13. $a_n = (-1)^n$

$a_1 = (-1)^1 = -1$

$a_2 = (-1)^2 = 1$

$a_3 = (-1)^3 = -1$

$a_4 = (-1)^4 = 1$

$a_5 = (-1)^5 = -1$

17. $a_n = 4n + 11$; $a_{15}$

$a_{15} = 4(15) + 11 = 60 + 11 = 71$

21. $a_n = (n + 1)(2n + 3)$; $a_8$

$a_8 = (8 + 1)(2 \cdot 8 + 3)$

$= (9)(19)$

$= 171$

25. $a_n = 3n^2(4n - 5)$; $a_5$

$a_5 = 3(5)^2(4(5) - 5)$

$= 3(25)(15)$

$= 1125$

29. $a_n = \frac{2.6431n}{n + .8725}$; $a_7$

$a_7 = \frac{(2.6431)(7)}{7 + .8725}$

$= \frac{18.5017}{7.8725}$

$= 2.3502$ (rounded)

33. 4, 8, 12, 16, ... can be written as 4(1), 4(2), 4(3), 4(4), ... so

$a_n = 4n.$

37. At the end of 1 year: $\frac{4}{5}(10,000)$

At the end of 2 years:

$\frac{4}{5}[\frac{4}{5}(10,000)]$

$= (\frac{4}{5})^2(10,000)$

At the end of 5 years:

$(\frac{4}{5})^5(10,000)$

$= 3276.8$

The value of the car is \$3276.80.

**Section 12.2 (page 486)**

1. $\sum_{i=1}^{6} 2i = 2(1) + 2(2) + 2(3) + 2(4) + 2(5) + 2(6)$
$= 2 + 4 + 6 + 8 + 10 + 12$
$= 42$

5. $\sum_{i=0}^{4} \frac{i}{i+1} = \frac{0}{0+1} + \frac{1}{1+1} + \frac{1}{2+1} + \frac{3}{3+1} + \frac{4}{4+1}$
$= 0 + \frac{1}{2} + \frac{2}{3} + \frac{3}{4} + \frac{4}{5}$
$= \frac{163}{60}$

9. $\sum_{i=1}^{5} (1)^i = 1^1 + 1^2 + 1^3 + 1^4 + 1^5$
$= 1 + 1 + 1 + 1 + 1$
$= 5$

13. $\sum_{i=1}^{5} (-1)^i = (-1)^1(1) + (-1)^2 2 + (-1)^3 3 + (-1)^4 4 + (-1)^5 5$
$= -1 + 2 - 3 + 4 - 5$
$= -3$

17. $\sum_{i=1}^{6} x^i = x^1 + x^2 + x^3 + x^4 + x^5 + x^6$
$= x + x^2 + x^3 + x^4 + x^5 + x^6$

21. $\sum_{i=2}^{6} \frac{x+i}{x-i}$
$= \frac{x+2}{x-2} + \frac{x+3}{x-3} + \frac{x+4}{x-4} + \frac{x+5}{x-5} + \frac{x+6}{x-6}$

Several different answers may be possible in Exercises 25 and 29.

25. $1 + 3 + 5 + 7 = \sum_{i=1}^{4} (2i - 1)$

29. $x + 1 + \frac{x+2}{2} + \frac{x+3}{3} + \frac{x+4}{4}$
$= \sum_{i=1}^{4} \frac{x+i}{i}$

33. $\bar{x} = \frac{8 + 11 + 14 + 9 + 3 + 6 + 8}{7}$
$= 8.43$ (rounded)

37. $\bar{x} = \frac{-6.104 + 2.503 + 4.272 - 1.383}{4}$
$= -.178$

41. $\sum_{i=1}^{n} k\, x_i = k^n \cdot \sum_{i=1}^{n} x_i$

Let $n = 3$, $k = 5$:

$5x_1 + 5x_2 + 5x_3 = 5^3(x_1 + x_2 + x_3)$

Let $x_1 = 1$, $x_2 = 2$, $x_3 = 3$:

$5(1) + 5(2) + 5(3) = 5^3(1 + 2 + 3)$
$30 = 125(6)$ *False*

Since the statement is not true in this example, it is not true in general.

45. $(\sum_{i=1}^{n} x_1)^2 = \sum_{i=1}^{n} x_i^2$

Let $n = 3$, $x_1 = 1$, $x_2 = 2$, $x_3 = 3$:

$(x_1 + x_2 + x_3)^2 = x_1^2 + x_2^2 + x_3^2$
$(1 + 2 + 3)^2 = (1)^2 + (2)^2 + (3)^2$
$(6)^2 = 1 + 4 + 9$
$36 = 14.$ *False*

This statement is not true in general.

**Section 12.3 (page 492)**

1. 1, 2, 3, 4, 5,...
$d = 3 - 2 = 1$

5. -10, -5, 0, 5, 10,...
$d = 10 - 5 = 5$

9. $-\frac{5}{3}, -1, -\frac{1}{3}, \frac{1}{3}, \ldots$

$$d = -1 - (-\tfrac{5}{3})$$
$$= -\frac{3}{3} + \frac{5}{3}$$
$$= \frac{2}{3}$$

13. $2, 4, 6, \ldots; a_{24}$

$$a_n = 2 + 2(n - 1)$$
$$a_{24} = 2 + 2(24 - 1)$$
$$= 2 + 46 = 48$$

17. $a_{12} = -45, a_{10} = -37; a_1$

$$d = \frac{1}{2}(a_{12} - a_{10})$$
$$= \frac{1}{2}(-45 - (-37))$$
$$= -4$$
$$a_n = a_1 + d(n - 1)$$
$$-37 = a_1 - 4(10 - 1) \quad \textit{Substitute}$$
$$-37 = a_1 - 36$$
$$-1 = a_1$$

21. $a_1 = 2, d = 5$

$$a_n = a_1 + (n - 1)d$$
$$= 2 + (n - 1)(5) \quad \textit{Substitute}$$
$$= 2 + 5n - 5$$
$$= 5n - 3$$

25. $-3, 0, 3, \ldots$

$$a_1 = -3$$
$$d = 0 - (-3)$$
$$= 3$$
$$a_n = -3 + 3(n - 1) \quad \textit{Substitute}$$
$$= -3 + 3n - 3$$
$$= 3n - 6$$

29. $\frac{3}{4}, 3, \frac{21}{4}, \ldots, 12$

$$a_1 = \frac{3}{4}, a_n = 12$$
$$d = 3 - \frac{3}{4} = \frac{9}{4}$$
$$a_n = \frac{3}{4} + \frac{9}{4}(n - 1) \quad \textit{Substitute}$$
$$12 = \frac{3}{4} + \frac{9}{4}(n - 1)$$
$$\frac{45}{4} = \frac{9}{4}(n - 1)$$
$$5 = n - 1$$
$$n = 6$$

33. $a_1 = 6, d = 3$

The formula $S_n = \frac{n}{2}(a_1 + a_n)$ requires values for $n$, $a_1$, and $a_n$. So for $S_6$, $n = 6$, $a_1 = 6$, and

$$a_6 = a_1 + (n - 1)d$$
$$= 6 + (6 - 1)(3)$$
$$= 6 + 5(3)$$
$$= 21.$$

$$S_6 = \frac{6}{2}(6 + 21) \quad \textit{Substitute}$$
$$= 3(27)$$
$$= 81$$

37. $a_n = 4 + 3n$

$$a_1 = 4 + 3(1) = 7$$
$$a_6 = 4 + 3(6) = 22$$
$$n = 6$$
$$S_6 = \frac{6}{2}(7 + 22) \quad \textit{Substitute}$$
$$= 87$$

41. $\sum_{i=1}^{10} (8i - 5)$

$= S_{10}$

$= \frac{10}{2}(a_1 + a_{10})$ *This is an arithmetic series*

$a_n = 8n - 5$

$a_1 = 8(1) - 5 = 3$

$a_{10} = 8(10) - 5 = 75$

$\sum_{i=1}^{10} (8i - 5) = \frac{10}{2}(3 + 75)$ *Substitute*

$= 5(78)$

$= 390$

45. $\sum_{i=1}^{20} (2i - 5) = S_{20} = \frac{20}{2}(a_1 + a_{10})$

$a_n = 2n - 5$

$a_1 = 2(1) - 5 = -3$

$a_{20} = 2(20) - 5 = 35$

$\sum_{i=1}^{20} (2i - 5) = \frac{20}{2}(-3 + 35)$ *Substitute*

$= 10(32)$

$= 320$

49. We want $\sum_{n=1}^{750} a_n$ where $a_n = n$.

So $a_1 = 1$, $a_{750} = 750$, and

$\sum_{n=1}^{750} a_n = S_{750}$

$= \frac{750}{2}(1 + 750)$

$= 281,625$

53. Your salaries at six-month intervals form an arithmetic sequence with $a_1 = 1600$ and $d = 50$.

After 5 years, since your salary is increased every 6 months, or $2(5) = 10$ times, your salary will be term $a_{11}$.

$a_{11} = a_1 + (n - 1)d$

$= 1600 + 10(50)$

$= 1600 + 500$

$= 2100$

Your salary will be $2100 a month.

**Section 12.4 (page 500)**

1. 4, 8, 16, 32,...

$r = \frac{8}{4} = 2$

5. 1, -3, 9, -27, 81,...

$r = \frac{-3}{1} = -3$

9. $1, -\frac{1}{2}, \frac{1}{4}, -\frac{1}{8}, \frac{1}{16}, \ldots$

$r = \frac{-1/2}{1} = -\frac{1}{2}$

13. $\frac{1}{9}, \frac{1}{3}, \ldots$

$a_n = a_1 r^{n-1}$. Now $a_1 = \frac{1}{9}$, so

$a_n = \frac{1}{9}r^{n-1}$. Now $a_2 = \frac{1}{9}r$, so

$\frac{1}{3} = \frac{1}{9}r$

$r = 3.$

Thus, $a_n = \frac{1}{9}(3^{n-1})$ or $\frac{3^{n-1}}{9}$.

17. 2, 10, 50,..., $a_{10}$

$r = \frac{10}{2} = 5$ and $a_1 = 2$

$a_n = a_1 r^{n-1}$

$a_{10} = 2(5)^{10-1}$

$= 2(5)^9$

21. $a_1 = 1$, $r = 5$; $a_8$

$$a_n = a_1 r^{n-1}$$
$$= 1(5)^{8-1}$$
$$= 5^7$$

25. $a_3 = \frac{1}{2}$, $a_7 = \frac{1}{32}$; $a_{25}$

To find r, first find $a_1$ using the formula $a_n = a_1 r^{n-1}$.

$$a_3 = a_1 r^{3-1} \qquad a_7 = a_1 r^{7-1}$$
$$\frac{1}{2} = a_1 r^2 \qquad \frac{1}{32} = a_1 r^6$$
$$\frac{1}{2r^2} = a_1 \qquad \frac{1}{32r^6} = a_1$$

The expressions for $a_1$ are equal.

$$\frac{1}{2r^2} = \frac{1}{32}r^6$$

Solve for r.

$16r^4 = 1$ *Multiply by least common denominator $32r^6$*

$r^4 = \frac{1}{16}$ *Divide by 16*

$r = \pm\frac{1}{2}$

Now find $a_1$ using $a_3 = a_1 r^{3-1}$

$$\frac{1}{2} = a_1 (\pm\tfrac{1}{2})^2$$
$$\frac{\frac{1}{2}}{\frac{1}{4}} = a_1$$
$$2 = a_1$$

Use $a_n = a_1 r^{n-1}$ with $a_1 = 2$ and $r = \pm\frac{1}{2}$ to find $a_{25}$.

$$a_{25} = 2(\tfrac{1}{2})^{25-1}$$
$$= 2(\pm\tfrac{1}{2})^{24}$$
$$= 2(\tfrac{1}{2})^{24} \quad \text{or} \quad (\tfrac{1}{2})^{23}$$

29. $\frac{1}{3}, \frac{1}{9}, \frac{1}{27}, \frac{1}{81}, \frac{1}{243}$

$S_n = \frac{a_1(1 - r^n)}{1 - r}$ where $a_1 = \frac{1}{3}$, $r = \frac{1}{3}$, and $n = 8$.

$$S_8 = \frac{(1/3)[1 - (1/3)^5]}{1 - 1/3}$$
$$= \frac{(1/3)[1 - (1/3)^5]}{2/3}$$
$$= \frac{1 - (1/3)^5}{2}$$
$$= \frac{242/243}{2}$$
$$= \frac{121}{243}$$

33. $\frac{-4}{3}, \frac{-4}{9}, \frac{-4}{27}, \frac{-4}{81}, \frac{-4}{243}, \frac{-4}{729}$

Use $S_n = \frac{a_1(r^n - 1)}{r - 1}$ with $a_1 = -4/3$, $r = \frac{-4/9}{-4/3} = -\frac{4}{9}(-\frac{3}{4}) = \frac{1}{3}$, and $n = 6$.

$$S_6 = \frac{(-4/3)[(1/3)^6 - 1)]}{1/3 - 1}$$
$$= \frac{(-4/3)[(1/3)^6 - 1]}{-2/3}$$
$$= 2[(\tfrac{1}{3})^6 - 1]$$
$$= 2[\tfrac{1}{729} - 1]$$
$$= 2[\tfrac{1}{729} - \tfrac{729}{729}]$$
$$= 2[\tfrac{-728}{129}]$$
$$= -\frac{1456}{729}$$

37. $\sum_{i=1}^{8} 5(\frac{2}{3})^i = 5\sum_{i=1}^{8} (\frac{2}{3})^i$

Use $S_n = \frac{a_1(1 - r^n)}{1 - r}$ with $a_1 = 2/3$, $r = 2/3$, $n = 8$.

$$5(S_8) = 5\frac{(2/3)[1 - (2/3)^8]}{1 - 2/3}$$

$$= \frac{(10/3)[1 - (2/3)^8]}{1/3}$$

$$= 10[1 - (\frac{2}{3})^8]$$

$$\sum_{i=1}^{8} 5(\frac{2}{3})^i = \frac{63050}{6561}$$

41. $a_1 = 6$, $r = \frac{1}{3}$

Since $|r| = \frac{1}{3} < 1$, the sum $a_1/(1 - r)$ exists.

$$\frac{a_1}{1 - r} = \frac{6}{1 - 1/3}$$

$$= \frac{6}{2/3} = 9$$

45. $a_1 - 1{,}000$, $r = \frac{-1}{10}$

Since $|r| = |-\frac{1}{10}| < 1$, the sum exists.

$$\frac{a_1}{1 - r} = \frac{1000}{1 + 1/10}$$

$$= \frac{1000}{11/10}$$

$$= 1000 \cdot \frac{10}{11}$$

$$= \frac{10000}{11}$$

49. $\sum_{i=1}^{\infty} \frac{12}{5}(\frac{5}{4})^i$

Since $\frac{5}{4} > 1$, the sum does not exist.

53. 0.125 125 125 ... can be written

0.125 + 0.000125 + 0.000000125 + ⋯.

This represents the sum of the geometric sequence with $a_1 = 0.125$ and

$$r = \frac{0.000125}{0.125}$$

$$= 0.001.$$

Since $|r| = 0.001 < 1$, the sum exists.

$$\frac{a_1}{1 - r} = \frac{0.125}{1 - 0.001}$$

$$= \frac{0.125}{0.999}$$

$$= \frac{125}{999}$$

57. The problem suggests the following sequences:

$10,\ 10 \cdot \frac{3}{5},\ (10 \cdot \frac{3}{5}) \cdot \frac{3}{5}, \ldots$ or

| At start | After first bounce | After second bounce | |
|---|---|---|---|
| ↓ | ↓ | ↓ | |
| $10(\frac{3}{5})^0$, | $10(\frac{3}{5})^1$, | $10(\frac{3}{5})^2$, | ... |

So, the height after the fourth bounce is given by

$$10(\frac{3}{5})^4 = 10(\frac{81}{625})$$

$$= \frac{810}{625}$$

$$= 1.3.$$

The height is $10(\frac{3}{5})^4$ or about 1.3 feet.

61. After n years, $(1.02)^n(1.1)$ billion units are used.

(a) After 5 years,

$$(1.02)^5(1.1) = 1.21$$

billion units are used.

Consumption doubles when

$$2.2 = (1.1)(1.02)^n$$
$$2 = (1.02)^n$$

$$\log 2 = n \log 1.02$$
$$\frac{\log 2}{\log 1.02} = n$$
$$35 \approx n$$

or at about 35 years.

**Section 12.5 (page 506)**

1. $(m + n)^4$

$$= m^4 + \frac{4!}{3!1!}m^3n + \frac{4!}{2!2!}m^2n^2 + \frac{4!}{1!3!}mn^3 + n^4 \quad \textit{Binomial theorem}$$
$$= m^4 + 4m^3n + 6m^2n^2 + 4mn^3 + n^4$$

5. $(2x + 3)^2$

$$= (2x)^3 + \frac{3!}{2!1!}(2x)^2(3) + \frac{3!}{1!2!}(2x)(3)^2 + (3)^3 \quad \textit{Binomial theorem}$$
$$= 8x^3 + 36x^2 + 54x + 27$$

9. $(mx - n^2)^3$

$$= (mx + (-n)^2)^3$$
$$= (mx)^3 + 3(mx)^2(-n)^2 + 3(mx)(-n^2)^2 + (-n^2)^3$$
$$= m^3x^3 - 3m^2n^2x^2 + 3mn^4x - n^6$$

13. $(3x - y)^{14}$

$$= (3x + (-y))^{14}$$
$$= (3x)^{14} + \frac{14!}{13!1!}(3x)^{13}(-y) + \frac{14!}{12!2!}(3x)^{12}(-y)^2 + \frac{14!}{11!3!}(3x)^{11}\cdot(-y)^3 + \ldots$$
$$= 3^{14}x^{14} - 14(3)^{13}x^{13}y + (91)(3)^{12}x^{12}y^2 - (364)(3)^{11}x^{11}y^3$$

17. $(2m + n)^{10}$ has fourth term

$$\frac{10!}{(10-3)!3!}(2m)^{10-3}(n)^3$$
$$= \frac{10!}{7!3!}2^7m^7n^3$$
$$= (120)(2^7)m^7n^3$$

21. $(k - 1)^9$ has third term

$$\frac{9!}{(9-2)!2!}k^{9-2}(-1)^2 = \frac{9!}{7!2!}k^7$$
$$= 36k^7$$

25. $(0.999)^{20}$

$$= (1 + (-0.001))^{20}$$
$$= 1^{20} + \frac{20!}{19!1!}(1)^{19}(-0.001) + \frac{20!}{18!2!}(1)^{18}(-0.001)^2 + \ldots$$
$$1 - 0.02 + 0.00019 = 0.980$$

**Chapter 12 Review Exercises (page 507)**

1. $a_n = 2n - 3$

$$a_1 = 2(1) - 3 = -1$$
$$a_2 = 2(2) - 3 = 1$$
$$a_3 = 2(3) - 3 = 3$$
$$a_4 = 2(4) - 3 = 5$$

5. $a_n = (n + 1)(n - 1)$

$$a_1 = (1 + 1)(1 - 1) = 0$$
$$a_2 = (2 + 1)(2 - 1) = 3$$
$$a_3 = (3 + 1)(3 - 1) = 8$$
$$a_4 = (4 - 1)(4 - 1) = 15$$

9. $\sum_{i=1}^{6} 2^i$

$= S_6$ *This is a geometric series*

$= \frac{a_1(1 - r^n)}{1 - r}$

$a_1 = 2^1 = 2,\ a_2 = 2^2 = 4,\ r = \frac{4}{2} = 2$

$\sum_{i=1}^{6} 2^i$

$= \frac{2(1 - 2^6)}{1 - 2}$ *Substitute*

$= 2(2^6 - 1)$

$= 126$

13. -6, -2, 2, 6, 10, ...
Arithmetic sequence

$d = -2 - (-6) = 4$

17. 64, 32, 16, 8, ...
Geometric sequence

$r = \frac{32}{64} = \frac{1}{2}$

21. 1, 7, 13, ...; $a_{40}$

$d = 7 - 1 = 6,\ a_1 = 1$

$a_n = 1 + 6(n - 1) = 6n - 5$

$a_{40} = 6(40) - 5 = 235$

25. 7, 10, 13, ..., 49

$a_1 = 7,\ d = 10 - 7 = 3$

$a_n = 49 = a_1 + d(n - 1)$

$= 7 + 3(n - 1)$

$= 4 + 3n$

Solve for n.

$49 = 4 + 3n$

$45 = 3n$

$n = 15$

29. $\sum_{i=1}^{8} (3i + 1)$

This is an arithmetic sequence.

$a_1 = 3 \cdot 1 + 1 = 4$ *Let $i = 1$*

$a_8 = 3 \cdot 8 + 1 = 25$ *Let $i = 8$*

$S_8 = \frac{8}{2}(4 + 25) = 116$

33. $a_1 = 1,\ r = -2;\ a_9$

$a_n = (-2)^{n-1}$

$a_9 = (-2)^{9-1}$

$= (-2)^8$

$= 2^8$

$= 256$

37. $2, \frac{2}{3}, \frac{2}{9}, \frac{2}{27}, \frac{2}{81}$

Use $S_n = \frac{a_1(1 - r^n)}{1 - r}$ with

$r = \frac{2/3}{2} = \frac{1}{3},\ a_1 = 2$, and $n = 5$.

$S_5 = \frac{2(1 - (1/3)^5)}{1 - 1/3}$

$= \frac{2(1 - 1/243)}{2/3}$

$= 3(\frac{242}{243})$

$= \frac{242}{81}$

41. $5.040404 = 5 + 4(.01) + 4(.01)^2 + 4(.01)^3 + \ldots$

$= 5 + 4[\frac{.01}{1 - .01}]$

$= 5 + \frac{4}{99}$

$= \frac{499}{99}$

45. Since 100% - 3% = 97% = .97, the population
after 1 year is .97(50,000)
after 2 years is .97[.97(50,000)]
or $(.97)^2(50,000)$
after n years is $(.97)^n(50,000)$.
After 6 years, the population is

$(.97)^6(50,000) \approx 41648$
$\approx 42,000$ (rounded).

49. The fourth term of $(3a + 2b)^{19}$ is

$\frac{19!}{16!3!}(3a)^{16}(2b)^3$ *Binomial theorem*

$= 7752(3)^{16}a^{16}b^3$.

**Chapter 12 Test (page 509)**

1. $a_n = 4 + n$
$a_1 = 4 + 1 = 5$
$a_2 = 4 + 2 = 6$
$a_3 = 4 + 3 = 7$
$a_4 = 4 + 4 = 8$
$a_5 = 4 + 5 = 9$

2. $a_n = 2(n + 3)$
$a_1 = 2(1 + 3) = 8$
$a_2 = 2(2 + 3) = 10$
$a_3 = 2(3 + 3) = 12$
$a_4 = 2(4 + 3) = 14$
$a_5 = 2(5 + 3) = 16$

3. $a_n = (-1)^n + 1$
$a_1 = (-1)^1 + 1 = 0$
$a_2 = (-1)^2 + 1 = 2$
$a_3 = (-1)^3 + 1 = 0$
$a_4 = (-1)^4 + 1 = 2$
$a_5 = (-1)^5 + 1 = 0$

4. $a_n = (n + 1)(n + 2)$
$a_1 = (1 + 1)(1 + 2) = 6$
$a_2 = (2 + 1)(2 + 2) = 12$
$a_3 = (3 + 1)(3 + 2) = 20$
$a_4 = (4 + 1)(4 + 2) = 30$
$a_5 = (5 + 1)(5 + 2) = 42$

5. $a_1 = 4$, $d = 2$
$a_1 = 4$
$a_2 = a_1 + d = 4 + 2 = 6$
$a_3 = 6 + 2 = 8$
$a_4 = 8 + 2 = 10$
$a_5 = 10 + 2 = 12$

6. $a_3 = 5$, $d = 4$
Use $a_n = a_1 + (n - 1)d$ with $n = 3$
$5 = a_1 + (3 - 1)(-4)$
$13 = a_1$
$a_2 = 13 + (-4) = 9$
$a_3 = 9 + (-4) = 5$
$a_4 = 5 + (-4) = 1$
$a_5 = 1 + (-4) = -3$

7. $a_1 = 2,\ r = -3$

$a_1 = 2$

$a_2 = a_1 r = 2(-3) = -6$

$a_3 = -6(-3) = 18$

$a_4 = 18(-3) = -54$

$a_5 = -54(-3) = 162$

8. $a_4 = 6,\ r = \frac{1}{2}$

$a_4 = 6$

$a_5 = \frac{1}{2}(6) = 3$

$\frac{1}{2}a_3 = a_4$

$a_3 = 2(a_4)$

$= 2(6) = 12$

$a_2 = 2(12) = 24$

$a_1 = 2(24) = 48$

9. $a_1 = 6,\ d = -2$

$a_n = a_1 + (n - 1)d$

$a_4 = a_1 + (4 - 1)d \quad n = 4$

$= 6 + (3)(-2)$

$= 6 - 6$

$= 0$

10. $a_8 = 3,\ d = -5$

$a_n = a_1 + (n - 1)d$

$3 = a_8 = a_1 + (8 - 1)(-5)$

$= a_1 - 35$

$a_1 = 38$

Now,

$a_n = 38 + (n - 1)d$

$= 38 + (n - 1)(-5)$

$= 43 - 5n$

so

$a_4 = 43 - 5(4) = 23.$

11. $a_2 = 11,\ a_5 = 38$

Since $a_3$, $a_4$ lie between $a_2$ and $a_5$, then

$$38 - 11 = 3d$$
$$27 = 3d$$
$$d = 9.$$

Therefore, $a_4 = a_2 + 2d$

$= 11 + 2(9)$

$= 11 + 18$

$= 29.$

12. $a_1 = 5,\ r = 2$

$a_n = a_1 r^{n-1}$

$= 5(2^{n-1})$

$a_4 = 5(2^{4-1})$

$= 5(8) = 40$

13. $a_2 = 9,\ a_3 = 4$

$$r = \frac{a_3}{a_2} = \frac{4}{9}$$

$a_4 = a_3 r$

$= 4(\frac{4}{9})$

$= \frac{16}{9}$

14. $a_6 = 12,\ a_7 = 9$

$$r = \frac{a_7}{a_6} = \frac{9}{12} = \frac{3}{4}$$

Since $a_6 = a_4 \cdot r \cdot r = r^2 a_4$, we have

$$a_4 = \frac{a_6}{r^2}$$
$$= \frac{12}{(3/4)^2}$$
$$= \frac{12}{9/16}$$
$$= 12 \cdot \frac{16}{9}$$
$$= \frac{64}{3}.$$

15. $a_1 = 8$, $a_5 = 12$

$$S_5 = \frac{5}{2}(a_1 + a_5)$$
$$= \frac{5}{2}(8 + 12)$$
$$= \frac{5}{2}(20)$$
$$= 50$$

16. $a_2 = 12$, $a_3 = 15$

$d = a_3 - a_2 = 15 - 12 = 3$

$a_1 = a_2 - 3 = 12 - 3 = 9$

$a_5 = a_4 + 3 = a_3 + 6 = 15 + 6 = 21$

$$S_5 = \frac{5}{2}(9 + 21)$$
$$= 75$$

17. $a_2 = 9$, $a_3 = 18$

$$r = \frac{a_3}{a_2} = \frac{18}{9} = 2$$
$$a_2 = a_1 r$$
$$9 = a_1(2)$$
$$a_1 = \frac{9}{2}$$
$$S_5 = \frac{(9/2)(2^5 - 1)}{2 - 1}$$
$$= \frac{9}{2}(32 - 1)$$
$$= \frac{9}{2}(31)$$
$$= \frac{279}{2}$$

18. $a_5 = 4$, $a_6 = 2$

$$r = \frac{a_6}{a_5} = \frac{1}{2}$$

Use $a_6 = a_1 \cdot r^5$ to get $a_1 = 64$.

$$S_5 = \frac{64(1 - (1/2)^5)}{1 - 1/2}$$
$$= 128(1 - \frac{1}{32})$$
$$= 128(\frac{31}{32})$$
$$= 124$$

Exercises 19 and 20 illustrate two methods of evaluating arithmetic series.

19. $\sum_{i=1}^{5} (2i + 8)$

$= [2(1)+8] + [2(2) + 8] + [2(3) + 8] + [2(4) + 8] + [2(5) + 8]$

$= 10 + 12 + 14 + 16 + 18$

$= 70$

20. $\sum_{i=1}^{6} (3i - 5)$

Use the formula $S_6 = \frac{6}{2}(a_1 + a_6)$.

$a_1 = 3 \cdot 1 - 5 = -2$

$a_6 = 3 \cdot 6 - 5 = 13$

$S_6 = \frac{6}{2}(-2 + 13) = 3(11) = 33.$

21. $\sum_{i=1}^{500} i$

This is an arithmetic sequence with $a_1 = 1$, $a_{500} = 500$, and $d = 1$. Use the formula for $S_{500}$.

$$\sum_{i=1}^{500} = \frac{500}{2}(a_1 + a_{500})$$
$$= 250(1 + 500)$$
$$= 125,250$$

22. $\sum_{i=1}^{3} \frac{1}{2}(4^i) = \frac{1}{2} \sum_{i=1}^{3} (4^i)$

$$= \frac{1}{2}[4 + 16 + 64] = 42$$

23. $\sum_{i=1}^{\infty} (\frac{1}{4})^i$

This is an infinite geometric series with $a_1 = \frac{1}{4}$ and $r = \frac{1}{4}$.

Since $|r| = \frac{1}{4} < 1$, the sum is $a_1/(1 - r)$.

$$\sum_{i=1}^{\infty} \left(\frac{1}{4}\right)^i = \frac{a_1}{1 - r} = \frac{1/4}{1 - 1/4} = \frac{1/4}{3/4} = \frac{1}{3}.$$

24. $$\sum_{i=1}^{\infty} 6\left(\frac{2}{3}\right)^i = 6 \sum_{i=1}^{\infty} \left(\frac{2}{3}\right)^i = 6\frac{2/3}{1 - 2/3} = 6\frac{2/3}{1/3} = 12$$

The sum is a geometric series with $a_1 = 2/3$ and $r = 2/3$.
Since $|r| = \frac{2}{3} < 1$, the sum is $a_1/(1 - r)$.

25. $a_1 = 4$, $d = 7 - 4 = 3$, $n = 18$

$$a_n = a_1 + d(n - 1)$$
$$a_{18} = 4 + 3(18 - 1) = 4 + 3(17) = 55$$

26. For the sequence -82, -74, -66,...:

$$a_1 = -82$$
$$d = -74 - (-82) = -74 + 82 = 8$$
$$a_{23} = a_1 + (23 - 1)d = -82 + 22(8) = 94.$$

$$S_{23} = \frac{23}{2}(a_1 + a_{23}) = \frac{23}{2}(-82 + 94) = \frac{23}{2}(12) = 23(6) = 138$$

27. The fifth term of $(2x - \frac{y}{3})^{12}$ is

$$\frac{12!}{8!4!}(2x)^8\left(-\frac{y}{3}\right)^4 = \frac{14080}{9}x^8y^4.$$

28. $$(3k - 5)^4 = (3k)^4 - \frac{4!}{3!1!}(3k)^3(5) + \frac{4!}{2!2!}(3k)^2(5)^2 - \frac{4!}{1!3!}(3k)(5)^3 + (5)^4$$
$$= 81k^4 - 4(27k^3)(5) + 6(9k^2)(25) - 4(3k)(125) + 625$$
$$= 81k^4 - 540k^3 + 1350k^2 - 1500k + 625$$

29. At the beginning of July, there are 20 insects.
After 1 week: 3(20) insects
After 2 weeks: 3[3(20)]
or $3^2(20)$ insects
After n weeks: $3^n(20)$ insects
So at the end of September, after 12 weeks, there are $(3)^{12}20$ insects.

30. The amounts of unpaid balance during 15 months form an arithmetic sequence in reverse order,

300, 280, 260, ..., 40, 20

which is the sequence with $a_1 = 20$, $a_n = 300$, and $n = 15$.
Thus, the sum of these balances is

$$S_{15} = \frac{15}{2}(20 + 300) = 2400.$$

Since 1% interest is paid on this total, the interest paid is 1% of \$2400 or \$24.
Since the sewing machine costs \$300 (paid monthly at \$20), the total cost is \$300 + \$24 = \$324.